Modern Aspects of Colloidal Dispersions

Modern Aspects of Colloidal Dispersions

Results from the DTI Colloid Technology Programme

edited by

Ronald H. Ottewill

School of Chemistry,
University of Bristol,
Bristol, U.K.

and

Adrian R. Rennie

Cavendish Laboratory,
University of Cambridge,
Cambridge, U.K.

KLUWER ACADEMIC PUBLISHERS
DORDRECHT / BOSTON / LONDON

A C.I.P. Catalogue record for this book is available from the Library of Congress.

ISBN 0-7923-4819-2

Published by Kluwer Academic Publishers,
P.O. Box 17, 3300 AA Dordrecht, The Netherlands.

Sold and distributed in the U.S.A. and Canada
by Kluwer Academic Publishers,
101 Philip Drive, Norwell, MA 02061, U.S.A.

In all other countries, sold and distributed
by Kluwer Academic Publishers,
P.O. Box 322, 3300 AH Dordrecht, The Netherlands.

Cover photograph from P. Bartlett, R.H. Ottewill and P.N. Pusey (1992),
Physical Review Letters 68, 3801 with permission.

Printed on acid-free paper

Printed in the Netherlands.

TABLE OF CONTENTS

Preface

This book contains the papers presented at a meeting sponsored by the Colloid and Interface Science Group of the Faraday Division, Royal Society of Chemistry, which was held at Wills Hall, University of Bristol from the 14th - 16th April 1997.

The purpose of the meeting, which was entitled Colloidal Dispersions, was to discuss the subject of concentrated colloidal systems including, dispersions, emulsions and powders, in order to emphasize recent advances in experimental and theoretical understanding of these systems and how these advances could be applied to practical utilisation in the wide range of industries which are involved with colloidal systems.

The papers presented at the meeting were given by the principal participants in a 5 year project on Colloid Technology, which started on the 1st August 1992, and was funded by the Department of Trade and Industry (DTI) of the U.K. and a consortium of industries which was composed of ICI, Schlumberger, Unilever and Zeneca. The academic centres involved were, the Universities of Bristol, Cambridge, Edinburgh and Imperial College, London.

Each of the papers published in this volume formed the focus for a discussion on that topic so that each subject was discussed in some depth by the participants.

Jean Proctor and Meg Staff have been tremendously helpful as secretaries at Bristol and Cambridge respectively throughout the project. Also, their help with the various meetings and with the production of this volume was invaluable. We thank them most warmly for their very able assistance.

We also thank both David Friday and Shirley Bolton of the DTI for their help and enthusiasm during the course of the project.

July 1997

Ronald Ottewill, Bristol
Adrian Rennie, Cambridge

THE DTI COLLOID TECHNOLOGY PROGRAMME

ALEX LIPS
DTI Programme Manager,
Unilever Research, Colworth House, Bedford,
Beds. MK44 1LQ, U. K.

1. Introduction

This volume describes the conference proceedings of a meeting on Concentrated Dispersions held at Bristol University, 14-16th April 1997, organised by the Colloid and Interface Science Group of the Faraday Division, Royal Society of Chemistry (RSC). The meeting was an open discussion forum intended as a final review and dissemination of the research carried out under a major UK collaborative research initiative known as the 'DTI Colloid Technology Programme'. By way of introduction it is helpful for the reader to have some understanding of the background to this recently completed initiative and its original aims.

'Colloid Technology' was launched in 1992 as a five year programme of pre-competitive research by academic and industrial research laboratories with a total research budget of £7.12M. Half the funding was provided by the UK Department of Trade and Industry (DTI) and the rest by a consortium of large UK industries through a combination of cash and in-house research. The collaborating academic groups were the University of Cambridge (mainly the Cavendish Laboratory but also Chemical Engineering, the Department of Applied Mathematics and Theoretical Physics (DAMTP) and the Melville Laboratory), Bristol University (School of Chemistry), Imperial College (Department of Chemical Engineering and Chemical Technology), and, increasingly during the course of the project, Edinburgh University (Department of Physics and Astronomy). The collaborating industrial research laboratories were Unilever Research (Colworth and Port Sunlight Laboratories), Schlumberger Cambridge Research, and, at the start, the then ICI Corporate Colloid Science Group. Subsequent to the de-merger of ICI in 1993/94, both the new ICI and Zeneca continued to contribute to the project, respectively from the Colloid & Rheology Unit at ICI Technology, Wilton and the Jeallot's Hill and the Alderley Edge laboratories in Zeneca. The effort breakdown is shown in Table 1.

R. H. Ottewill and A.R. Rennie (eds.), Modern Aspects of Colloidal Dispersions, 1–13.

TABLE 1. Financial Details

Collaborating Group	Research Effort	Cash Funding
Cambridge University	£ 3,000,000	
Bristol University	£ 1,000,000	
Imperial College	£ 400,000	
Edinburgh University	£ 140,000	
Project Management	£ 80,000	
Unilever	£ 1,100,000	£ 560,000
ICI/Zeneca	£ 900,000	£ 250,000
Schlumberger	£ 500,000	£ 250,000
DTI		£ 3,560,000

At the time of planning the project in 1990/91, and still today, many reasons were apparent for creating a strong alliance between academic and industrial colloids research and for proposing strategic government funding. 'Colloid Technology', linking colloid science, polymer science and process science/engineering has almost ubiquitous industrial relevance and is an essential enabling activity for a wide range of industries, large and small. In e.g. foods, pharmaceuticals, new materials design, catalyst utilisation, detergents, novel cosmetics, paints, fine chemicals, advanced oil recovery, new fuels etc, it is colloids expertise which has a pivotal role in facilitating product innovation, often providing a rich source of product innovation in its own right. Not surprisingly, therefore, industry has a strong need for high quality trained manpower in this area and for directed research on major scientific and technological challenges. It was both these requirements, and particularly the latter, which provided the case for the initiative and which shaped an ambitious research strategy for the consortium.

Several factors were especially recognised:

the commonality of challenging fundamental science needs between the collaborating industries and across a wider spectrum of UK industry

the complexity and diversity of colloidal phenomena encountered in industrial problems

the difficulty of some of the scientific challenges which could not be tackled by industry alone the need for generic research initiatives in 'classical' colloid science to be steered towards real commercial applicability

the distinctly multidisciplinary nature of colloid technology concerned with the need to bridge molecular understanding to materials behaviour

the increasing need to match other countries' investments (Germany, Japan) in research on colloidal systems

the small amount of research activity on colloidal dispersions in British Universities

an industrial demand for postgraduates trained in colloid science which could not be met by UK Universities

It was a deliberate choice to stimulate the development of colloidal physics and physical chemistry under this programme with a major emphasis on theory in the condensed matter physics of soft materials and supportive new measurement science. The priority for new science was a fundamental understanding of the nature of concentrated dispersions which generated a programme in three main areas: *Flow and Structure of Concentrated Dispersions, Composite Colloids, and Colloidal Forces.*

2.Flow and Structure of Concentrated Dispersions

2.1 NEW UNDERSTANDING

The industrial collaborators brought a wide range of challenges for the academic groups in measurement science, theoretical insights and simulation. Industrial colloids are typically multiphasic, concentrated disperse systems in the form of powders, pastes, dispersions, emulsions, structured liquids or foams. An understanding of the mechanical and flow behaviour of these materials is of paramount technological importance for the design of products for their intended 'processes of use' (e.g. spreading, dispersibility, lubrication, fracturing, eating etc.) and for their performance in unit operations during manufacture (e.g. pumping, emulsification, packing). The disperse phases can be solid particles, as for example in paints, detergent powders, drilling muds, filled polymers or the abrasive in toothpastes. Deformable disperse phases are frequently encountered in e.g. pharmaceutical and personal products (emulsions, creams), and a wide range of food products such as creams, spreads, and concentrated dispersions of granular starch or plant cells. Ice cream presents just one example in which both solid (ice, fat droplets) and deformable (air) inclusions are present.

It was our judgement that the flow behaviour of concentrated dispersions of even well-defined solid particle systems was far from adequately understood despite a long history of scientific endeavour. The challenges were not simply in new mathematical and computational capabilities or innovative experimentation but more fundamentally in the development of a proper understanding of the local physics of dynamic contacts between dispersed particles. This has been a major preoccupation for the group at the Cavendish. Work on a range of simulation techniques has been progressed by several groups involving Lattice Boltzman and dissipative particle dynamics methods at Edinburgh and Unilever, co-moving mesh techniques at the Cavendish, and simulations for anisotropic particles at Schlumberger and Cambridge in the Physico-Chemical Hydrodynamics group at DAMTP. Light scattering and small angle neutron scattering measurements of structure in flowing concentrated dispersions (Bristol and Cambridge), and experimental rheological work at ICI and Unilever, has provided a test bed for the new theories and simulation.

Much of this emphasis was prompted by our previous lack of understanding of the technologically important phenomenon of shear thickening (dilatancy) in concentrated dispersions. This is now understood in principle as a result of the theory/simulation work at the Cavendish and arises from 'soft potential' features between interacting particles. To illustrate, a hard core/soft corona simulation code is now available which for the first time can model measured flow curves for concentrated suspensions of starch granules and microgel particles. Dynamic, as opposed to just static, colloidal forces then govern the flow of concentrated dispersions at just below close packing. These 'lubrication forces' can be of polymeric or, more simply, electrostatic origin. At high shear, these operate on molecular length scales; rheology is then extremely system specific reflecting on the detailed lubrication interactions. The new insights represent an important scientific advance and a platform for future work in foams, high internal phase emulsions and soft solids with highly deformed particles. In such systems the colloids are above critical packing density and their behaviour in shear reflects a complex interplay of particle deformation and lubrication forces from soft coronas. Quite some progress has been achieved already in code development for concentrated systems of deformable particles/emulsions (Cavendish, DAMTP), interactively with experimental rheology at Bristol, Unilever and Zeneca. Modelling of dispersions of anisotropic particles and of deformable droplet dispersions was contributed by Schlumberger.

The colloidal physics of disordered materials has been another important theme for this programme with contributions from all the collaborating groups. Industry faces many scientific challenges related to both the design and manufacture of such structures. Examples where progress has been achieved include reactive powder mixtures (ICI), aggregate compaction in filter cakes (Schlumberger), metastable food colloids and cosmetics with instability triggered by shear in use (Unilever). Scientific advance was stimulated on several fronts. ICI evaluated fractal scaling predictions for the sedimentation and linear elastic behaviour of flocculated model dispersions. Bristol and Unilever advanced the understanding of kinetics of shear induced gelation in metastable concentrated dispersions. Experimental rheology at Bristol, Cambridge Chemical Engineering, ICI and Schlumberger provided detailed characterisation of constitutive properties of reversibly as well as strongly aggregated colloidal systems covering continuous shear rheology and viscoelasticity. These studies included phenomenological descriptions of yield-to-flow transitions. Characterisation of floc structure was undertaken by Bristol and Unilever. Supporting these experimental studies was a substantial effort in theory and simulation (Cavendish, Edinburgh, Schlumberger, Unilever) for flow curves, structure and connectivity in concentrated disordered systems.

Powder physics initiatives were successfully progressed at the Cavendish which embraced the search for unifying theories for powder like colloids and new measurement approaches for powder packing and high speed flow visualisation. Fluidisation, segregation and stress propagation in granular materials has seen considerable progress. From the industrial perspective, the chemomechanic behaviour (colloid structure coupled to chemical reactions) of e. g. cements and clay compacts was studied at Schlumberger, and ICI developed models for reactivity in pyrotechnic powders. Dense packing of colloids in concentrated dispersions was progressed in Bristol and ICI with emphasis on the densification of particulate gels and by Schlumberger in the compaction of filter cakes. Fundamental experimental studies on the packing of model mixed colloids at Bristol provided an important test for theories. Bristol also studied the phase behaviour of weakly aggregated concentrated model colloids including model mixtures. Edinburgh advanced new insights into the physics of kinetic pathways for phase separating polymer-colloid mixtures. An advance of wide scientific interest has been the detailed characterisation of particle dynamics in model concentrated dispersions which was made possible by the establishment of the technique of two-wavelength dynamic light scattering at Edinburgh.

2.2 EXPERIMENTAL DEVELOPMENT

There has been substantial investment in new techniques (ca. one third of the funding to the Universities) to enable novel studies to be undertaken in the microstructure of concentrated dispersions. Environmental scanning electron microscopy (ESEM) established at the Cavendish found wide use in problems (e.g. paints, cement, ice cream, cement) of the industrial members of the consortium and in generating new science on film drying and direct characterisation of the structural behaviour of wet colloids. Investment in electroacoustics and image analysis at Bristol, and in flow visualisation techniques at Cambridge, have similarly proved valuable. In Chemical Engineering at Cambridge we have developed rheo-optical methodology and a new rheometer based on multi-pass capillary flow which circumvents the complication of wall effects of conventional rheometers. Bristol have successfully exploited small angle neutron scattering to study structure in concentrated model colloids including those in flow. This was also studied in Cambridge based on the development of a 2-d small angle light scattering apparatus coupled to a shear cell. Unilever contributed with the development of new methodology based on small angle laser light scattering enabling new studies to be carried out on the characterisation of floc structure and kinetics of floc growth. Establishment of the technique of two-wavelength dynamic light scattering at Edinburgh was another important advance enabling fundamental studies of particle dynamics in concentrated dispersions and the characterisation of colloid structure in optically turbid media. State of the art ellipsometry at Cambridge has already been well exploited in studies of polymer adsorption and in generating new science on glassy behaviour in thin films.

3. Composite Colloids

Rheological characterisation of linear and non-linear viscoelastic behaviour of complex composite colloids was carried out in Chemical Engineering at Cambridge. New theoretical understanding has been generated at the Cavendish on the equilibrium and dynamic behaviour of liquid-crystal fluids containing colloidal inclusions. In particular, the role of topology has been shown to be important for the stability of fluid inclusions in liquid crystal phases, in principle, enabling the design of thermodynamically stable 'macro' emulsions, as was confirmed by experiment. ICI has studied the rheology of concentrated dispersions containing disssolved associating polymers. Unilever has characterised the self-assembly and rheology of complex milk protein systems and the mechanical behaviour of phase separated biopolymer colloids. The rheology of clay-polymer systems, as model drilling fluids, was investigated by Schlumberger. Studies on interaction between suspended particles and emulsion droplets in concentrated 'suspoemulsions' were contributed by Zeneca.

The formation of composite colloids in thin films from phase separating polymer mixtures was investigated at the Cavendish with support from ICI. ESEM and ellipsometry proved important for the characterisation of colloidal size domains in drying films which has enhanced our understanding of the physics of film formation from polymer latex systems.

4. Colloidal Forces

The programme sponsored two initiatives in direct measurement of colloidal forces by probe microscopy, both at the Chemical Engineering and Chemical

Technology Department of Imperial College. An apparatus was built to measure the deformability of individual deformable particles. It was shown useful for a wide range of particles including biological cells. The results obtained for model microgel particles provided an important input to the modelling of viscoelasticity and rheology of concentrated suspensions of deformable particles. The second investment was in modified atomic force microscopy (AFM) methodology which was succesfully exploited in studies of steric, electrostatic, and van der Waals' forces between colloid particles.

At Bristol, image analysis techniques have been used to investigate the interaction of macroscopic oil drops with a smooth solid surface. New information has been gained on wetting films, both oil-wetting and water-wetting, in such systems in the presence of surfactants.

Interactions of polymers at interfaces have been studied at Bristol and Cambridge. The Bristol group have concentrated on the measurement (neutron reflection and small angle scattering) and modelling of block copolymer adsorption at fluid-fluid interfaces. At the Cavendish, research was targeted at polymer dynamics in thin films on solid substrates. Using spectroscopic ellipsometry it was demonstrated that confinement of polymers to mesoscale dimensions can drastically affect dynamics. Neutron reflection was also used to study both adsorbed and grafted polymer layers.

Theory development has been concerned with polymer-mediated forces, at Cambridge and Edinburgh, directed primarily at the dynamics of polymer coats. We now have more direct evidence that this is crucial to many kinetic colloidal phenomena such as the rheology of rapidly sheared colloids outlined above. Also a more refined theory has been developed for polymer depletion effects.

5. Management of Programme

Co-ordination of a programme of this size and range of activities was not a trivial matter. Having first defined the overall plan an appropriate set of work plans known as tasks were developed. Several of these tasks were linked, and in many there was input from more than one of the collaborating groups. It was a deliberate decision to integrate the academic and complementary industrial research within the task structure and to appoint an academic staff member from the University with the largest commitment to the task to lead it. The task portfolio so derived is given below; for convenience this is clustered under Cambridge-, Bristol-, and Imperial College-led tasks.

5.1 CAMBRIDGE-LED TASKS

Theories for non-Newtonian flow in suspensions (E.J. Hinch)
Unifying theories for powder-like colloids (S.F. Edwards)

Theories for the mechanical and flow behaviour of concentrated dispersions (R.C. Ball)
Theories of flow-induced flocculation, metastability (M.E. Cates)
Dynamic properties of powders (J.M. Huntley and J.E. Field)
New approaches to rheological characterisation of colloidal dispersions (M.R. Mackley)
Rheological characterisation of gels, composite colloids and mesophases (M.R. Mackley)
Novel structural characterisation of concentrated colloids (A.M. Donald)
Structural characterisation of gels and composite colloids (A.R. Rennie and A.M. Donald)
Theories for phase equilibria and kinetics of composite colloids (M. Warner)
Densification of polymer latex dispersions (R.A.L. Jones)
Interfacial phenomena in polymer composites (R.A.L. Jones)
Theories for dynamics of adsorbed polymer layers (M.E. Cates)
Conformation of polymers at interfaces with inorganic materials (R.A L. Jones and A.R. Rennie)

5.2 BRISTOL-LED TASKS

Conformation of polymers at interfaces (T. Cosgrove)
Structure of model dispersions in defined flow fields (R.H. Ottewill and A.R. Rennie)
Rheology of model dispersions of deformable particles (J.W. Goodwin)
Shear-induced floc formation in concentrated dispersions (J.W. Goodwin)
Characterisation of particle packing (R.H. Ottewill, P. Bartlett and P.N Pusey (Edinburgh))
Heterosteric stabilisation in polymer solutions and melts (B. Vincent)
Thin liquid film studies (R.H. Ottewill)

5.3 IMPERIAL COLLEGE-LED TASKS

Direct measurement of long range forces between individual particles (P.F. Luckham)
Micro-deformability of individual particles (B.J. Briscoe)

The task structure needed only minor change during the course of the project and we succeeded in creating a culture which enabled each collaborating group to share a common interest in all the tasks and to have the opportunity to influence the direction of the total programme.

Overall co-ordination rested with the Programme Co-ordination Committee,

Prof. Sir Sam Edwards, Cambridge University (Chairman)
Prof. R.H. Ottewill, Bristol University
Prof. B.J. Briscoe, Imperial College
Dr. A. Lips, Unilever
Dr. R. Buscall, ICI
Dr. J. Stageman, Zeneca
Dr. G. Maitland, Schlumberger
A DTI Project Officer

who met at regular intervals to review specific targets, overall progress and budget issues. Technical progress reports and a programme of specialist scientific and technical meetings were the main mechanisms for reviewing and communicating progress.

We held major reviews restricted to members of the consortium at Unilever (Colworth) in 1993 and at Cambridge in 1994 and these were complemented by a series of smaller meetings on selected topics during the course of the project. In March 1995 at Bristol we held our first open meeting which was on Concentrated Dispersions organised, like the present conference, with the Colloid and Interface Science Group of the Faraday Division, RSC. The core of that major international meeting was the progress in our research at that time. There were two reasons for holding the meeting. The first was to ensure wider dissemination of some of our research findings which was a commitment made by the consortium to the DTI. The second was to provide an occasion to invite top international scientists from abroad to enable the DTI and ourselves to obtain international feed-back on the progress achieved. The DTI commissioned reports from a number of the invited international scientists who all expressed positive views, some highly complimentary, on the quality of our science and the aims of the project.

6. Progress against Objectives

The consortium had agreed a set of testable objectives with the DTI. It is useful here to summarise how we have performed against these.

6.1 DISSEMINATION

We have substantially exceeded our target of 100 publications. As summarised in this volume we have achieved in excess of 300 publications of which some have won awards. A substantial number of papers are also in press and in preparation. Our two open meetings, one a major international conference, were widely attended. Top international scientists have endorsed our research.

We have strengthened the long term position of colloid science in the UK with investment in the existing centres at Bristol and Imperial College and a new centre of colloid science at Cambridge. From 1930 to 1967 Cambridge had a distinguished Colloid Science Department led initially by Sir Eric Rideal and then by F.J.W. Roughton which trained several generations of colloid and polymer scientists. The Colloid Technology programme enabled the Tabor Laboratory to be established on the Cavendish site which brought to bear a wealth of research experience in condensed matter physics on colloidal problems. The ability of the Cavendish to attract high calibre graduates to colloid science has been a great boost of clear benefit to industry. As well as Cambridge, the Department of Physics at Edinburgh has developed a major centre for colloid physics and here the Colloid Technology Programme has been helpful partly through direct support but more importantly by providing an effective framework for linked research between the academic groups and also between academe and industrial research. The profile of academic colloid science has thus been raised significantly with both graduates and industry. Also, the network which was set up of individual scientists and academics will live on and be exploited for many years ahead.

6.2 TECHNICAL GOALS

The project has met its technical objectives:-

Generally exploitable theories and mathematical codes are now available for modelling the flow behaviour of concentrated colloidal dispersions. Here the work at the Cavendish represents a major breakthrough in scientific understanding and simulation capability. The project has generated a detailed understanding of the role of lubrication forces in dispersion rheology. We now know that this physics is a central aspect of the rheological behaviour of concentrated dispersions. Several initiatives in simulation have been successfully advanced together with an understanding of their complementarity and relative merit; we have also achieved progress in analytical treatments for some aspects of dispersion rheology. Experimental validation of theories has been actively pursued. Unilever has demonstrated the usefulness of these theories to its 'sheared hydrocolloid gel' technology for novel texturing of foods and skin cosmetics. ICI has found them helpful in the context of dispersing pigments and fillers in aqueous and other media.

For powders we have successfully progressed new theory and its validation through experiment.

Advance in theory/simulation for concentrated dispersions of deformable particles and non-spherical particles has also been substantial with new physics and code development. As a result of this work, the components are now in place to tackle the previously too difficult challenge of code development for the rheology of high internal phase colloids such as foams. Schlumberger has applied these new techniques to provide a sound basis for rheological models for both clay-based and emulsion-based drilling fluids.

The research in the physics of aggregated structures and the formation of such disordered materials has been of great value to the industrial partners in a number of product categories, e. g. food emulsions, drilling muds and paints providing a common framework of theory and new measurement approaches.

Though much still remains to be done there has been significant progress towards understanding aspects of stability, structure and rheology of composite colloidal systems. Experience was shared on many types of system reflecting the rich variety of industrial problems e.g. yield stress fluids with solid inclusions (drilling muds, toothpaste), surfactant mesophases with dispersed fluid phases (skin cosmetics, pharmaceuticals), structures with both solid and fluid dispersed phases (suspoemulsions, filter-cakes and ice cream). Theoretical work has elucidated not previously appreciated insights concerning emulsion stability in concentrated surfactant mesophases.

All the ambitions for new measurement science which this project has stimulated have been realised. Environmental scanning electron microscopy (ESEM) was the largest investment and has enabled the Cavendish to become one of the leading groups in this field. The consortium members have already derived considerable value from ESEM e.g. in the characterisation of drying of paints (ICI), in the setting of cements (Schlumberger), in characterisation of deposits on hair and skin and the structure of ice cream (Unilever). These and other successes have encouraged additional investment from another funding source. Electroacoustics, flow visualisation techniques and rheo-optical methodology have all been successfully commissioned and exploited in product-related problems, as has two-wavelength dynamic light scattering.

In the area of colloidal forces, our investment in building new apparatus has borne fruit enabling the micro-deformability of soft granular particles (starch, plant cells etc.) to be characterised and opening a wide range of opportunities for future research. A modified AFM apparatus has been successfully deployed in detailed measurements of colloidal forces and an application of image analysis has aided a study of colloid interactions in thin liquid films. New theory has been advanced for the dynamics of adsorbed polymer layers.

6.3 COMMERCIAL GOALS

The companies have benefited in several ways from this initiative and expect it to impact on their future commercial operations. We have demonstrated that systems as complex and diverse as 'mud, sludge and custard' can yield to deep scientific investigation, and share a common science in their mesoscale colloid physics. This science of 'soft materials', with its formidable challenges, needs strategic alliance between industry and academic research. The companies have been able to steer the academic research towards commercially important areas, and difficult generic scientific problems, which they could not have tackled on their own. The academics have benefited from programme continuity and resourcing at an appropriate scale.

All the research in this programme has been pre-competitive and the companies are still in a phase of assimilating some of the new science and concepts within their technologies. Nonetheless there are already indications that commercial advantage can be attributed in part to this project. For example, Unilever would not have been able to achieve rapid roll-out of a range of new food emulsion products without a strong science base in the colloid physics of disordered materials to which this project contributed in a major way. Also the science of shear thickening of concentrated dispersions is suggesting new approaches to formulating and processing food and cosmetic products with patentable opportunities. ICI has gained advantage in the rubber toughening of thermoplastics and in the formulation of a diverse range of products from delay fuses to surface coatings. For Schlumberger, the combination of new structural characterisation techniques and advances in our understanding of colloid-flocculated structures and colloidal compacts has stimulated and accelerated the development of a new generation of water-based drilling fluids poised to meet the technical and environmental challenges of oil-well construction over the next decade. New insight into the hydration mechanisms and structure of cement-based fluids is underpinning developments of more advanced well-bore construction materials. Zeneca will utilise the improved understanding of the rheology of concentrated particle plus soluble polymer systems in the design of specialist protective coatings and suspension formulations. The academic work to establish new techniques such as ESEM on a sound basis will directly encourage their use on problems of direct relevance to novel "effect" product design, e.g. stable rapidly redispersible solid formulations etc.

7. Concluding Remarks

All those involved in the project believe in its success and hope for its continuation in some form or other.

Industry has no shortage of major challenges for colloid scientists. The problem is not with the challenges, but with the organisation to meet these. Collaboration between industry and academic research will continue to be an essential need in the foreseeable future. We know that this view is not only held by the industries and academic institutes involved with this project but by a much larger range of UK industry and academe.

It is important to note, that our progamme, right from the start, was put to Government for support, not as a framework for linked research, but as a cohesive cluster of tightly targeted objectives with detailed analysis of research strategy involving both industry and academics.

Because we were able to define generic targets this has not restrained academic and industrial research; in fact, a substantial amount of very original research has resulted from the project.

Of course, it concerns us as a community of academic and industrial scientists how we can maintain our network and achieve an organisation providing continuity with the present type of approach which we believe to be an effective role model for meeting national strategic research needs.

8. Acknowledgements

Many people are owed a debt. I would like to thank especially Sir Geoffrey Allen and Sir Sam Edwards for their grand vision and crucial role in getting this initiative off the ground, and my colleague Peter Lillford for painting an even grander picture and pursuading me to take on the task of managing the programme. Richard Buscall, Geoff Maitland and John Stageman have managed the co-ordination of their companies' in-house research. I thank them warmly, and also Rodney Townsend, for their major inputs to the planning and execution of the overall programme and for ensuring our effective collaboration. Brian Kingsmill, David Friday, Maria Cody and Shirley Bolton from the DTI have always been a great help.

We are grateful to The Colloid and Interface Science Group of the Faraday Division for their assistance with our open meetings. Ron Ottewill, Adrian Rennie and Jean Proctor, organised the final meeting and spent many hours collating our inputs and shaping this volume.

Finally a special thanks to all the academics. You have been marvellously supportive in developing and sharing our vision, in nurturing a vibrant colloid science community, and in generating a wealth of technologically important new science.

DEFORMATION OF GEL PARTICLES

D.C. ANDREI, B.J. BRISCOE, P.F. LUCKHAM, D.R. WILLIAMS
Department of Chemical Engineering and Chemical Technology,
Imperial College of Science, Technology and Medicine,
Prince Consort Road, London SW7 2BY, UK.

This study describes an experimental procedure developed in order to quantify the mechanical and rheological properties of microscopic gel particles. The model material used was a cross-linked beaded form of dextran called "Sephadex" in a range of sizes between 60 and 600 μm diameter. Different cross-link densities for these hydrogel particles were supplied in an anhydrous form by Pharmacia, UK. A purpose built apparatus was constructed which allowed the compression of the individual particles between two parallel plates in two regimes: loading/unloading cycles and stress relaxation mode. A first order approximation of the Hertz theory for the quasi-elastic domain of low imposed strains onto the particles provided numerical solutions for the nominal zero strain elastic modulus. The material shows a remarkable change in the elastic modulus of the particles at the transition from the dry to the swollen state. Different restrictive aqueous swelling media were used and the subsequent mechanical responses of the particles analysed. The time-dependent behaviour of these gels and the related mechanisms are also presented. A special emphasis is placed upon the examination of the correlation between the stress relaxation behaviour and the migration of the liquid from the particles when they are subjected to compressive deformation.

1. Introduction

Gels form a large category of materials which mainly consist of a network structure which may or may be not polymeric in origin and significant quantities of imbibed liquid. An important class of gels according to their structure and the specific mechanisms of the gelation process is formed by chemically cross-linked polymeric gels [1]. Produced in chemical reactions like copolymerisation, polycondensation and vulcanisation of linear similar species, these polymeric gels usually exhibit a rubber-like consistency and have an extensive cross-linked structure surrounded by large quantities of entrained liquid [2-3]. As a consequence, the swelling and mechanical properties of these gels, are strongly dependent upon the interactions between the polymeric matrix and the solvent.

Numerous gels have microscopic dimensions and are dispersed in different liquid media (e.g. food products, pharmaceutical preparations, chromatographic columns, etc.). A specific value in studying single microscopic entities abstracted from these types of concentrated assemblies was shown recently [4-6]; the single particle deformation properties may be used to construct quantitative models for the rheology of the assemblies. To assess the mechanical properties of such systems, the experimental

R. H. Ottewill and A.R. Rennie (eds.), Modern Aspects of Colloidal Dispersions, 15–24.

method may make use of the conventional methods developed for macroscopic gel materials [7] as well as to accommodate the facilities specific to micro-manipulation techniques [8-9].

2. **Experimental procedures**

2.1 MATERIALS

The model material investigated was a cross-linked polysaccharide called "Sephadex" which is available in perfect spherical shapes (see Figure 1).

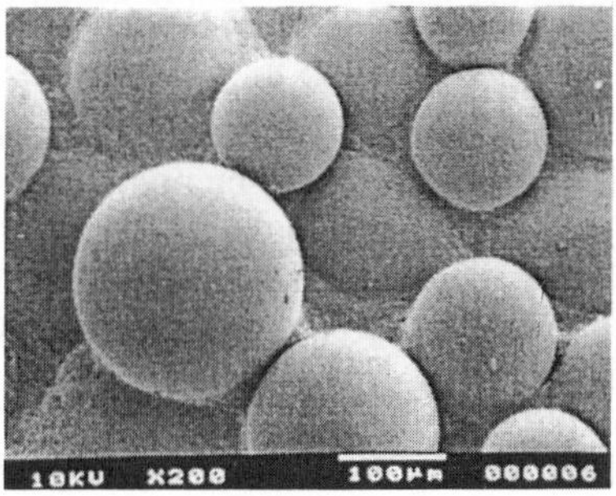

Figure 1 SEM micrograph of anhydrous "Sephadex" G50 particles (scale is shown).

Produced by Pharmacia as a gel filtration media, "Sephadex" particles possess a highly porous structure. The range of sizes of the pores are carefully controlled in the process of cross-linking and the matrix (dextran in this case) is chosen for its physical and chemical stability and inertness (lack of adsorptive properties). The pores of the gel matrix are usually filled in these applications with the liquid phase. In this study, "Sephadex" G50 and G25 were used for comparison purposes. G25 particles have a higher degree of cross-linking and implicitly a less porous structure. Figures 2 and 3 show the representation of the particles surface as detected using an AFM procedure. The particles show a pattern which seems more ordered in the case of G25 particles.

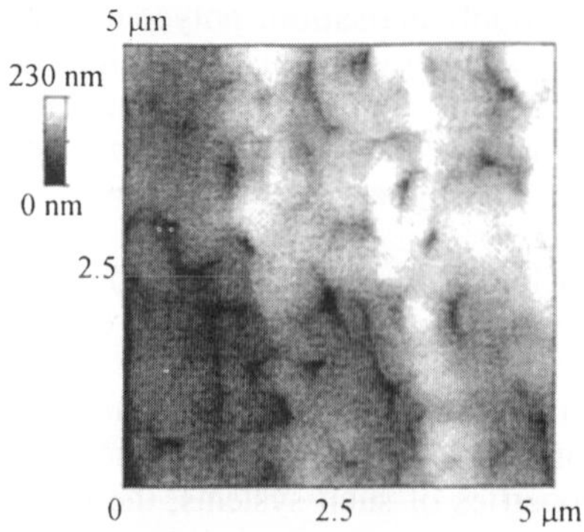

Figure 2 AFM micrograph of the surface of a "Sephadex" G50 particle.

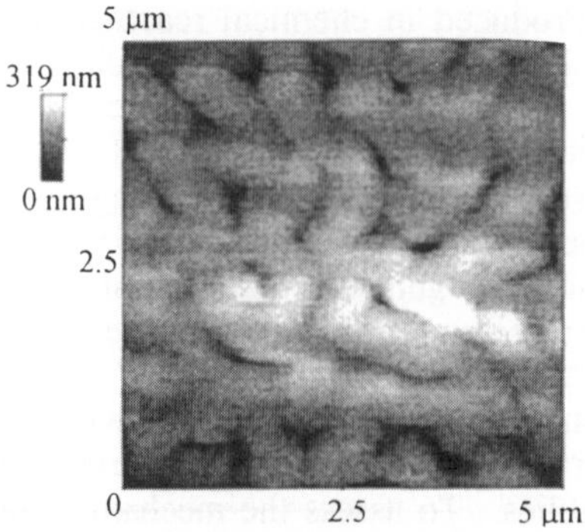

Figure 3 AFM micrograph of the surface of a "Sephadex" G25 particle.

At room temperature, these particles have a certain degree of moisture which may be detected by performing DSC analyses. Each set of particles was analysed twice. The first run consisted of heating from -40°C to 100°C (below the caramelisation temperature of ca. 120°C as found by Pharmacia) followed by an isothermal "hold" at this temperature for 15 minutes. A typical representation of the DSC curves obtained for the two calorimetric analyses is shown in Figure 4.

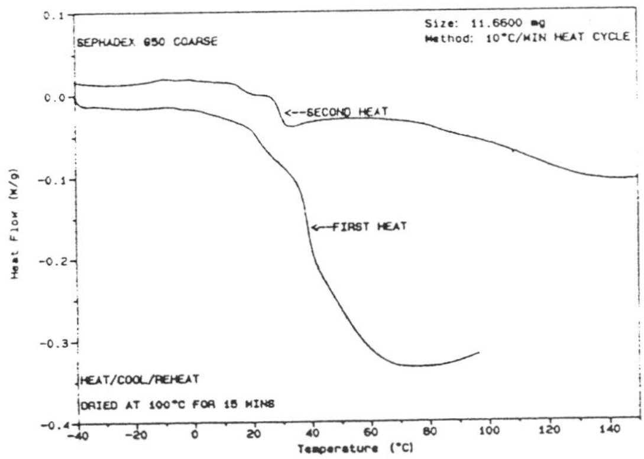

Figure 4 DSC curves for "Sephadex" G50 particles.

Due to the traces of water within the particles, the two curves which show only endothermic transitions differ greatly in shape. Moreover, the second curve which characterises the transitions of the polymer matrix alone contains three inflexion points in the ranges 10-20°C, 20-40°C and 50-90°C. These transitions may be interpreted using the micro-mechanical experiments performed on the particles.

2.2 APPARATUS

A detailed description of the developed micro-deformation apparatus is presented elsewhere [6]. Figure 5 presents a schematic representation of the most important components. The apparatus functions on the principle of the compression of the particle (3-Figure 5) between smooth parallel plates (between a polished brass platen (10-Figure 5) and a glass microscope slide (4-Figure 5)). A representation of the performance of the alignment system developed is shown in Figure 6.

The parameters measured continuously throughout the experiments are the responsive force F (measured with the cantilever transducer (11-Figure 5) of 10^{-5} N resolution) and the applied displacement h (monitored by the computer with a resolution of 40 nm) (see Figure 7). During the experiments, the particle is fully immersed in the media of interest (e.g. pure water, polymer solutions, aqueous alcoholic solutions). In all these cases, the components of the forces due to surface tensions, buoyancy or lateral drag effects were estimated or measured and found to be negligible in comparison with the responsive forces due to the particle deformation. Previously, it was found that the mechanical properties may vary for the same cross-link density from particle to particle [6]; therefore it is desirable that one single entity be studied in the course of a particular experiment (e.g. when studying the influence of the swelling ratio upon the modulus).

It is also known that the history of gel samples is important and more specifically the system reversibility with respect to modulus is not as good as to the degree of swelling [10]. Hence, the samples were subjected to a continuous variation of the imposed conditions (e.g. water activity of the medium) avoiding their complete or partial drying during the experiments. Visualisation of the particles during manipulation and deformation ensures that the particles were not damaged during the deformation.

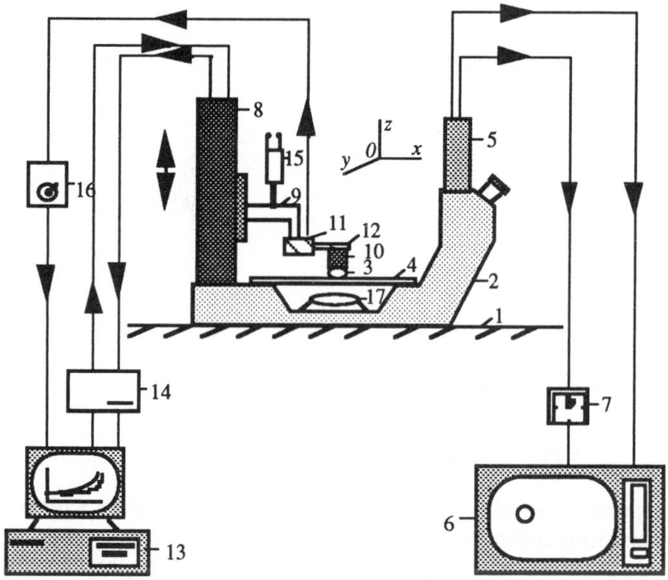

Figure 5 Schematic representation of the micro-deformation apparatus
1-antivibrational bench; 2-microscope; 3-particle; 4-microscope slide; 5-video camera; 6-video monitor; 7-time generator; 8-translational stage; 9-glass beam; 10-platen; 11-force transducer; 12-cantilever beam; 13-computer; 14-microstep driver; 15-position sensor; 16-transducer amplifier; 17-objective.

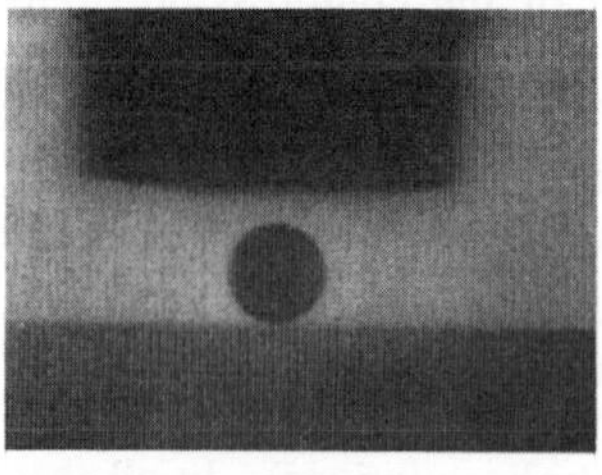

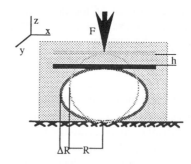

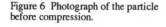

Figure 6 Photograph of the particle before compression.

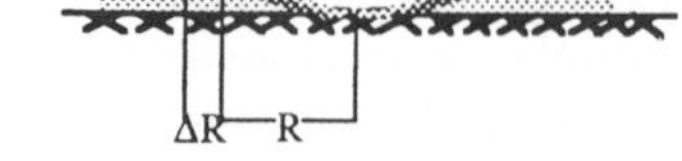

Figure 7 Schematic representation of the particle in compression
F-force; *h*-applied displacement; *R*-radius; *ΔR*-central expansion.

2.3 CURVE FITTING METHODS

Purpose written software was used to analyse the data from the two types of experiments: loading/unloading and stress relaxation. Schematic representations of the typical responses of "Sephadex" particles under the imposing of loading/unloading and stress relaxation deformations are represented in Figures 8 and 9. From the loading part of the compliance curves, a reduced elastic modulus may be computed using the Hertz theory for a smooth spherical surface in a non-adhesive contact with two flat surfaces:

$$F = \left[\frac{4}{3} \cdot \frac{R^{1/2}}{2^{3/2}} \cdot \frac{E}{1-\nu^2} \right] h^{3/2} \qquad (1)$$

where R is the initial radius of the sphere; E is the Young's modulus of the sphere; $E^* = E/(1-\nu^2)$ is the reduced elastic modulus of the sphere and ν is the appropriate choice for the Poisson's ratio.

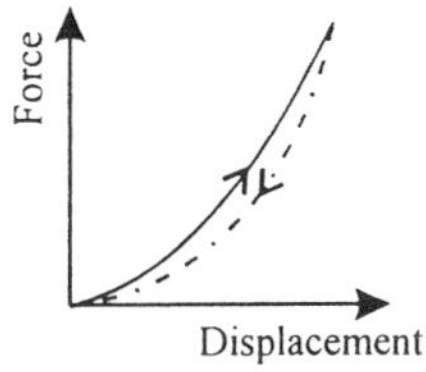

Figure 8 Typical response of a "Sephadex" particle during a loading/unloading test.

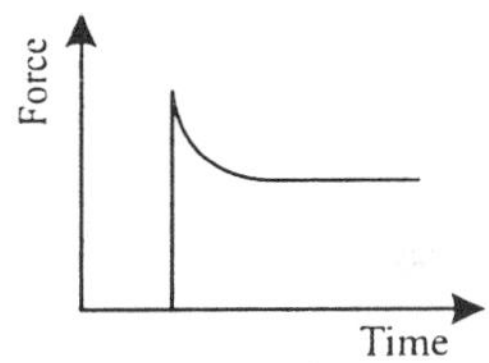

Figure 9 Typical response of a "Sephadex" particle during a stress relaxation test.

The proportionality between the force and the power "3/2" of the applied displacement was found valid up to 10% deformation. This regime corresponds to that of elastic deformations where time dependent effects are negligible. For higher strains, the material appears to show the phenomenon of strain hardening. Also, at this level, water transport phenomena from inside the particles may show a time dependent behaviour, i.e. a viscoelastic behaviour. Nevertheless, when the experiments are performed at a high velocity, quasi-elastic regimes may be obtained and theories specific to elastomeric materials successfully applied [11]. In these circumstances, stress relaxation experiments may provide information about the extent of the material flow and combined with the corresponding measurement of the particles dimensions may provide an estimate of the Poisson's ratio of the particles. An analytical relationship which provides a numerical solution for the Poisson's ratio of an elastomeric sphere compressed between two parallel plates up to 40% strain is given by:

$$\nu(\Delta R / R) = 81.163(\Delta R / R)^3 - 38.535(\Delta R / R)^2 + 6.989(\Delta R / R) + 0.04242 \qquad (2)$$

for $(\Delta R / R \geq 0.0485)$ where $(\Delta R / R)$ represents the dimensionless central lateral expansion of the sphere (see Figure 9).

3. Results and discussion

3.1 MICRO-MECHANICAL EXPERIMENTS OF THE PARTICLES IN THE DRY STATE

At room temperature and exposed to ambient humidity, "Sephadex" particles show a strong resistance to the applied deformation. Although the apparatus was designed to study soft particles and hence the machine compliance is sensed as a large factor of the response of the apparent particle deformation, an estimate of the modulus of the particle may be approximated. Hertz theory applied for ca. 5% effective strain in case of "Sephadex" particles irrespective of cross-link density provided an estimate for the reduced elastic modulus of the order of 10^9 Pa; this value is characteristic of glassy polymers. At room temperature, "Sephadex" particles are significantly below the glass transition temperature. Therefore, in the DSC curves, the domain which may be attributed to the transition between glassy to the rubbery state lies between 50°C and 90°C. The exact values for the inflection points for these transitions were 82°C and 81°C for "Sephadex" G50 and G25, respectively. The other transitions are believed to correspond to other "hindered" motions of the molecules, but the processes involved remain unclear. It may be that there is local ordering and the presence of crystalline domains. Dry "Sephadex" particles are not transparent. Also it is known that other polysaccharides such as starch have a partially crystalline structure [12].

3.2 MICRO-MECHANICAL EXPERIMENTS ON THE PARTICLES IN THE SWOLLEN STATE

On immersion in a "good" solvent such as water, "Sephadex" particles swell with the degree of swelling depending upon their cross-link density. The particles also become transparent. For a set of 10 particles, G50 and G25 coarse, the equilibrium swelling ratios defined as the ratios between the particle diameters in the swollen and the dry states were measured as 1.99 ± 0.03 and 1.60 ± 0.03, respectively. The slight variation in the value of the swelling ratio may be attributed mainly to the variations in their polymeric matrix cross-link densities. The difference in the cross-link densities was reflected in the values measured for the reduced elastic moduli; mean values of 2.2×10^5Pa and 1.7×10^6Pa were measured for the specimens mentioned above, respectively.

Stress relaxation experiments may generate, in conjunction with analyses of the geometry of deformation, useful information about the degree of compressibility of the polymeric matrix when it is swollen with liquid. Table 1 presents the variation of the normalised reaction force F/F_0 (where F is the force at time t and F_0 is the force at $t=0$) with time at a constant imposed 40% strain for a fully swollen "Sephadex" G50 particle, 418 μm diameter. The computed values of the Poisson's ratio for the particle using equation (2) are also presented. At the initial time of the experiment, the value of the Poisson's ratio was very close to 0.5 which is the value anticipated for incompressible materials such as rubbers. The decrease of the Poisson's ratio with time suggests that a migration of the liquid from the particle occurs during the deformation.

Table 1 Time dependent characteristics for a "Sephadex" G50 particle, 418 μm diameter.

t/s	4	10	15	30	60	90	120	150	200	400	600	900
F/F_0	.95	.90	.88	.82	.75	.72	.70	.69	.68	.67	.66	.66
ν	.487 ± .009	.487 ± .009	.478 ± .009	.478 ± .009	.454 ± .012	.440 ± .014	.440 ± .014	.424 ± .016	.424 ± .016	.424 ± .016	.424 ± .016	.424 ± .016

A characteristic time, τ, for the mass transport of water for "Sephadex" particles can be calculated using the relationship formulated by Monke *et al.* [13] as:

$$\tau = \frac{\left(d_p^{\infty}\right)^2}{4\pi^2 D} \tag{3}$$

where d_p^{∞} is the final radius of the particle in its fully state and D is the diffusion coefficient of the gel. This parameter was determined experimentally for "Sephadex" G75 particles by the same authors as 6.3×10^{-11} m^2/s. Using equation (3), a characteristic time for the mass transport of water for a "Sephadex" particle, 418 μm diameter is determined as ca. 70 s. This duration corresponds, in general terms, with the time associated with the significant decreases in the reaction forces and in the Poisson's ratio. Hence, the stress relaxation phenomena displayed by swollen "Sephadex" particles may be associated with the corresponding loss of the imbibed solvent into the ambient medium. A small amount of relaxation due to the presence of trapped entanglements which are specific to elastomeric networks may also be present.

The degree of swelling of "Sephadex" particles was controlled by dispersing different chemicals into the aqueous medium. Polymer solutions of dextran T2000 (provided by Pharmacia) of concentration of up to 12% weight were used. A maximum decrease of the swelling ratio was obtained for the G50 particles which was from ca. 1.99 down to a value of ca. 1.67. The mechanism of the restricting of the swelling has been attributed to the osmotic pressure induced in the system by the presence of dextran. This solute has a very large molecular weight (2,000,000) and therefore probably does not penetrate into the gel matrices. Nevertheless, no significant changes in the measured modulus were observed. It is believed that the high water content of these gels is not sufficiently altered to change the mechanical properties of the particles.

Water/n-propanol mixtures were found to be an effective medium for restricting the swelling of "Sephadex" particles; the lowering of the water activity in the mixtures contributes to the variation in the swelling of the particles. Figure 10 illustrates a "Sephadex" G50 particle in three different states: (a) dry state; (b) particle immersed in 15% vol. water/n-propanol solution and (c) extensively swollen particle. Due to the porous structure or semicrystalline nature of the particle, the dry state is characterised by an opaque appearance and also by a high reduced modulus, as mentioned previously.

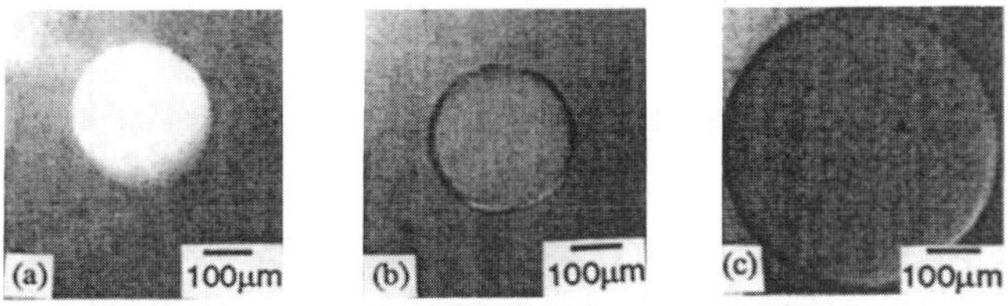

Figure 10 Photograph of a G50 particle on immersion in : (a) air; (b) 15% and (c) 80% vol. water/n-propanol solutions.

On immersion in the 15% vol. water/n-propanol solution, the particle's degree of opacity decreases and the measured reduced elastic modulus drops from ca. 10^9 Pa up to a value of ca. 5.7×10^5 Pa. The measured swelling ratio was ca. 1.02. This value corresponds to ca. 4.5% vol. of water inside the particle. The plasticisation of the amorphous matrix may have disrupted the ordered (crystalline) regions and/or closed the internal pore structure as a result of the swelling process. Upon increasing the water concentration of the environment to ca. 80% vol. water/n-propanol which corresponds to a swelling ratio of ca. 1.95, the particle becomes quite transparent and the modulus attains a monotonic value of ca. 1.7×10^5 Pa.

Figure 11 shows the variation in the shear modulus with increase in the swelling ratio for this "Sephadex" G50 particle. The shear modulus was calculated using the general relationship between moduli:

$$E = 2G(1 + \nu) \tag{4}$$

The choice for Poisson's ratio was 0.5 and corresponds to the short time quasi-elastic regime of the deformation.

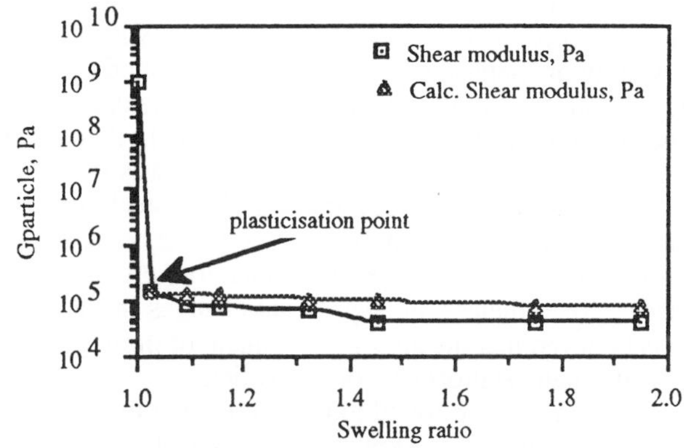

Figure 11 The variation of the shear modulus of a "Sephadex" G50 particle with the swelling ratio, ◻. The theoretical curve, Δ, is calculated using the theory of rubber elasticity.

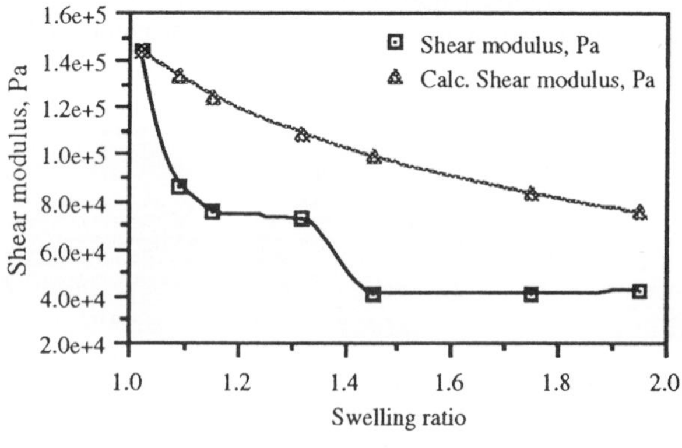

Figure 12 Detailed representation of the experimental-theoretical comparison for the plasticised domain.

If we assume that the particle in 15% vol. water/n-propanol (ca. 4.5% water volume fraction in the particle) corresponds to the plasticisation point of the particle (where the glass transition temperature is at the value of the ambient temperature), then conventional theories of rubber elasticity [14] may be used to interpret the mechanical properties of the particles. Figure 11 shows also the theoretical curve computed for the shear modulus by using the general relationship:

$$G' = G \cdot (1/s) \tag{5}$$

where G, G' are the shear moduli of the particle in the unswollen and swollen states, respectively and s is the swelling ratio as defined previously. The experimental curve shows a more pronounced decrease in the modulus than the theoretical one. The difference may be attributed to the arbitrary choice of the plasticisation point; Figure 7(b) also shows that the material of the particle is not highly homogeneous. This decrease in modulus obtained by the increase in the liquid content of the "Sephadex" G50 particles varies for different cross-link densities. For example, the modulus of the G50 particles which have a swelling ratio of ca. 1.99 was found to be ca. 8 times lower than that of G25 particles with a swelling ratio of ca. 1.6. Figure 12 does not show such a variation which suggests that the content of liquid alone cannot explain the mechanical behaviour of the particles; the response of polymeric matrix has also an important role.

4. Conclusions

An apparatus designed for the compression of microscopic gel particles (size range between 60 and 600 μm diameter), between parallel plates has been developed in order to quantify the intrinsic mechanical properties of the particles. Quasi-static compressive and stress relaxation deformations have been applied to the particles.

"Sephadex" particles have proven to be a model material for the compression studies due to their high degree of sphericity. The anhydrous particles appeared to be in a glassy state at 25°C, which is characterised by a comparatively high reduced elastic modulus (of the order of 10^9 Pa); at this stage the particles are opaque. Upon the immersion of the particles in pure water or aqueous solutions, a rapid plasticisation has been found to occur; all the particles become transparent. This significantly decreases the particles reduced elastic modulus to values of ca. 10^5-10^6 Pa; the glass transition temperature of the swollen particles was decreased below the ambient temperature. An initial plasticisation point has been estimated for the G50 particles when they are immersed in a 15% vol. of water/n-propanol solution which corresponds to a volume fraction of water of ca. 0.045. After reaching the rubbery state, the particles were found to effectively obey a relationship derived from the theory of rubber elasticity [14] when the degree of swelling is subsequently increased although the predictions are not in quantitative agreement with the experiment. The modulus of the particles depends upon the extent of the water comprising the granules as well as their polymeric matrices. The response of "Sephadex" particles under an imposed deformation depends quite markedly upon the time scale of the experiments and may be associated with the migration of water into and out of the polymer matrix.

5. Acknowledgements

The authors wish to thank the DTI Colloid Technology Research Programme for the provision of a research studentship for D.C.A. and a fellowship for D.R.W. We also thank Mr. Anders Meurk for the AFM photographs.

6. References

1. Flory, P.J. (1974) *Disc. Faraday Chem. Soc.*, **57**, 7.
2. Djabourov, M. (1991) *Polym. Intern.*, **25**, 135.
3. Osada, Y. and Ross-Murphy, S. (1993) *Scientific American*, **268**(5), 42.
4. Evans, I.D. and Lips, A. (1990) *J. Chem. Soc. Faraday Trans.*, **86**(20), 3413.
5. Patel, S.K., Rodriguez, F. and Cohen, C. (1989) *Polymer*, **30**, 2198.
6. Andrei, D.C., Briscoe, B.J., Luckham, P.F., Williams, D.R. (1996) *J. Chim. Physique*, **93**(5), 960.
7. Zrinyi, M. and Horkay, F. (1987) *Polymer*, **28**, 1139.
8. Duszyk, M., Schwab III, B., Zahalak, G.I., Qian, H. and Elson, E.L. (1989) *Biophys. J.*, **55**, 683.
9. Zahalak, G.I., McConnaughey, W.B. and Elson, E.L. (1990) *J. Biomech. Eng.*, **112**, 283.
10. Anbergen, U. and Oppermann, W. (1990) *Polymer*, **31**, 1854.
11. Briscoe, B.J., Liu, K.K, Williams, D.R. (1996) *J. Colloid and Interface Sci.*, (submitted).
12. Trommsdorff, U. and Tomka, I. (1995) *Macromolecules*, **28**, 6128.
13. Monke, K., Velayudhan, A., Ladisch, M.R. (1990) *Biotechnol. Prog.*, **6**, 376.
14. Treloar, L.R.G. (1975) *The physics of rubber elasticity*, Clarendon Press, Oxford.

THE RHEOLOGY OF AQUEOUS MICROGEL DISPERSIONS

J. W. GOODWIN & T. J. HUANG[1]
School of Chemistry, University of Bristol,Cantock's Close
Bristol BS8 1TS, UK
[1]Interfacial Dynamics Corporation, 17300 upper Boones Ferry Road.,
Portland, Oregon 97224, USA

Aqueous dispersions of particles of polyN-isopropylacrylamide, cross-linked with bis NN' acrylamide have been prepared and characterised. Electron and optical microscopy as well as dynamic light scattering have been utilised to measure the temperature dependence of size. The polymer has an LCST of ~32 °C and so the particles were highly swollen at 15 °C and in a compact form at 45 °C. Capillary viscometry was used to study the effect of temperature and electrolyte on the viscosity of dilute solutions. In most cases the intrinsic viscosity was less than the value for hard spheres indicating that flow was occurring inside the particles. Oscillatory rheometry on concentrated dispersions indicated that the systems behaved as elastic solids and the high frequency elasticity was recorded. A model is suggested in which the compliance of the swollen gel particles reduces the electrostatic repulsion by allowing the particles to deform. The model was capable of giving a good fit to the experimental observations.

1. Introduction

The rheological behaviour of concentrated dispersions of deformable colloidal particles is of major interest in studies in areas such as food processing, the manufacture of pharmaceutical / personal care products, coatings and drilling fluids for example. A number of experimental and theoretical studies have been made over recent years [1-15] but, due to the wide range of systems studied, a unified theoretical understanding is still lacking. Some of the work has been devoted to concentrated emulsions or foams [1-3] with volume fractions approaching unity. Other work has been on hard particles with relatively thick steric stabiliser layers [4-9], offering a degree of softness, whilst some work has dealt with non-aqueous microgel systems [8, 9].

A model microgel system that can be prepared in an aqueous environment in a monodisperse form is a cross-linked N-propylacrylamide particle dispersion [16-20]. This is the system that was used for the work described in this paper. The polymer has a lower critical solution temperature (LCST) of ~32 °C. It may be prepared as a latex using the usual range of water-soluble free radical initiators at temperatures around 70 °C. This provides a particulate system in which the swelling, and therefore the

R. H. Ottewill and A.R. Rennie (eds.), Modern Aspects of Colloidal Dispersions, 25–39.

hardness, can be varied simply by varying the temperature. Hence at room temperature the cross-linked particles form a highly swollen gel. The initiator fragments will be in the outer layer of the corona as they would have been located on the outer surface of the particle during polymerisation. The mutual repulsion of the charged groups can be expected to enhance the swelling of the outer corona and hence contribute to the softness by both intraparticle interactions and interparticle interactions. This polyelectrolyte nature of the outer layer has some parallels with protein stabilised colloidal dispersions.

In this paper we discuss some of the problems of characterising the hydrodynamic size of these systems through dynamic light scattering and capillary viscometry on dilute dispersions. Following this we describe experiments on concentrated dispersions which show viscoelastic behaviour and indicate how some of the results may be analysed.

2. Experimental

2.1 PREPARATION OF THE MICROGEL DISPERSIONS.

The preparations were carried out according to the method of Pelton and Chibante (16) and a brief outline is given below.

2.1.1 *Materials.*

The N,N'-methylene bisacrylamide (BA) cross-linking agent was supplied as an Electran Grade reagent from BDH in a 2% w/v solution, ammonium persulphate (APS) was AnalaR Grade from BDH and the water used was prepared from a Purite reverse osmosis system which included a de-ionising and filtration system to produce a quality at least equal to double distilled water. The monomer, N-isopropylacrylamide (NIPAM), was obtained from Eastman Kodak and purified as follows. The NIPAM was mixed with hexane and diethyl ether was added dropwise until complete dissolution was obtained. Activated charcoal was used to remove any yellow colouration, the solution was cooled and the white NIPAM crystals were filtered off and dried under vacuum.

2.1.2 *Polymerisation.*

The NIPAM was polymerised in the same manner as that used for the emulsifier-free polymerisation of styrene (16,21). The cross-linking agent BA was added to give a composition of 10%. A typical recipe is :-

NIPAM	100g
BA (2%soln)	500cm^3
APS	5g
Water	4400cm^3
Temperature	80 °C

At the end of the reaction, the latex was filtered hot through glass wool and immediately centrifuged at 12,000 r.p.m. in order to remove as much soluble polymer as possible. The latex was then redispersed in de-ionised water and dialysed until the conductivity of the dialysate was constant. at ~9×10^{-5} Sm^{-1}.

3. Characterisation of Microgel Particles

3.1 MICROSCOPY.

Particle sizing by transmission electron microscopy (TEM) was carried out using an Hitachi HS 7S electron microscope. The samples were prepared from dilute dispersions at room temperature by drying down onto a collodion covered grid and shadowing with Chromium at an angle of 20°. The diameter and height (from the shadow length) were measured using at least 500 particles. The mean radius, r_p, and height, h_p, were calculated as r_p = 600 ± 20 nm; h_p = 150 ± 20 nm, which corresponds to an equivalent spherical radius of 189 ± 10 nm for the dried particles. When the particles were prepared from dispersions in 0.5M sodium chloride solutions values of r_p = 200 ± nm and h_p = 150 ± nm giving an equivalent spherical radius of 182 ± 10nm clearly demonstrating that high concentrations of sodium chloride cause the particles to shrink back to almost hard particles.

Optical microscopy on aqueous dispersions at room temperature (20 °C) was carried out using a Nikon Microphot-SA microscope with a Microflex HFX-DX attachment with a FX-35DX camera. Brownian motion prevented very sharp images from being obtained, however sufficient quality was available to give a diameter in water of 0.5_7 μm and a radius in 0.5M sodium chloride solution of 0.4_8 μm.

3.2 DYNAMIC LIGHT SCATTERING.

Dynamic light scattering measurements were carried out using a Malvern instrument at various temperatures and sodium chloride concentrations. the results are plotted in Figure 1. In water and in dilute electrolyte at temperatures of 45 °C the particle radius has shrunk to ~208 nm. At room temperature and below the particle radii expand to more than twice this value. Here the particles are swollen until they contain ~90% solvent. There is a trend to larger particle size as the salt concentration is reduced which is indicative of the firstly the effect of electrostatic interactions on the swelling but secondly the change in the LCST with electrolyte. The rapid increase in diameter of the particles at 0.5M electrolyte at temperatures >25 °C is indicative of the electrostatic nature of the colloid stability of the particles at high temperatures. Comparison with the electron microscopy data indicates that there is still a significant amount of solvent associated with the particles even in their collapsed state.

3.3. PARTICLE CHARGE.

The surface charge of the particles was determined by conductometric titrations of aliquots of the dispersions after all the surface groups had been converted into the acid

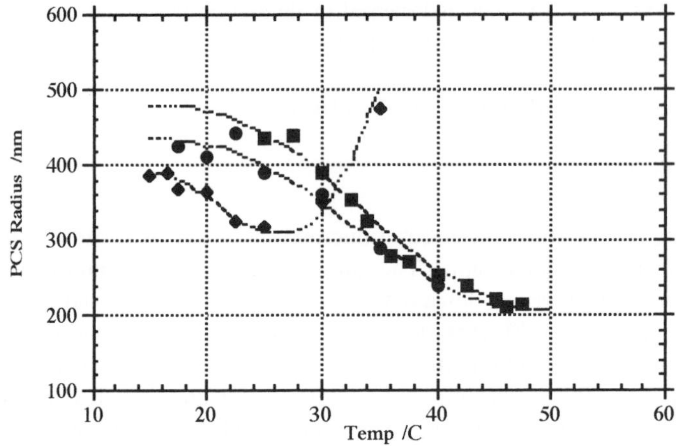

Figure 1. DLS data on poly-NIPAM particles as a function of temperature at various sodium chloride concentrations. The lines are to guide the eye.
■ Water; ● $5x10^{-3}$ M; ◆ $5x10^{-1}$ M Sodium Chloride.

form by treatment with a mixed-bed ion-exchange resin. The result gave a strong acid surface with an equivalence point at 7.5 μeq/g polymer. The density of the polymer in water was found to be $1.17x10^{3}$ kg m^{-3} using a Paar DMA60 Density meter. The surface charge densities were calculated as a function of temperature as illustrated in Figure 2

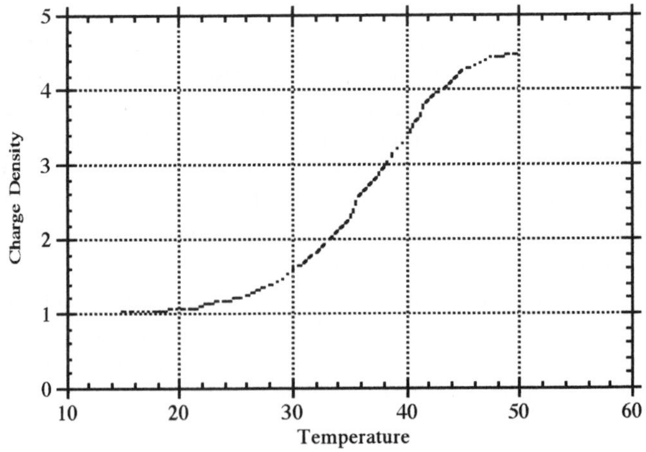

Figure 2. Surface charge density $\mu C\ m^{-2}$ as a function of temperature in $5x10^{-3}$ M sodium chloride for poly-NIPAM microgel particles.

where the surface charge density increases from 1μC m^{-2} to 4.5 μC m^{-2} as the temperature is increased from 15 °C to 50 °C.

3.4 CAPILLARY VISCOMETRY

Capillary viscometry measurements were carried out using a Canon-Fenske capillary viscometer. A good quality thermostat bath was used to control the temperature in the range was 15 °C to 45 °C. Electrolyte concentrations of 0, 0.5M and $5x10^{-3}$ M with respect to sodium chloride were studied. The microgels dispersions were made up by weight from a stock dispersion the concentration of which had been determined by dry weight measurements. As a result the weight fractions, w_p, of the measured samples were known accurately.

Now the volume fraction, φ_h, of the dispersions with the microgels in an hydrated state can be calculated from :-

$$\varphi_h = \left(\frac{a_h}{a}\right)^3 / \left(1 + \frac{\rho_p}{\rho_o}\left[\frac{1}{w_p} - 1\right]\right) \quad (1)$$

Here ρ is density with the subscript p indicating polymer and o water at the appropriate temperature and the radius of the collapsed particle is a. The hydrodynamic radius was taken from the DLS estimates and is given the symbol a_h. The experimental data are shown in Figures 3a-3c in which the relative viscosities are plotted as a function of volume fraction calculated using Equation 1.

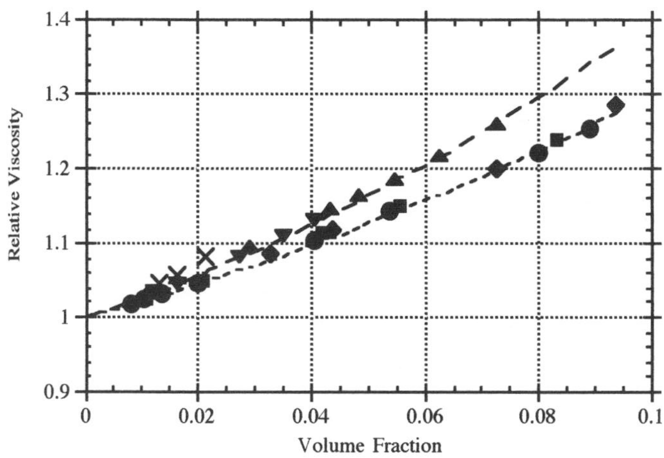

Figure 3a. Relative viscosity as a function of volume fraction of microgel dispersions in water. Experimental points and fitted curves.
■ 15 °C; ● 20 °C; ◆ 25 °C; ▲ 30 °C; ▼ 35 °C; × 40 °C.
.... $k_1 = 2.1, k_2 = 9$; – – – – $k_1 = 2.6, k_2 = 14$

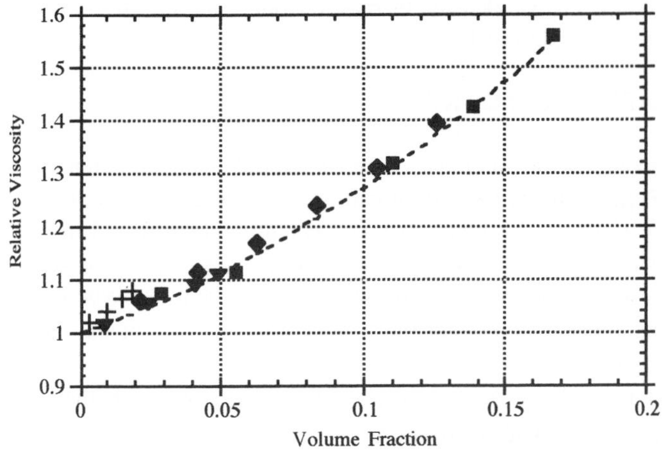

Figure 3b. Relative viscosity as a function of volume fraction of microgel dispersions in $5x10^{-3}$ M NaCl. Experimental points and fitted curves.
■ 15 °C; ◆ 25 °C; ▼ 35 °C; + 45 °C.
········ $k_1 = 1.8, k_2 = 9$; – – – – $k_1 = 2.1, k_2 = 9$

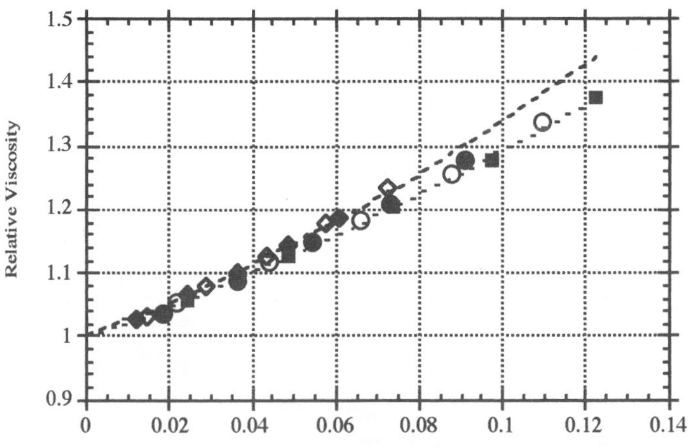

Figure 3c. Relative viscosity as a function of volume fraction of microgel dispersions in $5x10^{-1}$ M NaCl. Experimental points and fitted curves.
■ 15 °C; ○ 17.5 °C; ● 20 °C; ◇ 22.5 °C; ◆ 25 °C; ▲ 30 °C;
········ $k_1 = 2.3, k_2 = 6.25$; – – – – $k_1 = 2.5, k_2 = 9$.

The relative viscosity versus hydrodynamic volume fraction data were fitted to an equation of the form :-

$$\eta_r = 1 + k_1\varphi_h + k_2\varphi_h^2 \qquad (2)$$

and the values of the coefficients are given in the legends to the figures. Unfortunately the data at 40 and 45 °C were over to narrow a range of concentration but both sets seem to show a rather high intrinsic viscosity (k_1=[η]) with values >2.5 although the uncertainties are rather large. The remainder of the data was obtained over a much wider range and this enabled reasonable fits to be obtained and the values of the coefficients are given in Table 1 below. Even when water was used for dilution so no sodium chloride was added there will still be a finite concentration of ions present and so an electrolyte concentration of 10^{-5} M was taken as a representative value.

An alternative approach to the fitting procedure would have been to have plotted the relative viscosities versus the volume fraction and to have used the intrinsic viscosities so obtained to estimate the extent of the swelling. This procedure was utilised by Buscall [9] when analysing the data of Wolfe and Scopazzi [8]. The problem is that all changes in the intrinsic viscosity are then assigned to changes in particle volume. Although in some situations this may be a safe assumption, the additional size information gathered in this work meant that the intrinsic viscosity could be examined more fully.

TABLE 1. Coefficients of Equation 2 from plots of relative viscosity as a function of volume fraction.

NaCl Concentration (/M)	[η] = k_1		k_2	
	low T	high T	low T	high T
5×10^{-1}	2.3	2.5	6.3	9
5×10^{-3}	1.8	2.1	9	9
$\sim10^{-5}$	2.1	2.6	9	14

At low temperatures, i.e. at temperatures well below the LCST, the data at each electrolyte concentration was fitted by a single curve. With the particles in the most swollen state, the intrinsic viscosity at each electrolyte concentration was found to be less than 2.5, the hard sphere value [22]. The particles were charged and an increase of the intrinsic viscosity might have been expected, however, their diameters were large and their charge densities were low and so the primary electroviscous effect would have been small in each case. For example, calculation using the analysis of Booth [23] for an electrolyte concentration of a 10^{-5} M 1:1 electrolyte shows that the intrinsic viscosity would only be enhanced by 1% and so the primary electroviscous effect can be neglected here. Taylor (24) and Edwards and Schwartz [10-12] have considered the problem of deformable particles surrounded by a membrane. Taylor was studying the intrinsic viscosity of emulsion droplets and gave his well known result in terms of the viscosity of the continuous phase, η_o, the disperse phase, η_d:-

$$\eta_r = 1 + 2.5\frac{0.4\eta_o + \eta_d}{\eta_o + \eta_d} \tag{3}$$

If $\eta_o = \eta_d$ the value of the intrinsic viscosity is 7/4 which is the same value as obtained by Edwards and Schwartz [10-12].

Both treatments consider a membrane separating the internal from the external phases but across which the stress may be transmitted. For microgel particles there is no physical membrane producing this separation. However the particles deform along the principle axes of the stress as well rotate due to the vorticity of the field and hence flow (i.e. energy dissipation) will occur inside them giving some similarities to emulsion droplets. If we carry this analogy forward we would expect the intrinsic viscosities to be less than 2.5 and we may put a lower bound on the intrinsic viscosities of 1.75. It is tempting to press the analogy further and estimate an effective relative viscosity of the internal region of the particle. A value of 3 times the continuous phase viscosity of an emulsion drop would yield an intrinsic viscosity of 2.1 for example.

At high temperatures the intrinsic viscosity values are close to the hard sphere value as was expected for a collapsed configuration. The values of k_2 are generally higher than the 6.25 calculated by Batchelor [25] for hard spheres in shear flow. The secondary electroviscous effect will tend to increase this coefficient due to the electrostatic interactions producing an enhanced collision diameter. The shear rate dependence of second electroviscous effect has been analysed by Russel [26]. The shear rate in a capillary viscometer varies linearly from zero at the tube centre to a value of $O(10^3)$ at the wall and this makes it difficult to give a precise estimate of the coefficient. However, the value of $k_2 = 9$ at $5x10^{-3}$ M sodium chloride corresponds to a collision radius with an excluded shell of $3/\kappa$, which is typical of the values found at the phase transition of electrostatically stabilised colloidal particles. This confirms the observation of Buscall [9] that microgel particles are relatively hard in terms of collisions, i.e. much harder than might have been expected for a dilute solution of weakly cross-linked polymer.

4. Oscillatory Rheology of Concentrated Microgel Dispersions.

4.1 MEASUREMENTS

The linear viscoelastic response of the microgel dispersions were measured using a Bohlin VOR Rheometer with a cone and plate measuring geometry. Strain sweeps were performed on each sample in order to define a strain which ensured a linear response throughout the frequency range whilst maximising the measured signal. The samples were studied over a range of weight fractions at temperatures between 15 °C and 30 °C. The values of the storage (G') and loss (G'') moduli were then used for further analysis. In addition the wave rigidity modulus, $\tilde{G}$, was measured at 22.5 °C using a Rank Shearometer. In most cases the dispersions showed solid-like behaviour with phase angles below ~10° and frequency invariant storage moduli. A typical plot is shown in Figure 4 below which shows the experimental data at various temperatures. The addition of electrolyte reduces the storage modulus and results in a rapid rise in phase angle with increasing temperature. This is shown in Figure 5 for a microgel dispersion at a weight fraction of 0.113 in water and 0.5 M sodium chloride. Note how sensitive the moduli are to temperature and that at high added sodium chloride concentrations the modulus falls to zero and the phase angle rapidly increases at 24 °C. This is the point where the DLS experiment indicated aggregation due to loss of colloid stability.

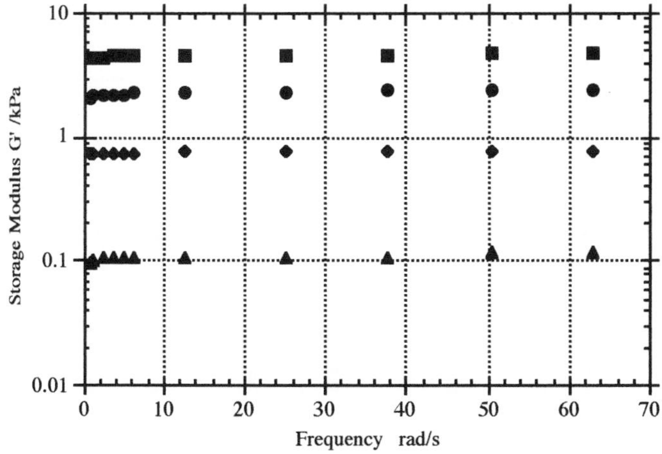

Figure 4. Storage Modulus as a function of radial frequency for a microgel dispersion in water at a weight fraction of 0.14.
■ 15 °C; ● 20 °C; ♦ 25 °C; ▲ 30 °C.

This paper will concentrate on data from experiments such as those illustrated in Figure 4. Here the systems were solid-like; indicating that the microgel particles were tightly packed together, and the moduli can be taken as the value of $G(\infty)$ for the structures. Figure 6 shows the variation of this high frequency limiting value of the storage modulus with concentration. The high frequency limit of the storage modulus

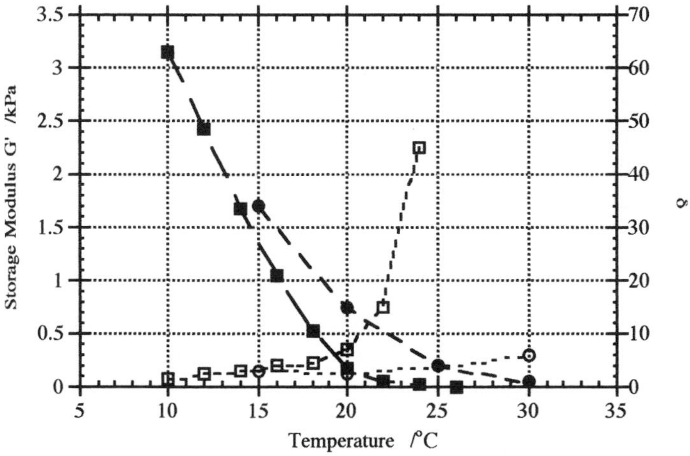

Figure 5. Storage modulus and phase angle, δ, as a function of temperature and added NaCl concentration.
■ G′ & □ δ in 5×10^{-1} M NaCl; ● G′ & ○ δ in 0 M NaCl.

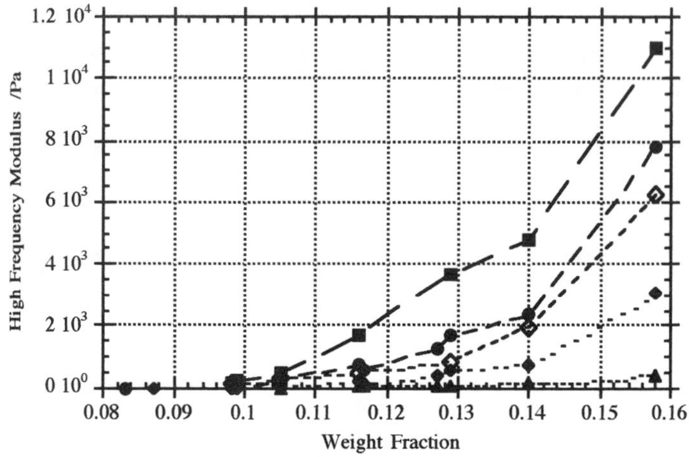

Figure 6. $G(\infty)$ as a function of weight fraction at various temperatures. ■ 15 °C; ● 20 °C; ◇ 22.5 °C; ◆ 25 °C; ▲ 30 °C.

increases rapidly with increasing weight fraction and with decreasing temperature i.e. with increasing particle interaction. Even at 15 °C, a weight fraction of 0.08 was required for the interactions to be strong enough to reduce the relaxation frequency to a value below the instrument range and hence enable a ready measure of $G(\infty)$ to be obtained.

4.2. MODELLING THE HIGH FREQUENCY LIMIT OF THE STORAGE MODULUS.

The value of $G(\infty)$ for a suspension of rigid interacting particles, but which are colloidally stable, can be adequately described in terms of the structural arrangement and the local elastic modulus via the curvature of the pair potential. The elasticity and the structural relaxation of the phases produced has been reviewed recently [25]. In general, the elasticity is given by the relationship :-

$$G(\infty) = N_p k_B T + \frac{2\pi N_p^2}{15} \int_0^\infty g(r) r^4 \frac{d^2V(r)}{dr^2} dr \tag{4}$$

Here N_p is the number density of the particles, V(r) is the interaction energy between two particles at a centre - centre separation of r, and g(r) is the pair distribution function of the system at solids loading N_p. For systems with strong, long range electrostatic repulsion acting between the particles, an ordered phase is produced which enables Equation 4 to reduced to that of a lattice cell :-

$$G(\infty)_{rigid} = \frac{\alpha}{R} \frac{d^2V(R)}{dr^2} \tag{5}$$

The nearest neighbour interactions dominate and the mean separation is R. The constant α is a function of the lattice type and the averaging procedure. For an fcc lattice the value of α can be taken as 0.83.

When the particles are deformable an additional effect arises at high concentrations. The charge layer is in the outer region of the corona and this is the origin of the electrostatic interaction. However the corona is deformable and strong interactions can be expected to produce a degree of flattening at the closest distance of approach. This will be opposed by both the modulus of the gel particle and the lateral interaction between charges on the surface. Although we note the lateral charge interaction here, we will neglect the contribution in the remainder of the model and only consider the electrostatic interaction normal to the surfaces with a surface - surface separation h, and the compliance of the gel particle. We can write Equation 5 for a pair of our interacting microgel particles in terms of the stress and the strain produced by the rapid deformation of the system as :-

$$\sigma = \gamma \frac{\alpha}{N_1 R} \frac{d^2V(h)}{dh^2} \tag{6}$$

N_1 is the co-ordination number of the lattice, i.e. 12 for fcc symmetry. This stress produces a deformation δa_p of the particles whose compliance is J_p :-

$$\frac{\delta a}{a} = \gamma_p = \sigma J_p \tag{7}$$

and the new particle surface - surface separation is :-

$$h' = h + 2\delta a \tag{8}$$

where $$2\delta a = 2aJ_p\sigma = 2aJ_p\gamma \frac{\alpha}{N_1 R} \frac{d^2V(h')}{dh'^2}$$

and so $$2\delta a = \frac{2a\gamma}{N_1} J_p G(\infty)_{rigid,h'} \tag{9}$$

Adding the compliances of the particle and the "electrostatic spring" gives the total compliance :-

$$J(R)_{soft} = \frac{1}{G(\infty)_{rigid,h'}} + J_p$$

and hence $$G(\infty)_{soft} = \frac{1}{J(R)_{soft}} \tag{10}$$

4.3. FITTING THE EXPERIMENTAL DATA.

The parameters that are needed to fit the experimental curves are the compliance of the microgel particles and the charge properties of the particle surfaces. Both of these represent a degree of uncertainty at present and they will be discussed in the light of the fits obtained to the experimental data. The experimental data for the values of $G(\infty)$ as a function of weight fraction at various temperatures are shown in figures 7 and 8 . For the sake of clarity data at 15 °C and 25 °C are plotted in Figure 7 whilst data for 20 °C and 30 °C are in Figure 8. The calculated curves are also shown for Equations 5 and 10 in both cases. A good fit to the data is found when the particle compliance is considered especially for the softer systems, i.e. those at 15 °C and 20 °C, where the curves for the soft and hard particles markedly diverge. The calculated curves only cover the range up to structures close to space filling as even more detail would be required in the model to go beyond this point. For the harder systems there is much less deviation from the hard sphere curve and this is of course mainly due to the lower volume fractions due to the smaller sizes thus giving weaker interactions. It is the relative magnitude of the compliances of the "polymer springs" and the "electrostatic springs" that controls the deviation.

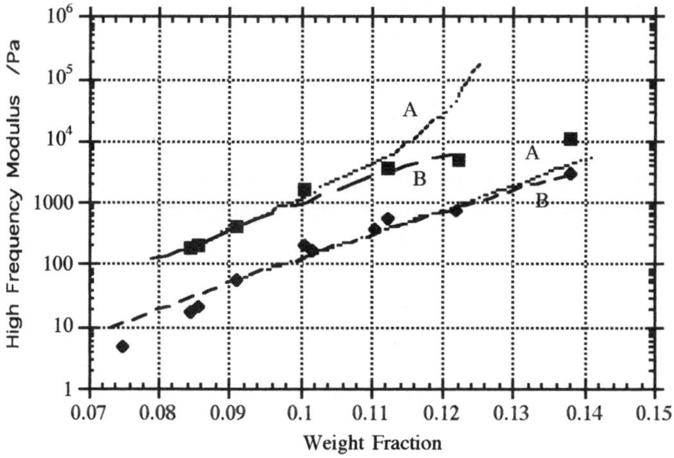

Figure 7. $G(\infty)$ as a function of weight fraction of polymer. ■ 15 °C; ◆ 25 °C; curve A hard spheres; curve B soft sphere.

4.4. THE VALUE OF THE PARTICLE COMPLIANCE.

As a first approximation the particles could be assumed to have a uniform structure and this facilitates the estimation of the particle compliance, J_p, from the elasticity of swollen elastomers (27) :-

$$J_p = 1/(N_x k_B T) \tag{11}$$

where N_x is the number density of cross-links in the swollen particle. This may readily be estimated from the cross-linker mole ratio at synthesis. However, the configuration of a uniform gel is not a likely one. The polymerisation conditions extant as a latex particle is being formed, combined with the electrostatic repulsion of the polymer end-groups, should result in a corona of decreasing density away from the central region. This would increase the compliance of the particle during the stages of the interaction investigated here. A detailed knowledge of the structure would be necessary for a full description and we will use the particle compliance as a fitting parameter here. The calculated particle compliances are plotted in Figure 9 below. It is worth noting that the curves are parallel indicating that the temperature dependence is just a swelling effect without any relative structural changes.

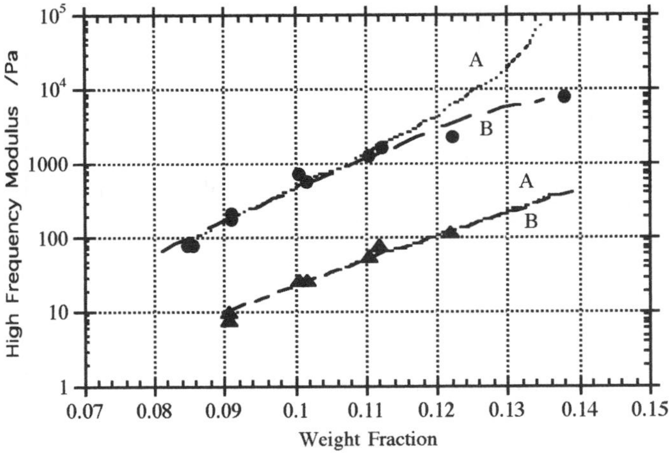

Figure 8. G(∞) a function of weight fraction of polymer. ● 20 °C; ▲ 30 °C; curve A hard spheres; curve B soft spheres.

3.5. THE VALUE OF SURFACE CHARGE DENSITY.

In order to calculate the pair potential we may neglect the attractive interaction from dispersion forces due to the heavily swollen nature of the particles. The electrostatic interaction arise from strong acid end-groups on the polymer chain and a constant charge model is the most suitable here [29] :-

$$V(h) = -2\pi\varepsilon a\psi^2 \ln[1 - \exp(-\kappa h)] \tag{12}$$

ε is the permittivity of the medium, ψ is the surface potential at infinite dilution and κ is the Debye-Huckel reciprocal decay length. For a situation with extensive particle distortion the radius a needs to be modified to reflect this. In a concentrated dispersion the value of κ must include the counter-ions from the particles :-

$$\kappa = \sqrt{\frac{N_A e^2}{\varepsilon k_B T} \cdot \frac{2c - \varphi\rho_p q}{1 - \varphi}} \tag{13}$$

where e is the fundamental unit of charge, c is the electrolyte concentration and q is the titratable charge on the particles. The surface charge density is simply calculated from the titratable charge, the particle size and the Faraday and then the surface potential at infinite dilution is related to the surface charge density, Φ, [30] :-

$$\Phi = \varepsilon\kappa \frac{k_B T}{e} \left[2\sinh\left(\frac{e\psi}{2k_B T} \right) + \frac{4}{\kappa a} \tanh\left(\frac{e\psi}{4k_B T} \right) \right] \tag{14}$$

A problem arises in utilising the titrated charge. Firstly, it is likely to be at a lower effective value due to counter-ion binding (Stern Adsorption), secondly, there will be a variable depth to the charged region and thirdly, there will be a counter-ion distribution inside the particle as well as outside. The degree of uncertainty introduced by these

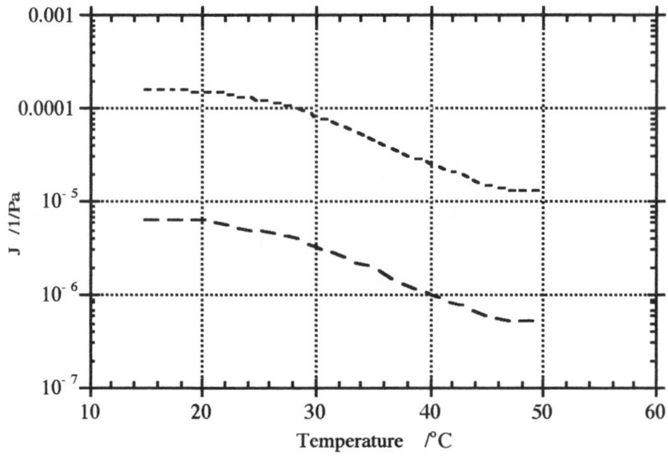

Figure 9. The calculated temperature dependence of the compliance of cross-inked poly-NIPAM microgel particles. Upper curve - data calculated from Equation 11; lower curve - values required to fit the experiments.

factors suggested that, like the particle compliance, the charge density is best treated as a fitting parameter at present. The value required to give a good fit to the data was equal to 1.2 μeq/g of polymer.

5. Acknowledgements

The authors are grateful for the financiacial support for this work which was carried out under the Department of Trade and Industry Project in Colloid Technology.

The authors also wish to thank the late Dr.Graham Johnson of the University of Bristol for performing the DLS experiments.

6. References

1. Princen, H. M. and Kiss, A. D. (1986) *J. Colloid & Interface Sci.*, **112**, 427.
2. Reinelt, D. A. and Kraynik, A. M. (1989) *J. Colloid & Interface Sci.*, **132**, 491.
3. Reinelt, D. A. (1993) *J. Rheology*, **37**, 1117.
4. Frith, W. J. and Mewis, J., (1987) *Powder Technology*, **51**, 27.
5. Mewis, J. and Frith, W. J., (1989) *AIChemE Journal*, **35**, 415.
6. Mewis, J. and D'Haene, P., (1993) *Makrmol. Chem. Macromol. Symp.*, **68,** 213.
7. Ploehn, H. J. and Goodwin, J. W., (1990) *Faraday Disc. Chem. Soc.*, **90**, 77.
8. Wolfe, M. S. and Scopazzi, C. (1989) *J.Colloid & Interface Sci.*, **133**, 265.
9. Buscall, R. (1994) *Colloids & Surfaces*, **83**, 33.
10. Schwartz, M. and Edwards, S. F., (1988) *Physica Acta* , **153**, 355.
11. Edwards, S. F. and Schwartz, M., (1990) *Physica Acta* , **167**, 595.
12. Schwartz, M. and Edwards, S. F., (1990) *Physica Acta* , **167**, 589.
13. Goddard, J. D. (1977) *J. Non-Newtonian Fluid Mech.*, **2**, 169.
14. Grimson, M. J. and Barker, G. C. !1990) *Molecular Physics*, **71**, 635.
15. Danov, K. D., Petsev, D. N., Denkov, N. D. and Borwankar, R. (1993) *J.Chem. Physics*, **99**. 7179.
16. Pelton, R. H. and Chibante, P. (1986) *Colloids & Surfaces*, **20**, 247.
17. Pelton, R. H., Pelton, H. M., Morphesis, A, and Rowell, R. L. (1989) *Langmuir*, **5**, 816.

18. Pelton, R. H. (1988) *J.Polymer Sci.: Part A : Polym. Chem.*, **26**, 9.
19. Tam, K. C., Wu, X. Y. and Pelton, R. H. (1992) *Polymer*, **33**, 436.
20. Wu, X. Y., Pelton, R. H., Hamielec, A. E., Woods, D. R. and McPhee, W. (1994) *Colloid & Polymer Sci.*, **272**, 467.
21. Goodwin, J. W., Hearn, J., Ho, C. C. and Ottewill, R. H. (1974) *Colloid & Polymer Sci.*, **252**, 464.
22. Einstein, A. (1908) *Ann. Physik*, **19**, 289.
23. Booth, F. (1950) *Proc. Roy. Soc.*, **A203**, 533.
24. Taylor, G.I., (1932) *Proc. Roy. Soc.*, **138**, 41.
24. Batchelor, G.K. (1977) *J. Fluid Mech.*, **83**, 97.
25. Russel, W.B., (1978) *J. Fluid Mech.*, **85**, 209.
26. Goodwin, J.W. and Hughes, R.W., (1992) *Adv. Colloid & Interface Sci.*, **42**, 303.
27. Green, M.S. and Tobolsky, A.V. (1946) *J.Chem. Phys.*, **14**, 80.
28. Wiese, G. and Healy, T. W. (1970) *Trans. Faraday Soc.*, **66**, 490.
29. Loeb, A. L., Wiersema, P. H. and Overbeek, J. Th. G. (1961) in "*The Electrical Double Layer Around a Spherical Colloid Particle*", MIT Press, Mass., p37.

STRUCTURAL STUDIES OF CONCENTRATED DISPERSIONS

A.M. DONALD, S. KITCHING, P. MEREDITH and C. HE
Polymers and Colloids Group
Cavendish Laboratory
University of Cambridge
Madingley Road
Cambridge CB3 0HE
UK

Abstract
Recent developments in environmental scanning electron microscopy (ESEM) have shown this technique to be well suited to the study of concentrated dispersions, with unique capabilities including the ability to follow processes such as aggregation in real time. This manuscript highlights some applications of ESEM to colloidal systems.

1. Introduction

Traditional methods of studying the structure of concentrated colloidal systems can broadly be divided into two categories - scattering and microscopic methods. The former approach has the advantage that it can sample wet systems, but the data is averaged over relatively large volumes and lacks spatial resolution. High resolution microscopies have traditionally not been able to image wet systems, although recent developments are beginning to overcome this problem. The environmental scanning electron microscope (ESEM) is one such approach. In this technique a system of differential pumping through a series of pressure limiting apertures permits the electron gun to be maintained under the necessary high vacuum, yet allow the sample chamber to contain gas at pressures up to about 15 torr. This means that samples can be kept fully hydrated during observation, if appropriate control of the sample temperature is also maintained. In addition, it is possible to change the state of hydration during the course of an experiment to permit either dehydration of a wet sample or hydration of a dry one, whilst imaging throughout if required. This clearly opens up the field of potential experiments to a much wider class than possible in a conventional SEM.

In this paper illustrative examples of both types of processes are presented. Aggregation of particles from aqueous dispersions (with and without salt) is followed, permitting a direct visualisation of the surface structure of the aggregates as they grow. Qualitative information on the probability of sticking upon a collision can also be obtained, together with the fractal dimension. The form of the final dry state of the 'film' that forms upon final drying of the dispersion can also be determined. The second example involves the hydration of dry ordinary Portland cement powder and its constituents. In this case it is

R. H. Ottewill and A.R. Rennie (eds.), Modern Aspects of Colloidal Dispersions, 41–50.

possible to interrogate the structure at different times after the onset of hydration by pulling off excess water from the surface and observing how the features on a particular grain have changed. The features observed can be related to the result of other measurements taken outside the ESEM, and has allowed us to test the hypothesis that a semi-permeable membrane forms on the surface of a grain leading to the familiar dormancy period during cement hydration.

2. Experimental

The work described here was all performed on an Electroscan E3 Environmental Scanning Electron Microscope.

2.1 AGGREGATION OF PARTICLES

The polymer latex dispersion was supplied by ICI plc at a concentration of 30% (w/w) polymer in water. A core-shell latex was used with a three layer structure. It consists of a PMMA (polymethyl methacrylate) core, a rubber middle layer and a thin PMMA outer layer. The total diameter was in the range of 200 nm-250 nm.

The samples were placed wet in the ESEM and the pump down sequence described by Cameron at al [1] was followed, to ensure that there was no premature evaporation of water which could favour aggregation initiating before imaging started. The temperature of the stage was maintained at around $3^{o}C$, which allows the sample to be moved easily between states of hydration and dehydration by small changes in temperature [2]. Care was taken to minimise beam damage by limiting exposure time (or increasing the scanning rate). A low gun voltage can also keep the level of beam damage down, but at the expense of image quality if the gun voltage is reduced too far. This effect may be due to the inability of the electrons to penetrate the thin layer of condensed water which covers the surface of the polymer latices when in a fully hydrated state. Because of this, a high gun voltage is required to penetrate this thin water layer should a good quality of image be desired. For the work described here a gun voltage of 18 keV was used, with a chamber pressure of 3-5 torr and temperature of 0-3°C.

For those samples dispersed in solutions containing $MgSO_4$ of different molarities, ranging from 0.016 -0.25 M, aggregation was studied by mixing the polymer latex and the salt solution (as a one to one mix) in a sample holder immediately prior to examination in the ESEM. Further details of the experimental approach can be found in [3].

2.2 HYDRATION OF TRICALCIUM SILICATE

One of the main ingredients of Ordinary Portland Cement (OPC) is tricalcium silicate (C_3S). For this work the tricalcium silicate samples were supplied by Construction Technology Laboratories (CTL), Skokie, Illinois- part no. TJC01 and contained rough grains in the size range 1-100μm. To assess the crystallinity, and identify the particular polymorph of the sample, X-ray diffraction measurements were carried out at Schlumberger Cambridge Research (SCR). The results of this characterisation indicated that the C_3S used was of the triclinic form and had a very low levels of impurity.

The C_3S powdered sample was put in a copper pot (~50ml capacity) and placed on the ESEM Peltier stage at a temperature of ~20°C. After appropriate pumpdown, the morphology of the "dry" sample was recorded for reference. The sample temperature was then reduced to ~5-7°C and condensation of water onto the surface of the powder allowed to occur. To follow the progress of the hydration, different samples were hydrated for periods varying from 30 seconds to 3 hours. At the end of the designated times, excess surface water was removed by raising the stage temperature by ~2°C, allowing the C_3S microstructure to be recorded. At this point, the grain surfaces were still partially hydrated, and as such the products remained in their natural state when micrographs were recorded. This method is estimated as giving a high water to solid ratio of ~10:1. Such a regime has previously been used in "wet" electron microscopy work by Jennings *et al.* [4] and also Double *et al.* [5], and in "wet" optical microscopy performed by Diamon *et al.* [6].

X-ray analysis was also carried out on OPC powder *in situ* using a Kevex X-ray detector which permitted mapping of different selected elements both before and after hydration had been carried out. OPC is a complex multiphase mixture of tricalcium aluminate, dicalcium silicate, tetracalcium aluminoferrite and gypsum, whose behaviour is to some extent dominated by the behaviour of the C_3S (see for instance [7]). For both C_3S and OPC it is known from conduction calorimetry that there is an initial rapid reaction (stage I), followed by an induction or dormancy period (stage II) when thermally the reaction seems to have ceased. This period typically extends from ~30-180 minutes, after which further reaction is observed to occur calorimetrically. For a further discussion of the stages of cement hydration in relation to ESEM studies see [2].

3. Results and Discussion

3.1 AGGREGATION STUDIES

Before any water is lost from the sample, the colloidal particles are seen to be moving about individually. Since it is not possible to see through a thick layer of water, much of the surface is featureless with only individual particles being seen sometimes protruding through the surface. The images of these particles are not very good, since the particles are moving under Brownian motion, and may blur the image during the time a micrograph is recorded. As water is allowed to evaporate, the particles start to aggregate. When no salt is present in the solution, these aggregates tend to be well-packed, and for the most part the collision of two particles does not lead to sticking. Figure 1 shows an example of structure at this stage, in which both hexagonal (Figure 1a) and square (Figure 1b) packing arrangements can be seen in the ordered region. However at this time there is still a significant fraction of particles that do not exhibit an ordered structure, as a fraction of the particles are still in motion.

As further evaporation of water occurs, the volume fraction of polymer latex increases still further, and finally reaches its critical concentration in the surface layer, so that all particles are involved in the close-packed arrays. Again both hexagonal and square regions of packing can be seen, but the former is much more common than the latter. It

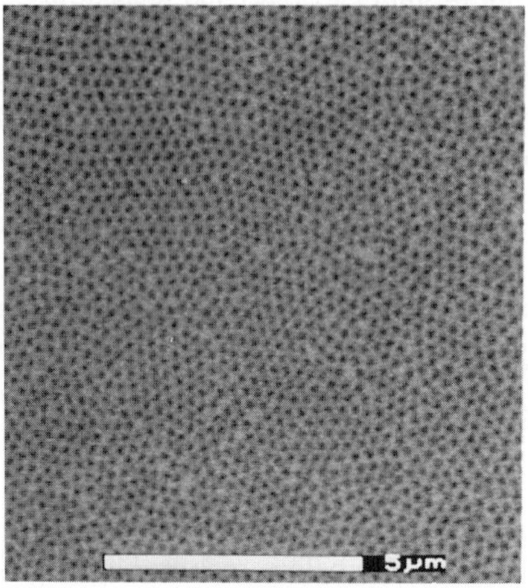

Figure 1a Hexagonal packing in the aggregates.

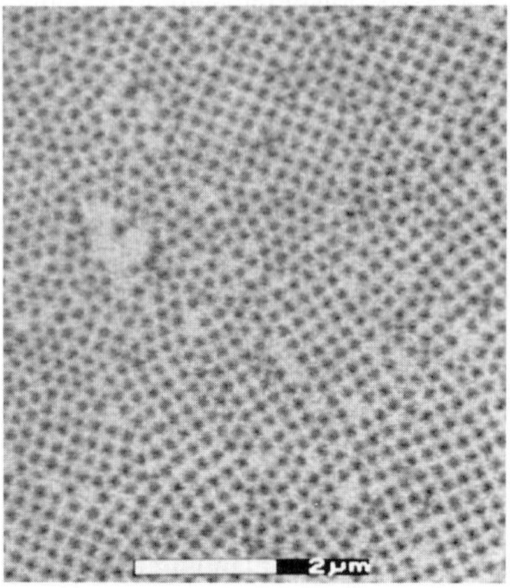

Figure 1b Square packing in the aggregates.

should be noted that it is the dark spheres in the micrographs that correspond to the polymer particles, whereas the bright edge is the boundary of the latex which, due to its hydrophilic nature, is covered with a thin layer of water. The origin of this contrast mechanism, in which hydrated latex particles exhibit bright edge contrast, has been described and explained elsewhere[8]. Figures 2 show the nature of aggregates in latices containing $MgSO_4$ at concentrations of 0.072M and 0.25M at a ratio 1:1 v/v. It can be seen that the packing is much less regular than in the no-salt case. The size of the flocs increases with the salt content.

In the absence of salt it appears that a colloidal crystal is forming at intermediate spacings, due to soft repulsive forces. For this, one may expect an equilibrium to be set up between the colloidal crystalline state and particles still in solution, with a local phase separation between the two states. As the concentration increases (due to loss of water in our experiments), the proportion of particles involved in the crystals will increase. However, many collisions do not lead to aggregation, with the particles subsequently moving apart again under their Brownian motion. This is because of the height of the potential barrier to the primary minimum. Those collisions which do cause sticking in the weak secondary minimum lead to the formation of the colloidal crystal, with well-organised packing. From the evidence the surface layer seems to favour hexagonal symmetry, although in places four-fold symmetry is seen. It is assumed that these differences simply reflect which plane of a close-packed crystal happens to form the surface layer.

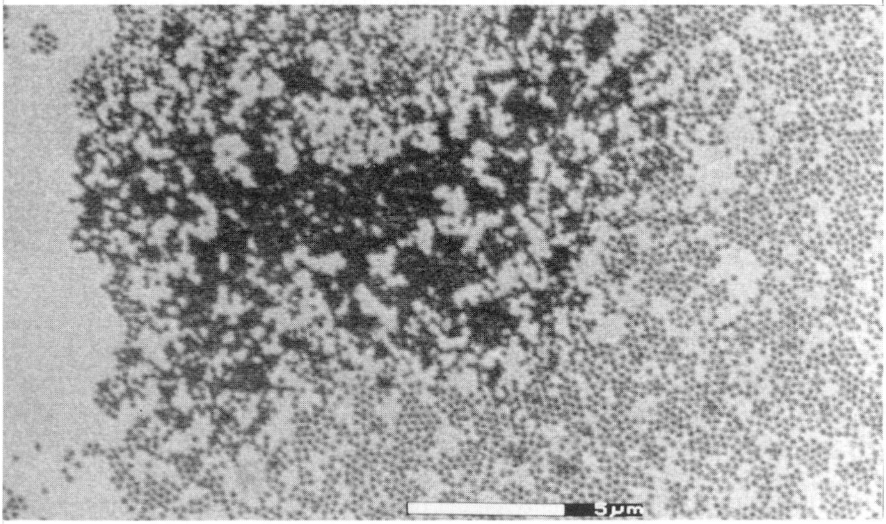

Figure 2a Aggregates from latices containing 0.25M $MgSO_4$.

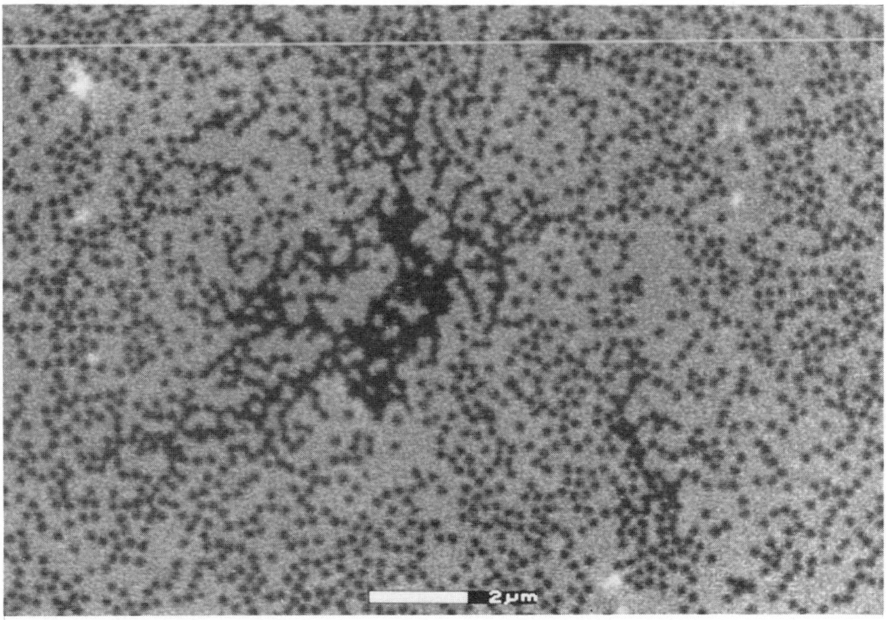

Figure 2b Aggregate from a latex containing0.072M $MgSO_4$.

The effect of adding salt is to reduce systematically the height of the potential barrier to the deep minimum of the pair potential, and also to increase the depth of the secondary minimum. In this case aggregation may occur with many particles sitting in the primary minimum. Most collisions may be expected to lead to this, and the structure will thus be more disordered. The aggregation will be irreversible[9]. In fact ESEM shows that the number of irreversible aggregate increases with an increase of salt concentration. Any collisions which lead to sticking only in the secondary minimum will on the contrary give rise to a reversible aggregate.

3.2 HYDRATION OF CEMENT AND TRICALCIUM SILICATE

Dry grains of C_3S grains are fairly featureless. They are irregularly shaped and highly polydisperse. Upon hydration a surface product was observed immediately upon contact with water. At early stages, for example after 1 minute, the layer appeared irregular and discontinuous. Hydration for longer times (t<20 minutes) had the effect of propagating and thickening this layer, whilst not producing any inter granular material. By ~20-30 minutes whole clusters of grains were seen to be completely covered by this product, and large pits on grain surfaces were covered and bridged by a self supporting layer. A comparison between the grain surface at zero time, i.e., before hydration, and after 25 minutes is shown in Figure 3 The area on the centre left shows clearly how an originally isolated grain becomes engulfed by the covering material. This material displayed very little distinct structure or crystallinity on the resolution scale of the instrument (~20nm under these conditions). This structure is consistent with that known for hydrated calcium silicate (C-S-H), and these results indicate that, using the terminology developed for thermal analysis, it continues to grow up till the end of stage I, after which time no further growth was noted.

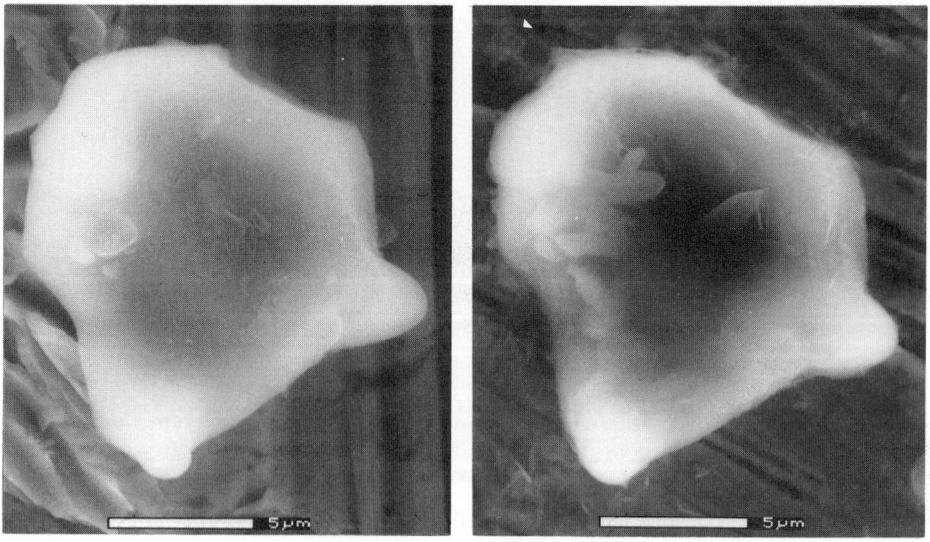

Figure 3 A grain of C_3S before and after 7 minutes of hydration.

During the early stages of hydration, calcium ions are rapidly released into solution [10] leading to an increase in pH to ~12 after a few minutes [11]. However, the dissolution of silicate ions is not as dramatic [12], and the resultant C-S-H surface product is therefore silicate rich. The ESEM evidence suggests that continued extension of this layer can occur until the entire surface of the grain has been covered. Hence, it is postulated that the well known dormancy period in the reaction of C_3S with water (induction) is associated with the exhaustion of easily available surface reactants. The rate of reaction is known to be critically dependent upon water to solid ratio and temperature, consistent with the exhaustion point varying according to hydration conditions and sample composition [10].

Further structural evolution did not occur during the period from 30 to 180 minutes of hydration, corresponding to this induction period. However, the samples exposed for ~3 hours displayed a dramatic change in morphology. Large, regular, often hexagonal crystals could clearly be seen in intergranular regions, and over the grain surfaces themselves. These crystals were in various degrees of development, but occurred throughout the sample. A clear example of such structures is shown in Figure 4.

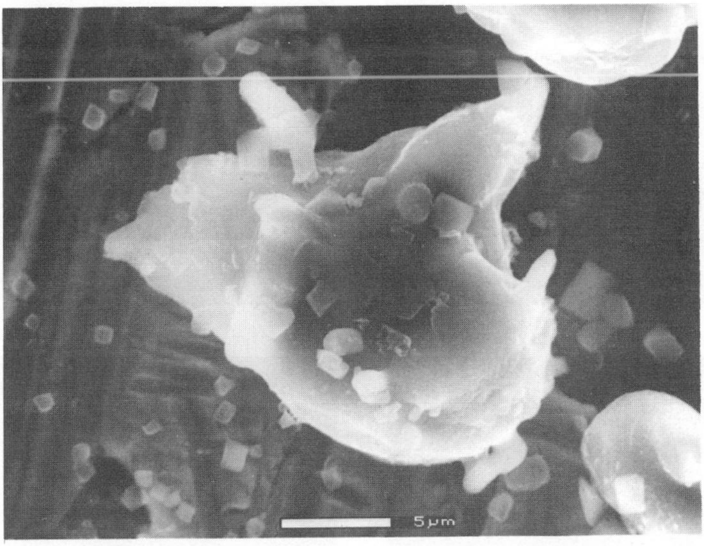

Figure 4 Calcium hydroxide (CH) crystals form after around 3 hours of hydration.

Such a structure is known to be associated with CH formed as a result of nucleation and growth from saturated or supersaturated calcium solution [13]. The presence of CH product has been noted in mature OPC pastes by other workers, although direct microstructural evidence for CH formation at the end of induction is rare. It has been suggested that during induction a critical ion concentration builds up, reaching the necessary levels at the end of stage 2 to cause crystallisation of CH. Such a clock reaction type process has in fact already been suggested by Billingham and Coveney

[14]], to explain the reaction of C_3A and gypsum to produce ettringite during OPC hydration.

However how the calcium ion concentration builds up in solution during induction is critically dependent upon the microstructure of the protective C-S-H layer that forms. During this period, anhydrous C_3S must continue to hydrate but the exact structural nature of the C-S-H coating is uncertain. The critical question as far as the continued limited stage 2 hydration of C_3S is concerned, is whether this coating is impermeable, semi-permeable or porous to water. The ESEM results presented here suggest that a protective layer of amorphous C-S-H forms on the grain surface which allows for the through transport of both hydrate and calcium ions by an osmotic mechanism [5]. At a later stage this structure will therefore not preclude the build up of the critical ion concentration required for the precipitation of crystalline calcium hydroxide.

Turning now to the case of OPC, X-ray mapping demonstrates how the distribution of the different elements in C_3S change upon hydration (Figure 5). The image shows that the

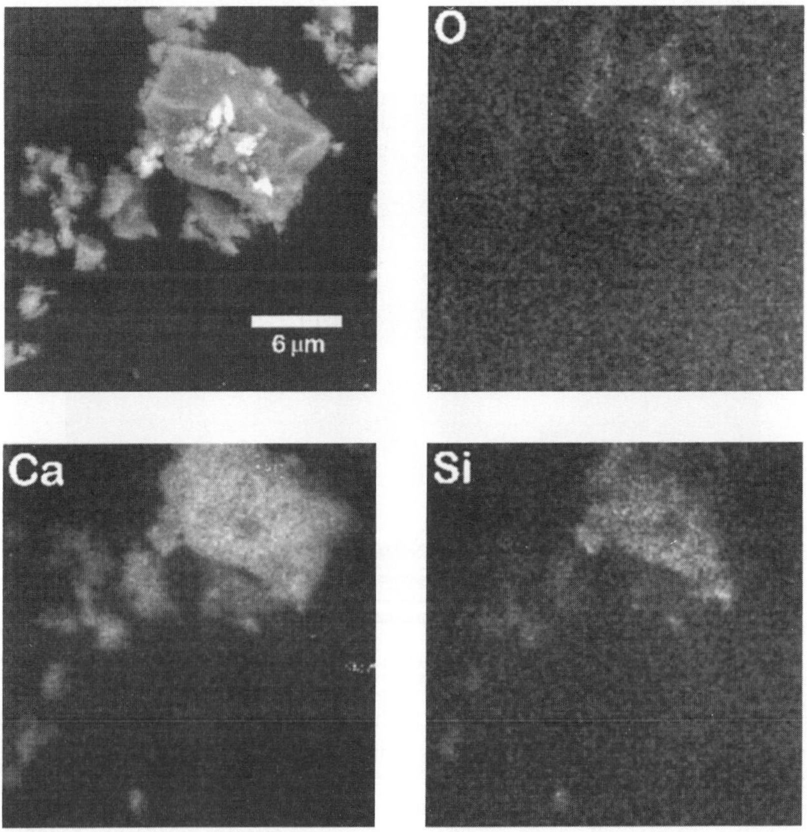

Figure 5a

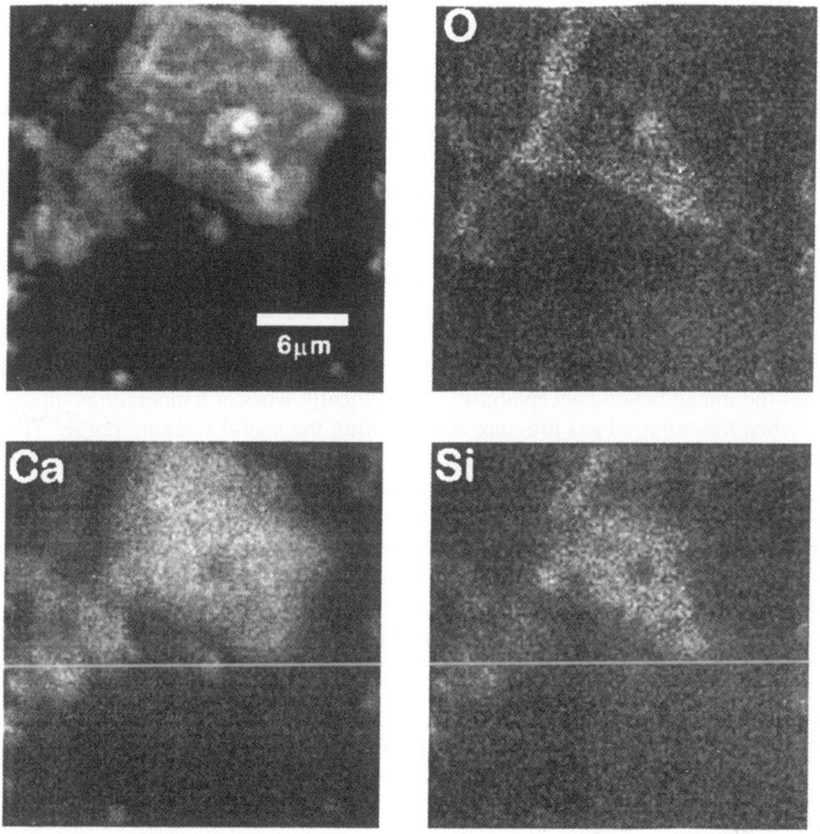

Figure 5b

Figure 5 Xray maps before (a) and after (b) 15 minutes hydration: image plus maps in the elements specified in each case

selected grain becomes covered with a layer of amorphous material, as in the case of C_3S referred to above. The X-ray maps of the dry grain show oxygen, silicon and calcium confined to the grains of the powder itself. Upon hydration the area from which significant signal for each element can be detected increases in line with the increasing size of the grain. Other elements such as iron and potassium (not shown) show very little difference in distribution before and after hydration. Sulphur shows some delocalisation. This X-ray mapping is therefore consistent with the idea that a layer of amorphous C-S-H covers the swelling cement grain.

4 Conclusions

In this paper only a few representative examples of how the ESEM may provide novel insights into the physical processes that may occur in dense colloidal dispersions are presented. The ability to image fully hydrated samples, and to follow hydration and dehydration experiments dynamically, clearly makes the ESEM ideally suited to follow a wide range of processes. It is to be expected that its use in this field will continue to spread. However the limitations of the technique must also be appreciated, and a full understanding of the electron-gas interactions that underlie the imaging process is still to be developed. These are factors that must be taken into account when designing experiments which can use the full potential of the instrument. One of the main limitations of the ESEM in its current form is that imaging can only take place under saturated water vapour at temperatures below ambient. Otherwise the high gas pressure degrades the image beyond acceptability. We typically work at temperatures in the range 3-7°C, when the required gas pressure is still within the useful working range. However, this implies that hydration at room temperature cannot be studied dynamically, which for some purposes may weaken any conclusions that can be drawn. For colloidal dispersions, the use of water as the imaging gas is likely always to be required. However for imaging sub-zero (as in the case of ice cream or freeze-drying studies) an alternative imaging gas will be required. In order to understand how these different gases affect the imaging process, we have developed an improved model for the electron-gas interactions. However there is probably still a long way to go before the instrument's development is exhausted.

Acknowledgements
The financial support of the DTI Colloid Technology Programme is gratefully acknowledged.

REFERENCES

1.Cameron, R. E. and Donald, A. M. (1994) *J Micros*, **173**, 227.
2.Meredith, P., Donald, A. M. and Luke, K. (1995) *J Mat Sci.*, **39**, 1921.
3.He, C. and Donald, A. M. (1996) *Langmuir*, **12**, 6250.
4.Jennings, H. M., Dalgleish, B. J. and Pratt, P. L. (1981) *J. of American Ceramic Soc.*, **64**, 567.
5.Double, D. D., Hellawell, A. and Perry, S. J. (1978) *Proc. R. Soc. Lond. A*, **359**, 435.
6.Diamon, M., Ueda, S. and Konde, R. (1971) *Cem. Conc. Res.*, **1**, 391.
7.Taylor, H. F. W. (1990) ,Chemistry of Cements, Academic Press , London, 123.
8.Meredith, P. and Donald, A. M. (1996) *J Micros*, **181**, 23.
9.Jeffrey, G. C. and Ottewill, R. H. (1988) *Coll Poly Sci*, **266**, 173.
10.Jawed, I., Skalny, J. and Young, J. F. (1983) in *Structure and Performance of Cements*, ed P Taylor, 237.
11.Mindess, S. and J F Young (1981) *Concrete*, , 86.
12.Tadros, M., Skalny, J. and Kalyonceu, R. S. (1976) *J. Amer. Ceram. Soc.*, **9**, 344.
13.Williamson, R. B. (1972) *Prog. Mat. Sci.*, **15**, 3.
14.Billingham, J. and Coveney, P. V. (1993) *J Chem Soc Faraday Trans.*, **89**, 3021.

FILM FORMATION OF LATICES

J. L. KEDDIE,[1] P. MEREDITH, [2] R. A. L. JONES AND A. M. DONALD
Polymer and Colloid Group, Cavendish Laboratory
Madingley Road, Cambridge CB3 0HE

Our studies of latex film formation over the last several years have resulted in several findings that are reviewed in this paper. Both the mechanisms and time-dependence of latex film formation are a strong function of polymer viscoelasticity and temperature. Depending on the temperature of the process, one of two steps can be rate-limiting in film formation: either water evaporation or particle deformation. When the latter is rate-limiting, microscopic air voids (with sizes below the wavelength of light) develop at about the same time as the onset of optical clarity. Sintering models can then be adapted to describe void shrinkage. In latex blends, clusters of non-film-forming particles enhance void formation and markedly slow down film densification.

1. Introduction

Latex films and coatings were put to practical use in the 1950s long before there was much understanding of the mechanism by which they are formed from their initial wet state. Mysteries of latex film formation - the process by which a colloidal dispersion of polymer particles in water is transformed into a continuous material - did not prevent their application in paints, coatings, adhesives, membranes, gloves, carpet backings etc. With recent heightened awareness of the environmental impact of solvent-borne coatings, there is now a driving force to use latex in applications that demand higher performance and an understanding of film formation mechanisms.

Latex film formation presents particular analytical challenges because the polymer colloid particles are soft and the process occurs in the presence of, and often under the action of, water. Many techniques employed in the past to investigate film formation are invasive. Consequently they cannot be used to study film formation as it occurs over time. Conventional electron microscopies require that the sample is held in high vacuum, and so they are not suitable for the study of wet latex. Neutron scattering techniques have made inroads in determining the cellular structure of latices [1], in measuring polymer diffusivity [2], and in studying water in the Plateau borders [3]. Non-radiative energy transfer techniques have been particularly effective in determining how factors

---

[1] Present Address: Department of Physics, University of Surrey, Guildford, Surrey GU2 5XH

[2] Present Address: Procter & Gamble (Health and Beauty Care) Ltd., Rusham Park, Whitehall Lane, Egham, Surrey TW20 9NW

R. H. Ottewill and A.R. Rennie (eds.), Modern Aspects of Colloidal Dispersions, 51–59.

such as molecular weight and the presence of plasticisers influence interdiffusion [2]. Great confusion is often associated with the transition from a concentrated dispersion to a dense array of particles, and this important but less-often studied problem is a primary interest of our work.

2. Experimental Techniques

We developed two non-invasive techniques - environmental-scanning electron microscopy (ESEM) and spectroscopic ellipsometry - that have some outstanding strengths in tackling problems in latex film formation:

1. *Sensitivity to void content.* The interstitial space between latex particles is not always immediately eliminated during film formation. The presence of voids will degrade film performance, especially permeability to solvents and gasses. Detection and characterisation of void content provides crucial insight into the process of film formation and the resulting film properties.

2. *Applicability to wet, drying and dry samples over a broad temperature range.* A latex passes through several stages, and ideally one would like analytical techniques that are suitable for each of these. High vacuum techniques are not suitable for the wet stages. Light scattering techniques are primarily applicable to dilute colloidal dispersions.

3. *Relatively fast data acquisition.* Analytical techniques must be suitable to the time-scales of the changes in latex microstructure. This time scale can be quite short, especially at elevated temperatures.

Environmental-SEM, which is described elsewhere in this volume, enables the imaging of latex particle clustering, deformation and coalescence under controlled conditions. There is no need to coat the sample surface with a conducting film, as required with conventional SEM. Via the adherence to proper pumpdown procedures, latex can be imaged in an environmental chamber without drying out the sample [4]. Manipulation of pressure and temperature in the chamber leads to the condensation of water that will decorate features of the sample's surface topography. A great deal of insight has been gained into the mechanisms of imaging and the use of water to enhance images [5].

Spectroscopic ellipsometry determines the optical constants and structure at interfaces in a sample by analysis of changes in the phase and amplitude of polarised light. We use it to monitor the optical constants of a latex during film formation, and we then correlate these with the evolving microstructure. Ellipsometry is used in one of two primary modes. Slowly-changing samples can be probed with detailed scans of incident angle or wavelength. Samples evolving rapidly allow data acquisition as a function of time only at a single angle-of-incidence and wavelength. With this second mode, just one pair of ellipsometric parameters (ψ and Δ) are obtained. These parameters are defined by the ratio of the Fresnel reflection coefficients (r_s and r_p)for the interface:

$$r_p/r_s = \tan(\psi)\exp(i\Delta), \tag{1}$$

as described elsewhere [15]. The two measured parameters are sufficient for finding two unknown values of the optical constants (real and imaginary component of the refractive index) when the sample is modelled as a semi-infinite medium. The end result of such a measurement, the pseudo-refractive index, may then be correlated with film microstructure.

The precise relationship between pseudo-refractive index and film microstructure is still a subject of investigation and debate. It is well-accepted that the real part of the refractive index is proportional to material density, as is given by the Lorentz-Lorenz equation [6]. Likewise, effective medium approximations, which describe a latex as a mixture of polymer and voids, predict that refractive index will increase as void content decreases. It is less clear how the imaginary component of refractive index, also known as the extinction coefficient, is best interpreted. In polymers that are not optically absorbing, the imaginary pseudo-refractive index must result from surface roughness, scattering from voids, or a combination of both. If both are present, as is most probably the case, then de-coupling the individual contributions is not possible when there are only two measured parameters. Work is currently underway to use ellipsometry to measure both surface roughness and void contents.

Armed with our non-invasive techniques, we set about answering the following questions:

1. What is the structure of a latex during the stages of film formation?
2. How does the viscoelasticity of a latex correlate with the mechanism and time-dependence of film formation?
3. What influence does the temperature of film formation have on the process?
4. In what way does the presence of non-film-forming particles in a latex blend affect the microstructure of a latex and the time-dependence of its film formation?

Each of these questions are addressed in the following sections.

3. Morphological Development

We used ESEM to isolate latex in various states: an aqueous dispersion, a random packing of particles containing interstitial water, deformed particles packed in a "honey-comb-like" structure, and a continuous film-formed layer [7]. Previous studies of the surfaces of films cast from a dilute monodisperse, "model" latex have found hexagonal arrays of particles [8]. In the more practical, concentrated and somewhat polydisperse latex systems that we studied, disordered packing emerged [7].

Although it has long been realised that an optically transparent latex can contain voids with sizes below the wavelength of light, ESEM provided some of the first clear images of their size, shape and extent. Optical measurements corroborated these findings by providing evidence for the development of surface roughness and density decreases with the evaporation of water. Low refractive indices of films were correlated with the presence of air voids. An important conclusion is that, in some cases, latex particles are not deformed to fill all available space upon the evaporation of water. Instead, interstitial voids develop, and they are eliminated long after the onset of optical clarity. By correlating microscopy with ellipsometry, we mapped out microstructure as a function of time, temperature and polymer viscoelasticity during the various stages of film formation [7,9]. We thus correlated refractive index with latex morphology under various conditions. Notably, in our study of harder film-forming latices, we reported the existence of a void-containing microstructure [7, intermediate to a concentrated dispersion and a densely-packed array of particles, and that had not been previously studied in detail..

4. Time-Dependence of Film Formation of Harder *versus* Softer Particles

We studied a series of latices based on a copolymer of methyl methacrylate and 2-ethyl hexyl acrylate [7]. The latices, which were synthesised at ICI Paints (Slough, UK) by a two-stage emulsion polymerisation process, had roughly the same particle sizes (about 300 nm diameter) and surface chemistries, but their glass transition temperatures ranged from -28 to 62 °C. The latices were stabilised with both nonionic and anionic surfactants. A cellulose ether was used as a protective colloid. At room temperature, the viscoelasticity of the latices thus spanned the range from being practically non-deformable to being highly "soft." Softer latices showed no evidence - neither in ESEM nor in ellipsometry - for void formation following evaporation of water. A concentrated dispersion of particles in water is immediately converted to a densely-packed array. With increasing polymer viscoelasticity (as indicated by an increased glass transition temperature (Tg) of the film-formed polymer), void formation becomes more extensive and is slower to be eliminated. An example of such a void structure is seen in Figure 1.

Such findings indicate that the driving force for film formation varies with the polymer. In particles with a high rigidity, all water should be evaporated from the latex before deformation is complete, and so the surface energy of the polymer/air interface can reasonably be considered to be a dominant driving force. In softer particles, capillary forces are expected to dominate the coalescence process.

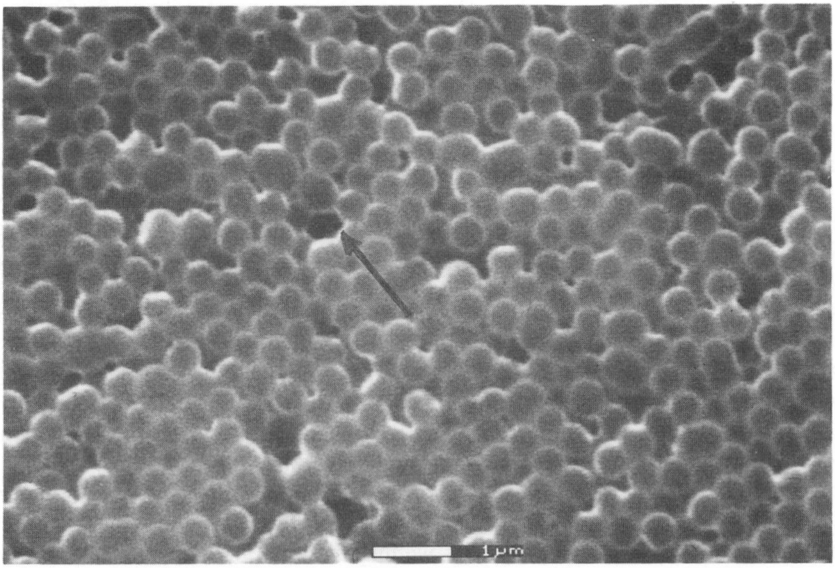

Figure 1. ESEM micrograph of void structure in a latex. Numerous interstitial voids are seen in this latex (T_g = 13 °C), which was film-formed in the microscope chamber at about 3 °C. Much more rarely seen are voids that apparently result from defects (vacancies) in particle packing (such as the hexagonal void indicated by the arrow).

5. Rate-Limiting Steps in Latex Film Formation

This concept of polymer viscoelasticity being related to the mechanism of film formation, led us to consider whether film formation has rate-limiting steps [9]. Sperry and co-workers [10] described the process of film formation as consisting of two steps - water evaporation and particle deformation - either of which might be rate-limiting. We compared two latices (one with a Tg of 13 °C and the other with a Tg of -5 °C) over a broad temperature range as a way of searching for the rate-limiting conditions. At low temperatures, evaporation rates are suppressed and polymer rigidity is enhanced, whereas at elevated temperatures evaporation is faster and polymers are much less viscous.

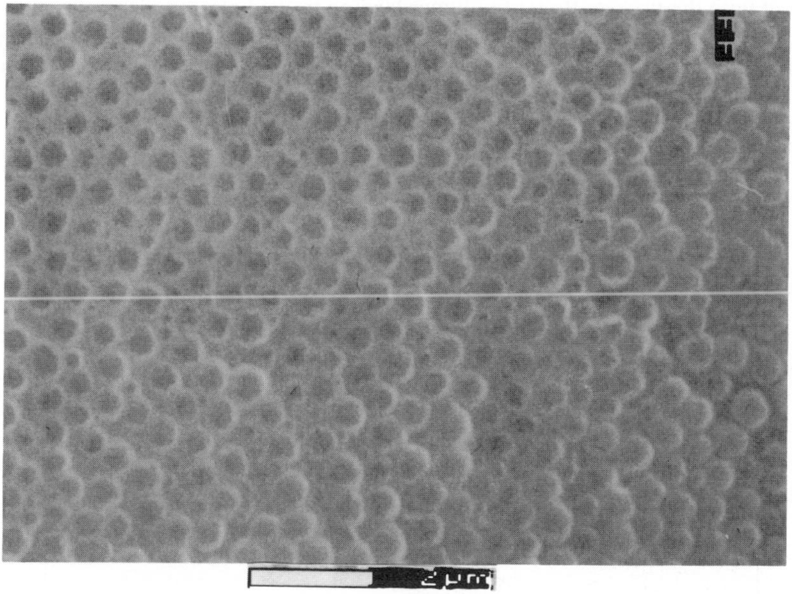

Figure 2. Evidence that film formation occurs immediately upon evaporation of water in a softer latex: an ESEM micrograph of a latex, with a T_g of -5 °C, during film formation at 3 °C. Particles emerge from the latex serum on the left; drier particles on the right have already deformed to create a dense array.

In the latices studied, a densely-packed, void-free film was formed concomitantly with the evaporation of water when film formation took place at temperatures about 20 °C or more above the polymer T_g. In this case, *evaporation* is rate-limiting, and water provides a driving force for film formation, probably via capillary forces [9].

With careful microscopy and a well-developed base of technical experience, we imaged concentrated dispersions in which particles began to emerge from the surface of the latex serum. In the soft latex (T_g = -5 °C) shown in Figure 2, we found rare evidence from microscopy that evaporation is rate-limiting. This figure shows the passage of a drying front in the plane of the film. On the left side, water fills the interstices between particles. On the right, water has evaporated during the time course of the observation, and particles are already seen to be deformed into a denser array. Within seconds of the image being obtained, a continuous film was formed without any apparent polymer/polymer boundaries remaining. When studied with ellipsometry at comparable temperatures, this latex showed an abrupt optical transition from a low refractive index (indicative of a wet, concentrated dispersion) to a high index that did not change much with time (indicative of a stable, continuous dense film).

At temperatures nearer to Tg, a drying front - distinct from a coalescence front in which particles are deformed and coalesced - is evident in the latex. That is, water evaporation creates air voids near the sample surface to define a drying front, and this front is separated in time from one that passes later in which the deformation of particles occurs. In this case, particle deformation is the rate-limiting step in film formation. In the absence of water, the driving force for particle deformation is likely to be the polymer/air surface energy. We measured the time interval between the passage of these fronts with ellipsometry. It increases with decreasing temperature in a manner that is consistent with the viscous flow of polymer leading to void closure, driven by the surface energy reduction of the polymer/air interface [9]. We based a model for this process upon an equation derived by Mackenzie and Shuttleworth [11] and upon the assumption that the optical clarity associated with the coalescence front occurs when the average void size decreases below a critical value. Figure 3 shows this result.

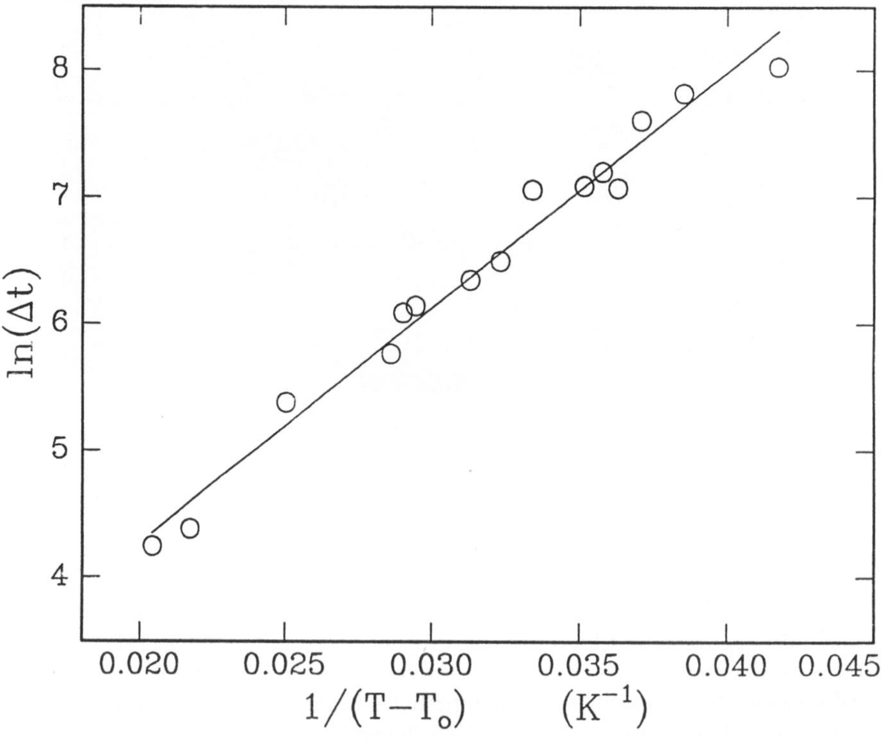

Figure 3. Application of a viscous flow model to interpretation of ellipsometry results. The time interval between the apparent passage of drying and coalescence fronts in the plane of a latex film is plotted against reciprocal temperature. The T_g of the latex is 286 K, and T_0 is fitted from the data to 257 K. (Time is in units of seconds.)

6. Latex Blends and the Application of Sintering Theories

In the absence of water, sintering models are applicable to latex film formation. These models equate the energy *expenditure* in particle deformation to the energy *savings* in the reduction of surface area. Tests of sintering models require precise measurements of density. These can be obtained easily for bulk materials. In thin coatings, in which gravimetric and dimensional measurements lead to large errors, optical properties can be feasibly correlated with density. In films cast from latex blends, we applied the Lorentz-Lorenz equation [6] to measured values of refractive index in order to calculate the relative density.

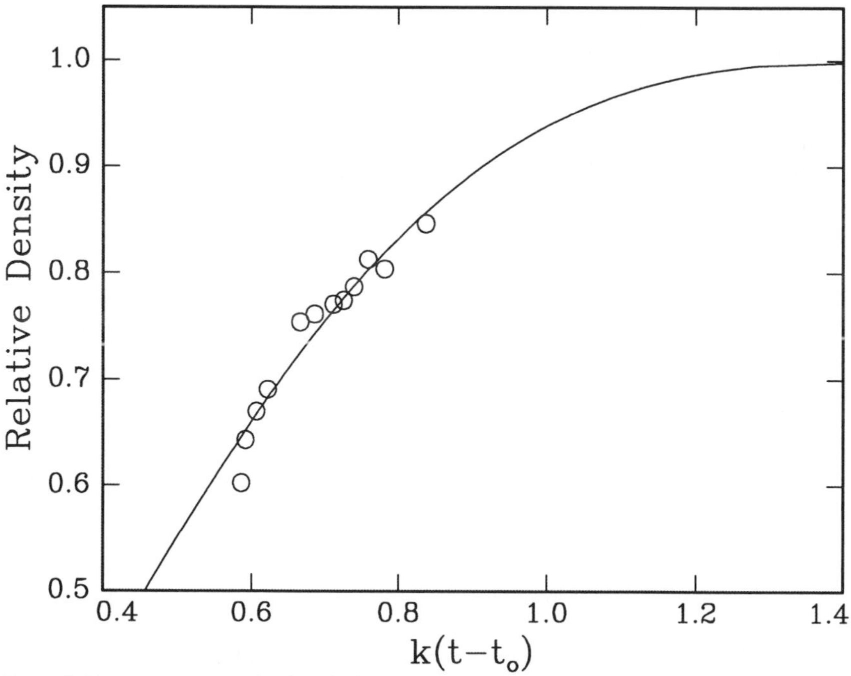

Figure 4. Room temperature density of a latex blend relative to the maximum density attainable, as a function of normalised time. The densities, inferred from optical measurements, follow the trend predicted by the M-S sintering model with viscosity fitted to $2 \cdot 10^{12}$ Pa s for the matrix of the blend (T_g = 13 °C).

We studied the room-temperature film formation of a series of latex blends consisting of increasing concentrations of non-film-forming (NFF) latex particles (Tg = 60 °C) in a film-forming (FF) latex (Tg = 13 °C) [12]. Both ESEM and ellipsometry reveal the pronounced effect that NFF particles have on the film formation of latex blends, even at low concentrations. With increasing concentration, the film density falls and the overall time required for film densification increases. This finding is explained by ESEM examination. Interstitial voids are found near NFF particles and especially near NFF clusters. There is a clear correlation between NFF particles and void formation. When the concentrations of NFF particles is as high as 40 weight percent, changes in optical constants - and thus the void size and film density - occur over a period of weeks. An important lesson can be learned from our study of polymer blends. Because contact between NFF particles creates voids, the density of a blend is increased by minimising the number of these contacts. Full dispersion of NFF particles is paramount to obtaining a void-free film.

The Mackenzie-Shuttleworth (M-S) model of sintering [11] predicts the density of a porous material as a function of time by assuming that the rate of void closure, dr/dt, is inversely related to the viscosity of the matrix, η, and to its relative density, $\rho(r)$:

$$\frac{dr}{dt} = -\frac{\gamma}{2\eta}\left(\frac{1}{\rho(r)}\right), \qquad (2)$$

where γ is the surface energy of the polymer/air interface. In a latex the sintering matrix is better described as being viscoelastic. Regardless of the description of the deformation and flow, however, sintering models predict an increase in density with time. Figure 4 shows how the M-S model, when applied to data obtained from a latex blend using fitted values of polymer viscosity, describes the observed changes in relative density [12]. The sintering constant, k, is defined as $\gamma z^{1/3}/\eta$, with γ defined as the surface energy of the polymer/water interface, z being the number of voids per unit volume, and η being the viscosity of the polymer matrix. A standard literature value was used for γ [12]; z was estimated from the particle size by assuming that a void was associated with each particle, and η was used as a fitting parameter. The fitted values of viscosity for a polymer blend are higher than those found for the corresponding single latex, consistent with predictions of viscosity for suspensions of hard spheres in a viscous matrix [13].

7. Concluding Remarks

Morphological development in latex is dependent on the ambient conditions and the particular viscoelastic properties of the polymer or polymers. By applying two non-invasive techniques in conjunction, we have explored several phenomena of latex film formation, particularly the development of voids. At the same time, we have initiated these techniques for application to problems in colloid science. Their development is ongoing.

The results of this research are placed in a broader scientific and historical context in a recent review of latex film formation [14]. Many challenges remain in this field. Much previous work has already addressed how the presence of surfactants and salts influences flocculation and ordering of polymer colloids. These effects should now be correlated with latex film formation and final film properties. Similarly, both particle ordering [16] and phase separation [17] can be induced in blends of particles with appropriate bimodal size distributions, and future studies need to determine how these effects influence film formation.

8. Acknowledgements

Funding for this work was provided by the Colloid Technology Programme. We are grateful to ICI Paints (Slough, UK) for providing the latices. We benefited from discussions with Drs. P. Sakellariou, D. Taylor and C. Meekings at ICI Paints and Dr. G. Meeten at Schlumberger Cambridge Research. JLK is grateful for funding for one year from the Oppenheimer Fund at the University of Cambridge.

References

1. Joanicot, M.; Wong, K.; Richard, J.; Maquet, J. and Cabane, B. (1993)*Macromolecules,* **26**, 3168.
2. Wang Y. and Winnik, M.A. (1993)*J. Phys. Chem.,* **97,** 2507.
3. Crowley, T.L.; Sanderson, A.R.; Morrison, J.D.; Barry, M.D.; Morton-Jones, A.J. and Rennie, A.R. (1992) *Langmuir,* **8**, 2110.
4. Cameron, R.E. and Donald, A.M. (1994) *J. Microsc. (Oxford),* **173**, 227.
5. Meredith, P. and Donald, A.M. (1996) *J. Microsc. (Oxford),* **181**, 23.
6. Born, M. and Wolf, E. (1975) *Principles of Optics*, Pergamon, New York, p. 87.
7. Keddie, J.L.; Meredith, P.; Jones, R.A.L. and Donald, A.M. (1995) *Macromolecules,* **28**, 2673.
8. Wang, Y.; Juhúe, D.; Winnik, M.A.; Leung, O.M. and Goh, M.C. (1992)*Langmuir,* **8**, 760.
9. Keddie, J.L.; Meredith, P.; Jones, R.A.L. and Donald, A.M. (1996) in *Film Formation in Waterborne Coatings*, ACS Symp. Ser. 648, American Chemical Society Washington, D.C., pp. 332-348.
10. Sperry, P.R., Snyder, B.S., O'Dowd, M.L. and Lesko, P.M. (1994) *Langmuir,* **10**, 2619.
11. Mackenzie, J.K. and Shuttleworth, R. (1949) *Proc. Phys. Soc.*, **62**, 838.
12. Keddie, J.L.; Meredith, P.; Jones, R.A.L. and Donald, A.M. (1996) *Langmuir,* **12,** 3793.
13. de Kruif, C.G.; van Iersel, E.M.F.; Vrij, A and Russel, W.B. (1986) *J. Chem. Phys.*, **83**, 4717.
14. Keddie, J.L., accepted for publication in *Journal of Materials Science and Engineering R: Reports.*
15. Azzam, R.M.A. and Bashara, N.M. (1987) *Ellipsometry and Polarised Light*, North-Holland, Amsterdam, p. 270.
16. Bartlett, P.; Ottewill, R.H. and Pusey, P.N. (1992) *Phys. Rev. Lett.*, **68**, 3801.
17. Steiner, U.; Meller, A and Stavans, J. (1995) *Phys. Rev. Lett.*, **74**, 4750.

FLOW INDUCED AGGREGATION OF COLLOIDAL DISPERSIONS

J. W. GOODWIN*, J. D. MERCER-CHALMERS
School of Chemistry, University of Bristol,
Cantock's Close, Bristol BS8 ITS, UK.

The aggregation of dispersions of polystyrene particles has been studied using the irreversible change of viscosity with shearing time. The volume fractions ranged from 0.2 to 0.5 and the salt concentrations were in the range 0.01 -0.1 M with respect to potassium chloride. A critical shear rate for aggregation was observed and the flocs formed were studied using scanning electron microscopy. The flocs were densely packed prolate ellipsoids with an axial ratio of ~1.5 : 1. The critical shear rates were predicted from a model which used the dispersion viscosity as an approximation for the multi-body hydrodynamic interactions in estimating the pair interaction and hence the critical shear rate. The size of the flocs produced was found to be a function of shear rate at which aggregation was produced. Bimodal latex blends were studied as a function of number ratio of small to large particles. A long delay time was observed, followed by a rapid viscosity change indicative of the aggregation. This delay time was shown to be a function of the total strain applied to the dispersion.

1. Introduction

The aggregation of colloidal particles in the presence of a flow field has been studied since the early part of this century. This long history is indicative of the importance of the process in many practical situations. For example, in many processing situations stability of a dispersion is highly desirable, but sensitivity to aggregation may be induced at high solids loading and at high shear rates. In other situations, such as in separation processes, the control of the aggregation process and morphology of the subsequent aggregates is of great importance to the efficiency of the processing.

An excellent discussion of most of the important aspects of particle aggregation has been given by Russel, Saville and Schowalter [1] and their nomenclature will mostly be used here. The aggregation of colloidal particles in the absence of a shear field and solely under Brownian motion is termed perikinetic aggregation. The kinetics of dilute dispersions has been described by many workers such as Smoluchowski [2] and Fuchs [3]. Smoluchowski [2] also studied the aggregation of dilute systems under conditions where a shear field dominated the Brownian component. This was termed orthokinetic aggregation but the effect of a repulsive barrier was not included. Fuchs [3] had considered the problem of a repulsive

R. H. Ottewill and A.R. Rennie (eds.), Modern Aspects of Colloidal Dispersions, 61–75.

barrier under perikinetic conditions. Both van de Ven and Mason [4] and Zeichner and Schowalter [5] studied the effect of the pair potential on the orthokinetic aggregation rate. The analysis was extended to include weak Brownian motion by Feke and Schowalter [6]. The bulk of the studies have been focusing on dilute systems and pair interactions. However, the later stages of the interactions and the morphology of the "processed" aggregates is often of interest. The morphology formed from perikinetic aggregation of strongly attractive particles is of fractal form typical of diffusion limited aggregation (see e.g. 7), but when such systems are subject to subsequent processing denser, more uniform structures are formed [8].

This paper describes a study of the aggregation of model systems consisting of polymer latices, both unimodal and bimodal, as a function of shear conditions. The volume fractions of the dispersions was in the range 0.2 - 0.5. The resulting morphology of the processed aggregates was studied by scanning electron microscopy (SEM).

2. Shear Induced Aggregation

2.1 DILUTE DISPERSIONS

In quiescent fluids, interparticle interactions are adequately described by the total potential energy from the theory of Derjaguin and Landau and Verwey and Overbeek (see for example ref.1). The pair potential is given as the sum of the electrostatic and the dispersion terms. The latex particles used in this study were stabilised by strong acid groups from initiator fragments and so a constant charge approximation of the repulsive interaction was used. The pair potential expression utilised was :-

$$V_T = -2\pi\varepsilon a q_\delta^2 \bullet \ln[1-\exp(-\kappa h)] - A_{12}a/12h \quad (1)$$

Here ε is the permittivity, κ is the Debye-Huckel parameter describing the decay of the diffuse layer potential, h is the surface - surface separation of the particles which have radius a, $A_{12} = 7.3\mathrm{x}10^{-21}$ J is the net Hamaker constant for polystyrene in water and q_δ is the diffuse layer charge density. This latter quantity can be calculated from the ζ-potential from the expression [10] :-

$$q_\delta = \varepsilon\kappa \frac{k_B T}{ze}\left[2\sinh\frac{ze\psi}{2k_B T} + \frac{4}{\kappa a}\tanh\frac{ze\psi}{4k_B T}\right] \quad (2)$$

where z is the ion valency and e is the proton charge. A typical pair potential curve is shown in Figure 1. The maximum in the pair potential acts as an activation energy barrier to coagulation.

In shear flow, the collision frequency is increased over that due to diffusion. The relative magnitude of shear to diffusion energy is given by the Péclet number :-

$$Pe = 3\pi a^3 \eta_o \dot{\gamma} / k_B T \quad (3)$$

where η_o is the viscosity of the continuous phase and $\dot{\gamma}$ is the shear rate. Clearly the combination of large particles and high shear rates favour the collisions due to the convective field. The hydrodynamic forces that act on a pair of colliding particles give a maximum compressive force followed by a maximum extensional force when the particle centres are oriented along the principle axes of shear, i.e. at $\pi/4$ to the flow

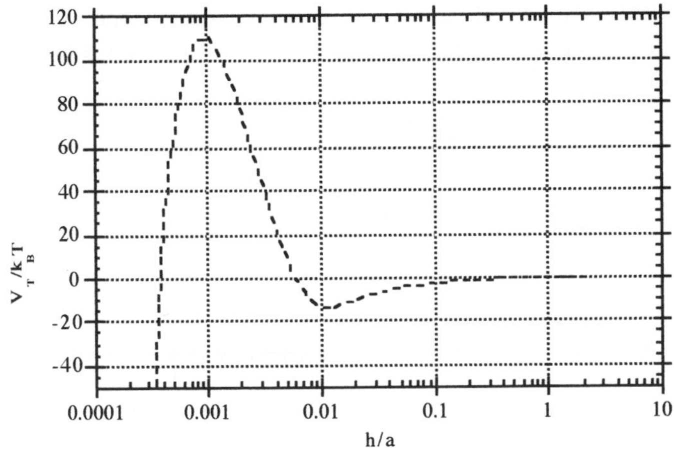

Figure 1. Pair potential for latex particles with a = 489 nm in 5×10^{-2} M Sodium Chloride solution with $q = 2.25\ \mu C\ cm^{-2}$.

direction. Due to the asymmetry of the pair potential, the important consideration is the compression. The magnitude of the compression from the hydrodynamic interaction compared to the colloidal term is :-

$$N_f = 6\pi\eta_o a^3\dot{\gamma} / \Delta V_T \tag{4}$$

Here ΔV_T is the height of the repulsive barrier taken from the value of the secondary minimum.

2.2 CONCENTRATED DISPERSIONS

In concentrated dispersions the rigorous trajectory analysis as carried out by Feke and Schowalter [6] for example, cannot be carried out. However, the feature which is the focus of this paper is the increasing ease of the induction of aggregation which occurs as the volume fraction is increased. The problem is not one of rate but of the threshold of stability which is indicated by $N_f > 1$. Equation 4 does not have any concentration dependence beyond that implied through changes in the pair potential due to the additional ion content when the counter-ion number become significant as compared to added electrolyte.

The feature that is absent from Equation 4 is the effect of multi-body hydrodynamic interactions. The straightforward solution is to use the suspension viscosity in Equation 4:-

$$N_f = \frac{6\pi a^3\dot{\gamma}}{\Delta V_T}\eta(\varphi,\dot{\gamma}) \tag{5}$$

The viscosity at moderate to high volume fractions is dependent on both solids loading and shear rate as illustrated in Figure 2. The range of the colloidal forces at large values of κa is sufficiently small that the viscosity of the suspensions will be close to that of hard spheres up to shear rates that produce coagulation. The variation of viscosity with volume fraction has been given by Krieger [11] and the high and low shear limits to the viscosity [12] as :-

$$\eta(0) = \eta_o\left(1 - \frac{\varphi}{0.54}\right)^{-1.35} \tag{6a}$$

$$\eta(\infty) = \eta_o\left(1 - \frac{\varphi}{0.605}\right)^{-1.51} \tag{6b}$$

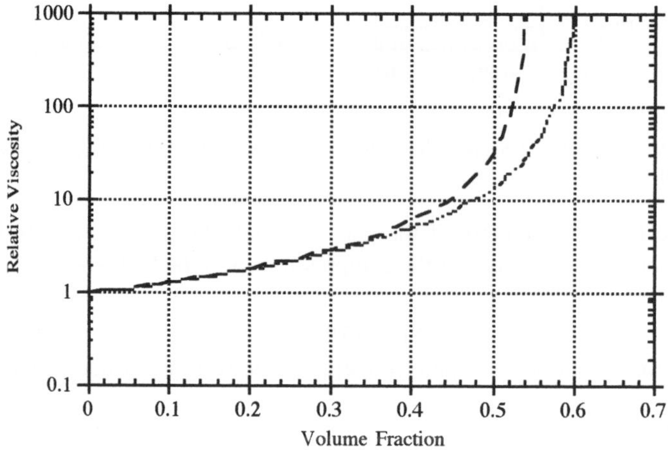

Figure 2. The relative viscosity as a function of volume fraction calculated for hard spheres from Equation 6.

The shear rate dependence can be described by, for example, the Cross equation [13] which for monodisperse spheres is :-

$$\eta = \eta(\infty) + \frac{\eta(0) - \eta(\infty)}{1 + \alpha\dot{\gamma}} \tag{7}$$

where α is the inverse of the shear rate at the mid-point of the curve, i.e. it can be identified from the Pe =1 condition.

By restricting the discussion to particles which are stable under quiescent conditions, only the high shear limit of the viscosity needs to concern us here. Equation 5 and Equation 6b can be combined to define a critical shear rate where coagulation under shear will be observed :-

$$\dot{\gamma}_{crit} = \frac{\Delta V_T}{6\pi a^3 \eta_o\left(1 - \frac{\varphi}{0.605}\right)^{-1.51}} \tag{8}$$

Figure 3 illustrates the marked dependence of the critical shear rate on the volume fraction of a dispersion, especially as $\varphi > 0.4$. Figure 4 shows the radius dependence of the critical shear rate for a value of $\Delta V_T = 50\ k_BT$.

3. Experimental

3.1 PREPARATION AND CHARACTERISATION OF THE LATEXES

A series of polystyrene latexes were synthesised as surfactant free systems [9]. After preparation the latexes were filtered while hot through glass wool and then extensively dialysed against mixtures of methanol and de-ionised water in order to remove electrolytes and other soluble materials. The water used was from a Purite system which consisted of a reverse osmosis de-ionisation followed by mixed bed ion-exchange treatment, further cleaning by means of an activated charcoal column and

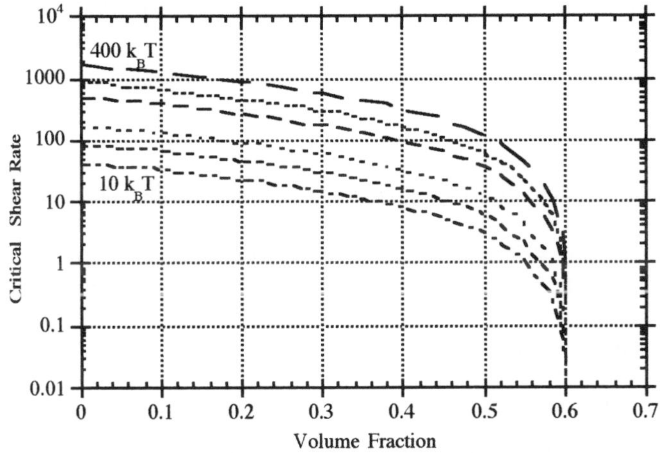

Figure 3. Calculated critical shear rates for a colloidal dispersion of 374 nm particles with ΔV_T = 10 k_BT; 40 k_BT; 120 k_BT; 200 k_BT; 400 k_BT.

filtration through a 0.2 μm filter. Initially a 1 : 1 mixture of water : methanol was used and this was followed by progressive dilutions to water. At least 25 changes of dialysate were made in order to ensure that the latexes were sufficiently clean. The latexes were stored in the refrigerator at 5 °C prior to use. The particle sizes of the particles were determined from micrographs prepared on an Hitachi HS 7S transmission electron microscope using a Carl Zeis TGZ3 particle size analyser. At least 500 particles were counted and a micrograph of a diffraction grating replica enabled the magnification to be determined. The particle sizes of the latices used in this study are given in Table 1.

TABLE 1. Particle size of polymer latexes..

	Diameter nm	Coef.of Variation %
plsk 9	266	5
jm 25	570	5
jm 17	747	3
jm 20	822	2
jm 28	978	5

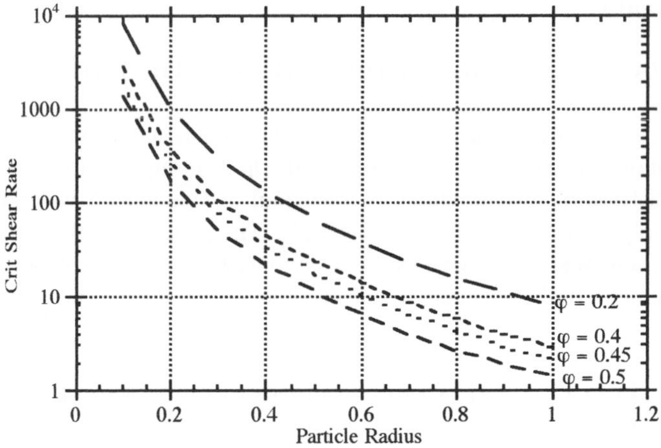

Figure 4. The calculated dependence of the critical shear rate on particle radius in μm with a $\Delta V_T = 50\ k_BT$.

Electrophoretic mobility measurements were made on the latexes in potassium chloride solutions using the PenKem System 3000. In addition latex samples jm 26 and jm 28 were studied in 40% glycerol solutions. The electrophoretic mobilities were converted to ζ-potential by means of the O'Brien and White computer program [13].

3.2 RHEOLOGICAL MEASUREMENTS

All the experiments were carried out using the viscometry mode of a Bohlin VOR Rheometer. The measuring geometry used was a C14 DIN standard cup and bob. Latex samples were prepared with the desired levels of sodium chloride either by dialysis or centrifugation and redispersion. Dialysis was the preferred method for the higher electrolyte concentrations. Experiments were carried out in two stages.

Firstly, a latex sample was introduced into the rheometer and the shear programme was set to vary the shear rate continuously up and then down. The upper limit was $\sim 10^3\ s^{-1}$, as beyond this rate the Reynolds Number was too high for a laminar flow field to be maintained and secondary flow patterns were introduced. At low volume fractions Newtonian behaviour was observed, but as the volume fractions were increased to values in excess of 0.3, shear thinning behaviour was observed. If the up and down curves of the shear stress, σ, versus shear rate, $\dot{\gamma}$, were superimposable, the latex was considered to be stable to the maximum shear used. The upper threshold was increased up to the usable instrument limit. If the latex was still stable, a different salt concentration and/or volume fraction was used. When instability occurred, thixotropy was seen, i.e. an hysteresis loop was observed with the return leg following a lower track than the up-sweep as illustrated in Figure 5. This defined a shear rate above which instability could be observed.

Secondly, a fresh sample of latex was placed in the rheometer and the shear stress was observed as a function of time at the shear rate corresponding to values above the onset of instability. This process covered an extended timescale and an environmental chamber was added to the instrument to control evaporation of the solvent phase. After shearing, the samples were removed and used for aggregate morphology studies employing scanning electron microscopy, SEM. Figure 6 shows a typical form of the viscosity time profiles obtained. If shearing was stopped and restarted, the viscosity values did not start from the original value but from the value reached when the shearing process was interrupted. This clearly indicated that an irreversible change had occurred to a new stable system. The data in Figure 6 is plotted against the dimensionless group, the reduced time, t_{red}, which is simply the product of the shearing time and the shear rate.

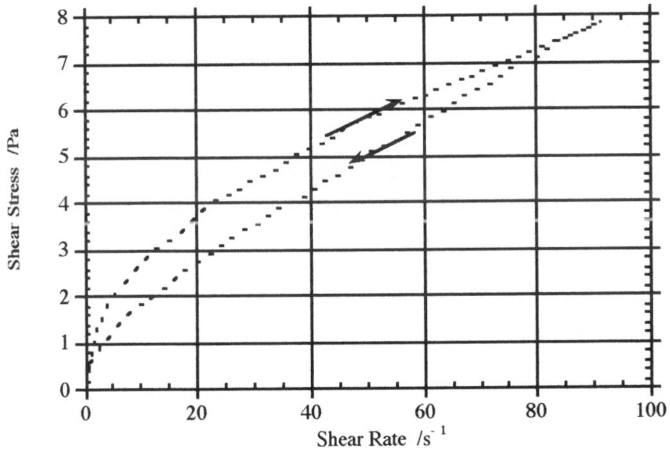

Figure 5. Typical rheogram for an unstable latex. Particle radius 379 nm in KCl at $5.5x10^{-2}$ M at $\varphi = 0.4$.

Latex dispersions were made up to contain 40% glycerol in the continuous phase in order to reduce Brownian motion due to the increased viscosity, and also to increase the shear forces in the system. Of course the glycerol will also change the permittivity and this will have altered the interparticle forces although in a predictable fashion. The lower permittivity of the systems containing glycerol changes the magnitude of the electrostatic interactions as well as the decay length of those interactions.

Finally, a series of blends of latexes with different particle sizes was prepared. The blends are detailed in Table 2 which gives the total volume fraction as well as the ratio of small to large particles in the bimodal systems. The viscosities of the systems were followed as a function of time, and systems were examined by SEM at the end of the runs.

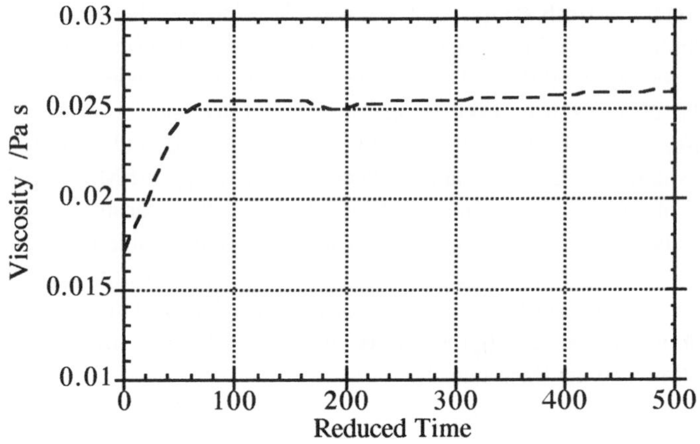

Figure 6. Typical viscosity versus reduced time profile for unstable latex. Particle radius 379 nm in KCl at 5×10^{-2} M at $\varphi = 0.4$ and sheared at 291 s-1.

3.3 COMPARISON WITH EXPERIMENT

The experimental data obtained for the onset of an irreversible shear thickening with shearing time, as illustrated in Figure 6 are given in Table 2. The calculated values are also given, as are the calculated values of ΔV_T. The agreement between the calculated and experimental values is satisfactory, especially when it is recalled that shear rates of up to two orders of magnitude higher would have been predicted if the effect of solids loading had not been accounted for. The experimental values can be expected to be tending towards higher values than the predictions as the threshold must be exceeded for the effect to be easily observed in an experiment.

The reason for the very high values of the critical shear rate seen with the experiments utilising 40% glycerol is less readily explained. The viscosity increase due the addition of glycerol to a level of 40% by weight in the continuous phase increases the viscosity by a factor of 3.6 which was expected to reduce the value of the critical shear rate. However the decrease in relative permittivity increases the interaction energy by reducing the value of κ. The rate of fluid drainage from between particles controls the timescale of the aggregation process, and if the drainage were very slow, then the onset of the viscosity rise would be delayed. The two glycerol data points tend to support this, as the system with the lower shear force, has the longer delay.

TABLE 2. Calculated and experimental values of the critical shear rate.

φ	Diam. /nm	NaCl /M	$\Delta V_T / k_B T$	$\dot{\gamma}_{crit}$ / s^{-1}	Expt. $\dot{\gamma}_{crit}$ /s^{-1}	t_{red}
0.4	374	0.05	95	77	291	2.0×10^4
0.45	374	0.03	135	72	41	3.7×10^4
0.5	411	0.10	40	9	15	1.9×10^4
0.5	411	0.08	65	15	23	-
0.35	285	0.018/ glyc	195	489	1459	3.6×10^5
0.4	285	0.018/ glyc	195	246	461	3×10^3

After shearing was completed, samples were removed for SEM analysis. The samples were air dried, mounted on an aluminium stub and then sputter-coated with gold. Photographs were then taken using an Hitachi SEM. Figure 7, 8 and 9 shows that the flocs produced were densely packed and ellipsoidal in shape. The flocs produced at a shear rate of 1459 s^{-1} were half the diameter of those produced at 461 s^{-1} i.e. the flocs would have experienced similar shear forces as $F_h \propto \dot{\gamma} a_{floc}^2$.

Figure 7. Flocs of jm25 formed after shearing latex in 40% glycerol and 0.03M KCl at 461 s^{-1}.

Figure 8. jm25 after shearing at 461 s^{-1},. $\varphi = 0.35$, 1,000X..

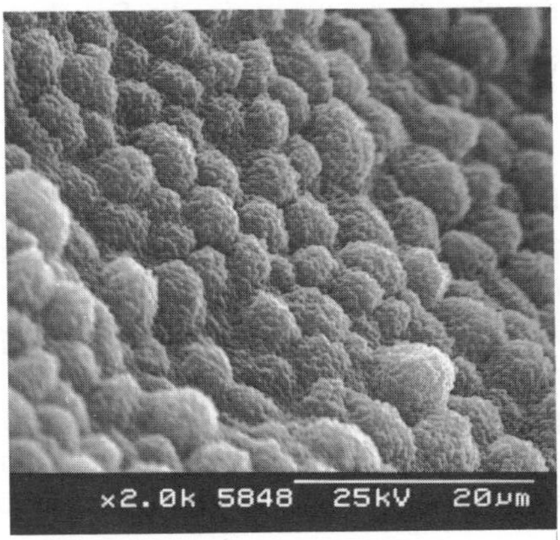

Figure 9. jm25 after shearing at 1459 s^{-1} $\varphi = 0.4$, 2,000X.

The shear forces acting on a pair of flocs can be calculated in a similar manner to that between normal particles. However in this case, we are concerned with the steady state aggregation/ dis-aggregation process. Figure 10 illustrates the model that we will consider. As aggregates collide, the shear forces firstly compress them together and then extend the unit until it separates into two parts. The figures show that densely packed spherical aggregates are produced by this "shear processing" action. The asymmetry of the pair potential means that it is easier to cause two particles to aggregate than to separate them i.e. if the energy available from the shear field is sufficient to result in separation, it is reasonable to model two flocs coming together and reorganising and then being pulled apart. as shown schematically in Figure 10. We model this by equating the cohesive energy of the intermediate spherical aggregate to the shear field energy whilst recognising that work done in rearranging the structure to

Figure 10. Schematic of the aggregation/dis-aggregation occurring during shear processing.

spheroidal units during this process is ignored. This reorganisation and subsequent break-up into two equal volumes will drive the system towards a narrow size distribution. Using dense random packing for the aggregates with a co-ordination number of 10, and the radius of the combined unit as $a_f 2^{1/3}$, we have the cohesive energy as :-

$$E_c = 7.63\Delta V_T'(a_f / a)^2$$

This is the work done by the shear field at break-up at a shear rate of $\dot{\gamma}_c$:-

$$E_h = 6\pi\eta a_f^3\dot{\gamma}_c$$

Equating these and using the Dougherty-Krieger equation for the viscosity of the suspension of aggregates gives :-

$$a_f = \frac{0.4\Delta V_T'(1 - \varphi / 0.39)}{a^2\eta_o\dot{\gamma}_c} \tag{9}$$

Here $\Delta V_T'$ is the energy barrier to dis-aggregation, i.e. the energy difference between the primary minimum and the primary maximum. The results from this calculation are illustrated in Figure 11. The energy barrier to dis-aggregation of the particles illustrated in Figures 8 is ~670 k_BT, if the closest approach is taken as 0.5nm. Of course this can only be taken as a guide as the reliability of Equation 1 at these distances must be very poor. However the sizes are of the correct order and a higher shear rate reduces them. In addition, it is important to note that ultra high shear rates would be required to reduce floc size to a value close to the particle size.

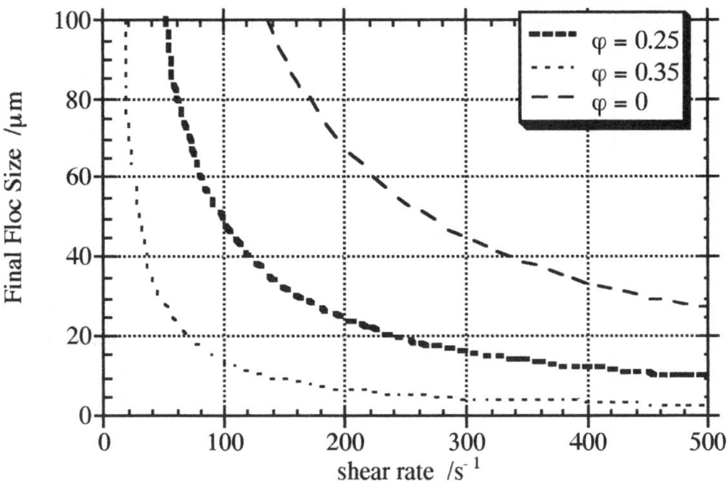

Figure 11. The radius of flocs predicted by Equation 9 as a function of volume fraction at $\Delta V_T' = 670\ k_BT$.

3.4 LATEX BLENDS

A series of bimodal systems were prepared by mixing two latexes with polystyrene particles of radii 489 nm and 133 nm. An electrolyte concentration of 1×10^{-2} M potassium chloride was used throughout these experiments. The stock latexes were concentrated initially by high speed centrifugation, but then ultrafiltration was used so that redispersion of a packed sediment was avoided. Aliquots of the latex concentrates were placed in the rheometer cup, mixed slowly with a plastic spatula, and the bob slowly lowered into the cup. The viscosity of the systems were then recorded whilst the systems were sheared at a constant shear rate. Table 3 gives the latex concentrations and the reduced shear times. These were calculated from the sharp rise in viscosity after shearing the sample as illustrated in Figure 12, which shows the typical behaviour of all the blends studied. Samples of the sheared blends were examined in the SEM, using the same procedure as that used for the samples in glycerol/water mixtures.

Figure 13 shows that the reduced time was approximately constant at 2×10^5, confirming that the aggregation time was simply dependent on the total strain. This in turn indicates that all the shear rates were in excess of the critical value for aggregation. The reduced time increased as the ratio of small particles to large particles was increased. Figure 14 shows a plot of the data and the linear curve fit obtained over the range studied at $\varphi \sim 0.2$. Comparison of run 123 with run 138, and also runs 121 and 137, indicates that the reduced time is decreased by at least a factor of 3 as the volume fraction is increased by a factor of 1.5 from 0.25 to 0.36. This was greater than the ratio of the relative viscosities where the increase was only a factor of 1.75.

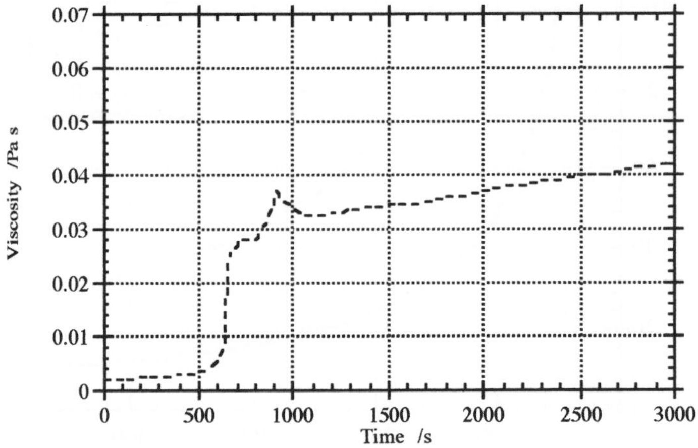

Figure 12 Viscosity as a function of shearing time for latex blend jmc 115 at a shear rate of 731 s^{-1}.

TABLE 3. Shear induced aggregation of latex blends as a function of composition and shear rate.

Run No.	φ	Small:Big	shear rate s^{-1}	t_{red} /x10^{-5}
115	0.22	19:1	731	4.4
116	0.24	12:1	731	2.9
117	0.186	31:1	731	9.5
118	0.22	20:1	731	5.8
119	0.22	20:1	731	14.6
120	0.24	16:1	731	4.0
121	0.24	16:1	581	4.6
123	0.25	14:1	581	4.4
126	0.36	19:1	116	2.5
129	0.36	19:1	146	1.3
130	0.36	19:1	146	1.5
131	0.36	19:1	184	1.5
132	0.36	19:1	231	1.7
133	0.36	19:1	291	2.9
134	0.36	19:1	367	2.9
135	0.36	19:1	461	1.4
136	0.36	18:1	461	3.5
137	0.36	18:1	581	1.5
138	0.37	14:1	581	1.3

The scanning electron micrographs again showed uniform flocs which were near spherical. The axial ratios of the prolate ellipsoidal shapes were ~1.5 : 1. Figure 15 and Figure 16 show the micrographs with the magnification shown by the 10μm bar. Both particle sizes are readily seen but fewer large particles are in evidence than small ones. The lower magnification picture in Figure 17 confirms the uniformity of the floc types found. The broad implication is that the majority of the large particles are inside of the flocs which have an outer layer rich in small particles.

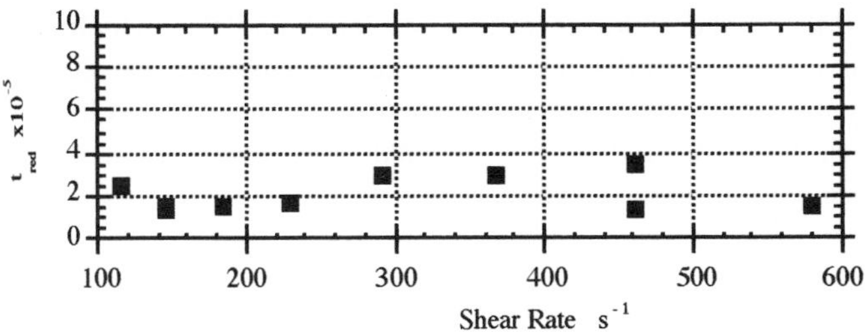

Figure 13. The shear rate dependence of the reduced time at $\varphi \sim 0.36$ and S:B = 14:1

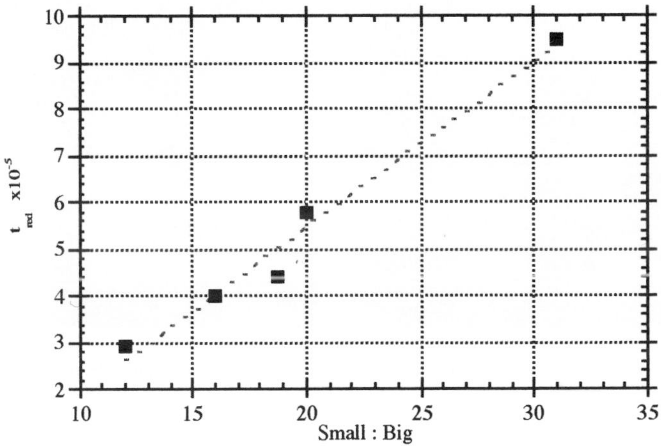

Figure 14. The reduced time for aggregation as a function of ratio of small particles to large particles in the dispersion at $\varphi = \sim 0.2$

The effect of blending particles to give a bimodal distribution is to induce instability at lower electrolyte concentrations than if a unimodal system was used. The ΔV_T values for the various combinations used are given in Table 4. It can be seen from this table that the large : large particle combination is the most shear sensitive although the energy barrier is highest. This would suggest that the aggregation would start with the large particles. The aggregates would then sweep up the smaller particles. This is confirmed by Figure 14 where the shortest reduced time is for the

TABLE 4. Energy barrier to aggregation of polystyrene particles with -60 mV on both particles and at 0.01 M KCl. The ratios of the energy maxima and critical shear rates for aggregation are also given.

sizes /nm	ΔV_T /kT	ratio of energies	ratio of crit. shear rates
489 : 489	1,000	1	1
489 : 133	450	0.45	3
133 : 133	270	0.27	12

systems with the higher number density of large particles which maximises the probability of favourable collisions. Aggregate break-up will be dominated by the lowest energy maximum and hence the aggregates can be expected to be have a core rich in large particles and a shell rich in small particles. This is confirmed by the micrographs shown in Figures 15 and 16.

Figure 15. Flocs from Run 115 Latex Blend at 5,000X

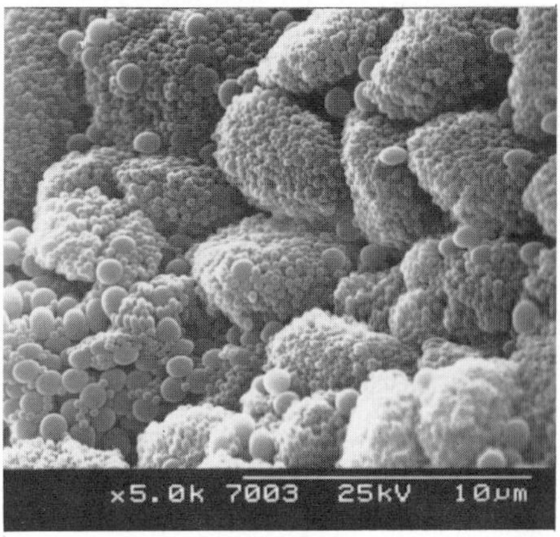

Figure 16. Flocs fron Run 116 Latex Blend at 5,000X

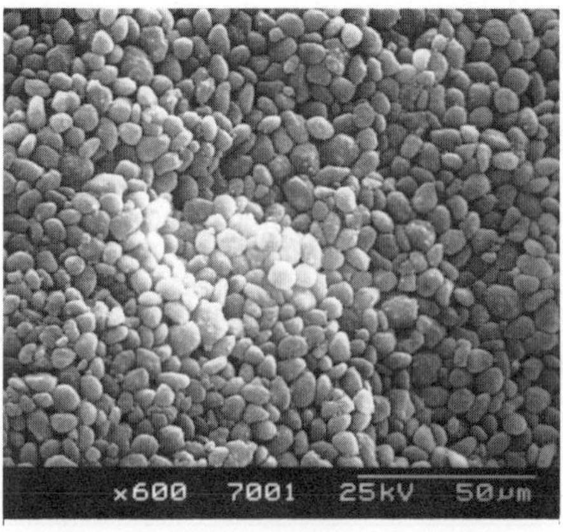

Figure 17 . Flocs from Latex Blend Run 115 after shearing at 731 s^{-1} at 600X.

4. Acknowlegements

We acknowledge with thanks support from the DTI Colloid Technology Project, including also, ICI, Schlumberger, Unilever and Zeneca.

5. References

1. Russel, W.B., Saville, D.A. and Schowalter, W.R. (1989) *Colloidal Dispersions*, University of Cambridge Press, Cambridge, U.K. Chapter 8.
2. von Smoluchowski, M. (1917) *Z. Phys. Chem.*, **92**, 129.
3. Fuchs, N. (1934) *Z. Phys.*, **89**, 736.
4. van de Ven, T.G.M. and Mason, S.G. (1976) *J. Colloid Int. Sci.*, **57**, 505.
5. Zeichner, G.R. and Schowalter, W.R. (1977) *A. I. Chem. E. J.*, **23**, 243.
6. Feke, D.L. and Schowalter, W.R. (1983) *J. Fluid Mech.*, **133**, 17.
7. Witten, T.A. and Sander, L.M. (1981) *Phys. Rev. Let.*, **47**, 1400.
8. Mills, P.D.A., Goodwin, J.W. and Grover, B.W. (1991) *Colloid Polym. Sci.*, **269**, 949.
9. Goodwin, J.W., Hearn, J., Ho, C.C. and Ottewill, R.H. (1974) *Colloid Polym. Sci.*, **259**, 464.
10. Loeb, A.L., Overbeek, J. Th. G. and Wiersema, P.H. (1961) *The Electrical Double Layer Around a Spherical Colloid Particle*, MIT Press, Cambridge, Mass., U.S.A.
11. Krieger, I.M. (1972) *Adv. Colloid Int. Sci.*, **3**, 111.
12. Goodwin, J.W. and Ottewill, R.H. (1991) *J. Chem. Soc. Faraday Trans.*, **87**, 357.
13. O'Brien, R.W. and White, L.R. (1978) *J. Chem. Soc. Faraday Trans. II*, **74**, 1607.

TWO-COLOUR DYNAMIC LIGHT SCATTERING STUDIES OF COLLOIDAL DISPERSIONS

Two-Colour Dynamic Light Scattering

P.N. PUSEY, P.N. SEGRÈ, S.P.MEEKER, A. MOUSSAÏD
and W.C.K. POON
Department of Physics and Astronomy
The University of Edinburgh
Mayfield Road, Edinburgh, EH9 3JZ, UK

Two-colour dynamic light scattering (TCDLS) is a relatively new technique which effectively suppresses multiple scattering, allowing accurate measurements to be made of the Brownian dynamics of turbid samples. It operates by cross-correlating scattered light of two different colours at exactly the same scattering vector. Using TCDLS we have measured the dynamic scattering functions $f(Q,\tau)$ of dispersions of "hard-sphere" poly-methylmethacrylate particles in cis-decalin over wide ranges of concentration and scattering vector Q. The decays of $f(Q,\tau)$ at short times show good agreement with computer simulation. At longer times the decays can be associated with "structural relaxation", the rearrangement of the relative positions of a particle and its cage of neighbours. We have found that the rate of structural relaxation shows exactly the same dependence on the concentration of the dispersion as its zero-shear-rate viscosity. We have also found an intriguing, as yet unexplained, scaling of the scattering functions which suggests that structural relaxation is controlled by the self diffusion of the individual particles. Recently we have exploited the fact that TCDLS provides a measurement of the ratio of single scattering to total (single + multiple) scattering to perform static light scattering measurements on turbid samples. In particular we have studied mixtures of colloidal particles and random-coil polymer, finding enhanced large-scale concentration fluctuations in "marginal" colloidal liquids.

1. Introduction

Over the past 25 years, dynamic light scattering (DLS) has proved to be a powerful technique for investigating the Brownian dynamics of a wide range of systems including colloidal suspensions. However, it is only if the sample is relatively transparent, so that single scattering dominates, that it is possible to relate the quantity measured by DLS to simple properties of the scattering medium (Equations (1) to (3)). The light scattered by optically turbid media contains contributions from both single

R. H. Ottewill and A.R. Rennie (eds.), Modern Aspects of Colloidal Dispersions, 77–87.

and multiple scattering. The relationships between the properties of doubly, triply etc., scattered light and those of the medium are complicated. Thus the application of DLS, in its usual form, is restricted to transparent media. In this paper we describe a relatively new technique, two-colour dynamic light scattering (TCDLS), which effectively suppresses multiple scattering and selects just the single scattering (see Section 3 below), allowing the study of optically turbid samples. The technique operates by constructing the time cross-correlation function of scattered light of two different colours at exactly the same scattering vector.

We review TCDLS studies of the dynamics of concentrated suspensions in a liquid of colloidal particles which interact like hard spheres. These experiments have yielded two unexpected findings. First, the rate of structural relaxation (defined below) of the suspensions shows the same dependence on suspension concentration as the inverse of their zero-shear-rate viscosity. Second, the intermediate scattering functions, measured by TCDLS, show an interesting scaling property which suggests that structural relaxation is controlled by self diffusion of the particles. Detailed descriptions of this work have been published recently. Thus this paper will give only a very brief summary which directs the reader to the literature (and follows Reference [1] quite closely).

Dynamic information about a sample is obtained by analysing the time-dependent part of the TCDLS intensity cross-correlation function. Useful information, namely the ratio of the single scattered intensity to the total scattered intensity, is also contained in the amplitude (the zero-time value) of the cross-correlation function. Thus it is possible to obtain the angular dependence of light scattered *singly* by a turbid sample by combining measurements of the angular dependences of the total scattered intensity *and* the cross-correlation amplitude. In Section 7 we describe the application of this new method to the determination of the static structure factors of the "liquid" phases of colloid-polymer mixtures in which the presence of free polymer induces an effective attraction between the particles by the depletion mechanism.

2. Background

Two-colour dynamic light scattering, and ordinary DLS for transparent samples, measures the normalised autocorrelation function of the amplitude of the singly-scattered light field (see Section 7 for further details). This quantity is equal to the normalised intermediate scattering function $f(Q,\tau)$ of the suspension,

$$f(Q,\tau) \equiv \frac{F(Q,\tau)}{F(Q,0)}, \tag{1}$$

where the intermediate scattering function $F(Q,\tau)$ is given by

$$F(Q,\tau)=\frac{1}{N}\left\langle \sum_{j=1}^{N}\sum_{k=1}^{N}\exp\left\{i\boldsymbol{Q}\cdot\left[\boldsymbol{r}_j(0)-\boldsymbol{r}_k(\tau)\right]\right\}\right\rangle, \tag{2}$$

and

$$F(Q,0)=S(Q) \tag{3}$$

where $S(Q)$ is the static structure factor. Here N is the number of particles, $\boldsymbol{Q}$ is the scattering vector and $\boldsymbol{r}_j(t)$ the position of particle j at time t. As can be seen from its definition, in general the intermediate scattering function measures a collective motion of the particles and can be recognised as the autocorrelation function of spatial Fourier components of the sample's density fluctuations of wavelength $2\pi/Q$.

In concentrated suspensions, where the fraction ϕ of the suspension's volume which is occupied by the particles may be 0.5 or larger, the static structure factors resemble those of simple atomic liquids, showing pronounced diffraction peaks at $2\pi/Q\approx 2R$, where R is the particles' radius. The dominant structure in the suspension, which gives rise to this peak, is the short-ranged ordering, or cage, of particles surrounding a given particle.

In a dilute suspension, where interactions between the particles can be neglected, the intermediate scattering function takes the simple form $f(Q,\tau)=\exp\left(-D_0 Q^2\tau\right)$, where D_0 is the free-particle (Stokes-Einstein) diffusion coefficient. In a concentrated suspension, due to both direct and hydrodynamic interactions between the particles, $f(Q,\tau)$ has a more complicated dependence on $Q^2\tau$, the slowest decay being found at the peak of $S(Q)$. Furthermore $f(Q,\tau)$ decays via a two-stage process: an initial exponential decay,

$$f(Q,\tau)=\exp\left[-D_S(Q)Q^2\tau\right], \qquad \tau \ll \tau_R, \tag{4}$$

where τ_R is the "structural relaxation time", essentially the lifetime of a particle's cage of neighbours; and a second, slower, approximately exponential decay at long times,

$$f(Q,\tau)\propto\exp\left[-D_L(Q)Q^2\tau\right], \qquad \tau \gg \tau_R, \tag{5}$$

with $D_L(Q)<D_S(Q)$.

3. Experimental

The basic idea of multiple scattering suppression in DLS, due to Phillies [2], is to use two illuminating laser beams and two detectors whose outputs are cross-correlated. The experiment is arranged so that, although the beam-detector pairs have different geometries, their associated scattering vectors are identical. Thus, for single scattering, each detector "sees" the same spatial Fourier component of the sample's density fluctuations. However for multiple scattering it can be shown that this degeneracy is broken. As a consequence, the time-dependent part of the measured intensity cross-correlation function (proportional to the square of the intermediate scattering function, Eq. (1)) reflects only single scattering, and multiple scattering contributes merely to the time-independent "baseline".

Phillies' original experiment in 1981 used counter-propagating laser beams of the same colour with detectors set at 90° on either side of the beams. While this experiment demonstrated clearly the suppression of multiple scattering, the arrangement could not be readily adapted to other scattering angles. In 1990 Schätzel and co-workers [3] proposed and demonstrated a more versatile equipment, based on the same principle, which by using laser beams of two different colours (the blue, 488 nm, and green, 514.5 nm, lines of the argon ion laser) can be operated over a range of angles, ~20°-140°, similar to that of conventional single-beam DLS equipment. This TCDLS equipment was subsequently developed in a collaboration between Schätzel and ALV, Langen, Germany into the commercial instrument used in the present work. A detailed description of the equipment and its operation is given in [4].

The suspensions consisted of sterically-stabilised particles of poly-methylmethacrylate in cis-decalin [see e.g. 5]. Due to a slight difference between the refractive indices of the particles and the liquid, these samples were quite turbid resulting in strong single scattering and significant multiple scattering. The multiple scattering was suppressed by the TCDLS technique, and the strong single scattering dominated that from dust and the sample cell walls, allowing the collection of accurate data. The polymer used in the experiments of Section 7 was polystyrene.

4. Short-Time Diffusion

The short-time diffusion coefficients $D_S(Q)$ describe the average motions of the particles over distances small compared to their radius and reflect both direct and hydrodynamic interactions between the particles. Extensive measurements [5] were made of $D_S(Q)$ as functions of both scattering vector Q and suspension volume fraction ϕ. These were compared with the predictions of theory and computer simulation, good agreement being found with the latter [5].

5. Long-Time Diffusion

The long-time diffusion coefficients $D_L(Q)$ describe motions of the particles over distances comparable to, or larger than, their radius. There is no complete theory to date of long-time diffusion. The new results outlined below may provide insights which will stimulate theoretical developments.

As noted above, in a dense fluid-like assembly of hard spheres the dominant structure, which gives rise to the main peak in $S(Q)$ at $Q = Q_m$, is the cage of particles surrounding a given particle. Thus it can be argued that the long-time decay of $f(Q_m, \tau)$, the intermediate scattering function measured at $Q = Q_m$, reflects the dominant structural relaxation of the system so that $D_L(Q_m)$ is a measure of the rate of structural relaxation. By comparing measurements of $D_L(Q_m)$ with measurements of the zero-shear-rate viscosity η of the suspensions we have found [6] that the rate of structural relaxation shows the same dependence on suspension concentration as the inverse of the viscosity over the whole range $0 < \phi < 0.50$, i.e. that

$$\frac{D_L(Q_m)}{D_0} = \frac{\eta_0}{\eta}, \tag{6}$$

where η_0 is the viscosity of the liquid in which the particles are suspended (see Figure 1). While one would certainly expect these two quantities to show similar

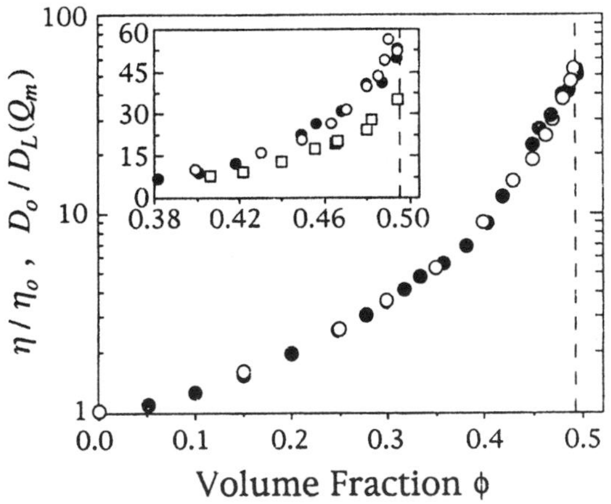

Figure 1. Relative viscosity η/η_0 (open circles) and inverse rate of structural relaxation $D_0/D_L(Q_m)$ (filled circles) versus volume fraction ϕ of suspensions of PMMA spheres (from [6], q.v. for an explanation of the data points represented by squares in the inset).

dependences on concentration - the processes of simple shear flow and structural rearrangement both involve the relative motions of neighbouring particles - the apparent identity found experimentally is surprising and remains to be explained by theory.

6. Scaling of the Intermediate Scattering Functions

As part of an attempt to understand better the mechanism of structural relaxation we made a second surprising discovery [7]. This was that for $QR > 2.7$, a range of scattering vector Q which encompasses most of the strong variation of structure factor $S(Q)$ including the main peak, plots of $\ln f(Q,\tau)/D_S(Q)Q^2$ against τ, measured at different values of Q, lay on a master curve. As can be seen from Equation (4), this way of plotting the data ensures that they superimpose at short times (since $\ln f(Q,\tau)/D_S(Q)Q^2 = -\tau$, for $\tau << \tau_R$). What is surprising is the additional

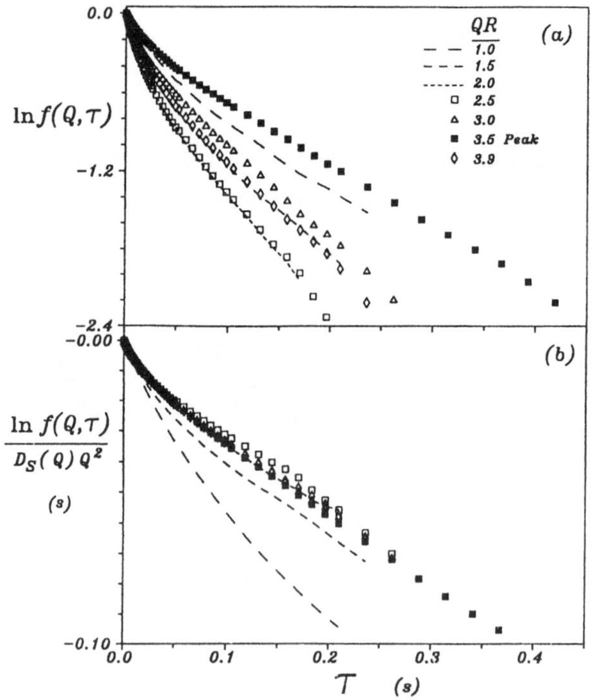

Figure 2. (a) Logarithm of normalised intermediate scattering functions $\ln f(Q,\tau)$ versus time τ for a PMMA suspension of volume fraction $\phi = 0.465$ for different values of QR as indicated. (b) Same data plotted as $\ln f(Q,\tau)/D_S(Q)Q^2$ versus τ, showing scaling for $QR > 2.5$ (from [5]).

superimposition of data at intermediate and long times, $\tau \geq \tau_R$ (see Figure 2). As noted in [7], this finding implies that, for $QR > 2.7$, the intermediate structure factor can be written

$$f(Q,\tau) \approx \exp\left(-\frac{D_S(Q)}{D_S(\infty)} \frac{Q^2}{6} \langle \Delta r^2(\tau) \rangle \right), \qquad (7)$$

where $\langle \Delta r^2(\tau) \rangle$ is the mean-square displacement of a single particle, suggesting that structural relaxation is controlled by self diffusion. Although previous work [8] has suggested a connection between structural relaxation and self diffusion, the detailed scaling implied by Equation (7) awaits a full theoretical explanation.

7. Static Light Scattering by Turbid Samples Using TCDLS

Two-colour dynamic light scattering measures $g_C^{(2)}(Q,\tau)$, the normalised time cross-correlation function of the total (single + multiple) scattered intensities, $I_B(Q,t)$ and $I_G(Q,t)$, of blue and green light, defined by

$$g_C^{(2)}(Q,\tau) \equiv \frac{\langle I_B(Q,0) I_G(Q,\tau) \rangle}{\langle I_B(Q) \rangle \langle I_G(Q) \rangle}. \qquad (8)$$

Under the usual conditions of operation, this quantity is related to the intermediate scattering function $f(Q,\tau)$ (Equation (1)) of the sample by [4]

$$g_C^{(2)}(Q,\tau) = 1 + \beta^2 \beta_{OV}^2 \beta_{MS}^2 [f(Q,\tau)]^2, \qquad (9)$$

where β, β_{OV} and β_{MS} are factors smaller than one. The first, β, is the usual quantity encountered in ordinary DLS, determined by the ratio of the area of the detector aperture to the coherence area of the scattered light. The "overlap" factor, β_{OV}, allows for the fact that the scattering volumes for the blue and green light are slightly different. The final factor, β_{MS}, is given by the ratios of the average intensities of single scattered light, $\langle I_B^S(Q) \rangle$ and $\langle I_G^S(Q) \rangle$, to those of the total scattered light:

$$\beta_{MS}^2(Q) = \frac{\langle I_B^S(Q) \rangle}{\langle I_B(Q) \rangle} \frac{\langle I_G^S(Q) \rangle}{\langle I_G(Q) \rangle}. \qquad (10)$$

When studying the dynamics of the medium, the intermediate scattering function $f(Q,\tau)$ is the quantity of interest and the β's are just unimportant "instrument constants". But it is apparent from Equation (10) that β_{MS} itself contains valuable information which can be used to allow interpretable static light scattering measurements to be made on samples turbid enough to produce significant multiple scattering.

Since $f(Q,0)=1$, (Equation (1)), the zero-time value of the intensity cross-correlation function can be written (Equation (9))

$$g_C^{(2)}(Q,0) = 1+\beta^2\,\beta_{OV}^2\,\beta_{MS}^2\,. \tag{11}$$

The factor β_{MS} can be determined experimentally by combining a measurement of $g_C^{(2)}(Q,0)$ for the sample of interest, giving the product $\beta^2\,\beta_{OV}^2\,\beta_{MS}^2$, with a measurement of $g_C^{(2)}(Q,0)$, under exactly the same experimental conditions, for a dilute sample for which there is no multiple scattering so that β_{MS} = 1, giving $\beta^2\,\beta_{OV}^2$. In general, the ratio of single to total scattering will depend on the wavelength of the light. However the wavelengths of the blue and green light used (488 nm and 514.5 nm) are close enough that it is a reasonable approximation to take

$$\frac{\left\langle I_B^S(Q)\right\rangle}{\left\langle I_B(Q)\right\rangle} \approx \frac{\left\langle I_G^S(Q)\right\rangle}{\left\langle I_G(Q)\right\rangle}, \tag{12}$$

so that Equation (10) becomes

$$\left\langle I_G^S(Q)\right\rangle = \beta_{MS}(Q)\left\langle I_G(Q)\right\rangle, \tag{13}$$

giving the single scattered intensity $\left\langle I_G^S(Q)\right\rangle$. (A more involved procedure which does not require the approximation of Equation (12) will be described elsewhere [9]; tests have shown that only small errors are introduced by the approximate approach used here.) Allowance must be made for the fact that both the intensity of light incident on the scattering volume and the intensity of light scattered singly by the volume are attenuated on passage through a turbid sample. In the geometry of our equipment, where the scattering volume is centred in a cylindrical scattering cell, the total attenuation is simply equal to the transmission T of light passing through the sample along a diameter of the cell,

$$T = \frac{I_T}{I_0}, \tag{14}$$

where I_T and I_0 are the transmitted and incident intensities. Thus we can write

$$\left\langle I_G^S(Q)\right\rangle_0 = \frac{\beta_{MS}(Q)\left\langle I_G(Q)\right\rangle}{T}, \tag{15}$$

where $\left\langle I_G^S(Q)\right\rangle_0$ is the intensity of light which would be scattered singly if there were no multiple scattering and no attenuation. It is apparent from Equation (15) that experimental determination of $\left\langle I_G^S(Q)\right\rangle_0$ requires measurements of the dependences on scattering vector Q of the factor $\beta_{MS}(Q)$ (by the procedure outlined above) and the total scattered intensity $\left\langle I_G(Q)\right\rangle$, combined with a single measurement of the sample's transmission T.

We have used this method to measure the static structure factors of the "colloidal liquid" phases [10] of three mixtures of PMMA colloids and polystyrene polymer. The polymer induces a depletion attraction between the particles. The strength of the attraction is determined by the polymer concentration and its range by the ratio ξ of the polymer's radius of gyration to the particle's radius. Previous work [10] has shown that when ξ is large enough, the mixture exhibits three equilibrium phases, colloidal gas, colloidal liquid and colloidal crystal. As ξ is reduced, corresponding to shorter-ranged attractions, the liquid region in the phase diagram becomes smaller and, for $\xi < 0.25$, no liquid phase is observed. Then the effect of adding polymer is simply to broaden the gas-crystal coexistence region of the pure hard-sphere system.

Figure 3 shows the structure factors $S(Q)$ of three colloidal liquids. The same colloidal particles, of radius R = 214 nm, were used in each case. Polymers of different molecular weights gave size ratios ranging from ξ = 0.62, giving a well-developed colloidal liquid phase, through ξ = 0.42, to ξ = 0.26, the last mixture showing just a "marginal" liquid (i.e. the liquid region in the phase diagram is vanishingly small). Samples were prepared in the three-phase (colloidal gas, liquid and crystal) coexistence regions of the phase diagrams [10], and the light scattering measurements were performed on the middle (liquid) phases. The transmissions T of the samples were between 0.2 and 0.4 and $\beta_{MS}(Q)$ ranged from 0.3 to 0.7, indicating significant multiple scattering. Structure factors were obtained from the intensities $\left\langle I_G^S(Q)\right\rangle_0$ by dividing by the particle form factor measured on a dilute suspension. In all cases the intensity of scattering by the polymer was negligible compared to the particle scattering.

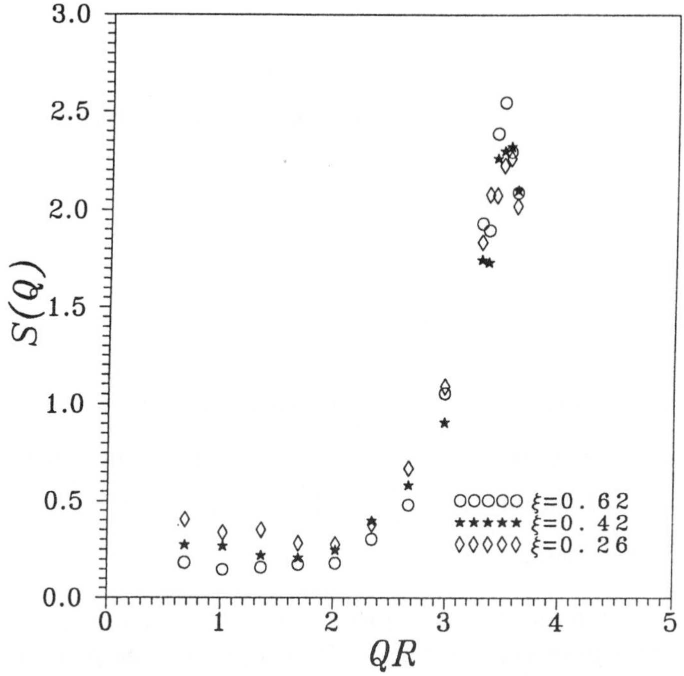

Figure 3. Static structure factors $S(Q)$ versus QR of the colloidal liquid phases of three mixtures of hard-sphere colloidal particles and polymer, for three different values of ξ, the ratio of polymer to particle radii.

The results of Figure 3 will be discussed in detail elsewhere [9]. Here we make three observations:

1. For all three liquids, the main peaks of $S(Q)$, which reflect the local structure in the liquid (a particle and its cage of neighbours), have similar amplitudes.
2. The amplitude of the structure factor at small Q, where density fluctuations of large spatial scale are probed, is much larger for the marginal liquid ($\xi = 0.26$) than for the well-developed liquid ($\xi = 0.62$). In the marginal colloid-polymer mixture, the gas-liquid critical point and the region of triple coexistence are quite close to each other in the phase diagram. Thus the observation of large long-ranged fluctuations could be regarded as a remnant of critical fluctuations, even in the triple-point liquid.
3. Even for the well-developed liquid ($\xi = 0.62$), the amplitude of $S(Q)$ at small Q is much larger than for atomic liquids such as argon [11].

These results illustrate clearly that TCDLS can be exploited to perform static light scattering on quite turbid samples.

8. Acknowledgements

The work reported here was supported by the DTI Colloid Technology project, by the Engineering and Physical Sciences Research Council, and by the Biotechnology and Biological Sciences Research Council.

9. References

1. Pusey, P.N., Segrè, P.N., Behrend, O.P., Meeker, S.P., Poon, W.C.K. (1997) *Progress in Colloid and Polymer Science*, in press.
2. Phillies, G.D.J (1981) *J. Chem. Phys.* **74**: 260; (1981) *Phys. Rev. A* **24**: 1939.
3. Drewel, M., Ahrens, J., Podschus, U. (1990) *J. Opt. Soc. Am.* **7**: 206; Schätzel, K., Drewel, M., Ahrens, J. (1990) *J. Phys.: Condens. Matter* **2**: SA393; Schätzel, K. (1991), *J. Mod. Optics* **38**: 1849.
4. Segrè, P.N., van Megen, W., Pusey, P.N., Schätzel, K., Peters, W. (1995) *J. Mod. Opt.* **42**: 1929-52.
5. Segrè, P.N., Behrend, O.P., Pusey, P.N. (1995) *Phys. Rev. E* **52**: 5070-83.
6. Segrè, P.N., Meeker, S.P., Pusey, P.N., Poon, W.C.K. (1995) *Phys. Rev. Lett.* **75**: 958-61; Poon, W.C.K., Meeker, S.P., Pusey, P.N. (1996) *J. Non-Newtonian Fluid Mech.* **67**: 179-89; Meeker, S.P., Poon, W.C.K., Pusey, P.N. (1997) *Phys. Rev. E*, in press.
7. Segrè, P.N., Pusey, P.N. (1996) *Phys. Rev. Lett.* **77**: 771-4.
8. de Schepper, I.M., Cohen, E.G.D., Pusey, P.N., Lekkerkerker, H.N.W. (1990) *Physica A* **164**: 12-27.
9. Moussaïd, A., Pusey, P.N., Poon, W.C.K., to be published.
10. Ilett, S.M., Orrock, A., Poon, W.C.K., Pusey, P.N. (1995) *Phys. Rev.* E **51**: 1344-52.
11. Yarnell, J.L., Katz, M.J., Wenzel, R.G., Koenig, S.H. (1973) *Phys. Rev. A* **7**: 2130.

CHARACTERISATION OF PARTICLE PACKING

G. D. W. JOHNSON+, R. H. OTTEWILL
School of Chemistry, University of Bristol, Bristol BS8 1TS, U.K.

A. R. RENNIE
Polymers & Colloids Group, University of Cambridge, Cavendish Laboratory, Madingley Road, Cambridge CB3 0HE, U.K.

\+ *This paper is dedicated to the memory of Dr G. D. W. Johnson who died in June 1996*

An investigation has been made of the structure of bimodal mixtures of large particles, A, and small particles, B, by means of small angle neutron scattering. At low number ratios, N_B/N_A, the results suggest that the smaller particles can fit into interstitial positions in a fluid lattice of larger particles. At higher ratios the smaller particles appear to form clusters both in the fluid systems and subsequently in the crystals formed.

1. Introduction

The packing of colloidal particles has an important bearing on many aspects of technology, for example, on the strength of ceramic materials [1], on the scattering of light by pigmented paint films [2], on the rheology of concentrated dispersions [3] and many others. In an industrial context the particulate dispersions are seldom monodisperse and frequently bimodal dispersions are used.

The aim of the present project was to investigate the behaviour of bimodal dispersions in an attempt to understand the role of particle-particle interactions which determine the properties of such systems in both aqueous and nonaqueous environments under both static and sheared conditions. In order to do this we have used binary blends of monodisperse latices thus building on the experience of recent research on the properties of single component systems.

The real inception of the need for well-defined monodisperse systems in order to understand the complexity of colloidal sytems can be traced back several decades. It really began in the 60's and 70's with the development of

R. H. Ottewill and A.R. Rennie (eds.), Modern Aspects of Colloidal Dispersions, 89–99.

synthetic methods for the formation of dispersions of spherical particles with a narrow distribution of particle sizes both in aqueous [4,5] and nonaqueous environments [6]. Following the availability of these dispersions early work uncovered many new phenomena which included interesting phase behaviour and the packing of particles into well-defined crystalline structures [7,8]. Moreover, with the development of photon correlation spectroscopy, PCS, and small-angle neutron scattering, SANS, it became possible to probe the structure of concentrated colloidal dispersions [9,10].

Since 1975 the field has expanded considerably with investigations of concentrated monodisperse systems under normal gravitational conditions [11], under time-average zero gravity conditions [12] and more recently in space [13].

In addition to work on monomodal systems there have also been developments in understanding the behaviour of bimodal systems. For example, the phase behaviour of mixtures of sterically stabilised particles, essentially hard spheres, has been examined using both light scattering and SANS [14,15,16,17]. Crystalline structures of composition AB_2 and AB_{13} were obtained from bimodal mixtures of large particles, A, and small particles, B, the exact structure depending on the particle size and volume fraction ratios [18]. An example of a crystalline array of AB_2 particles, formed under time-average zero-gravity conditions with slow evaporation of the medium, is shown in Figure 1, with A having a radius of 1810 Å and B a radius of 1050 Å. These results were linked with liquid-state theory through simulation [19] which showed regions of superlattice structures and fluid phases which were entirely consistent with the experimental observations.

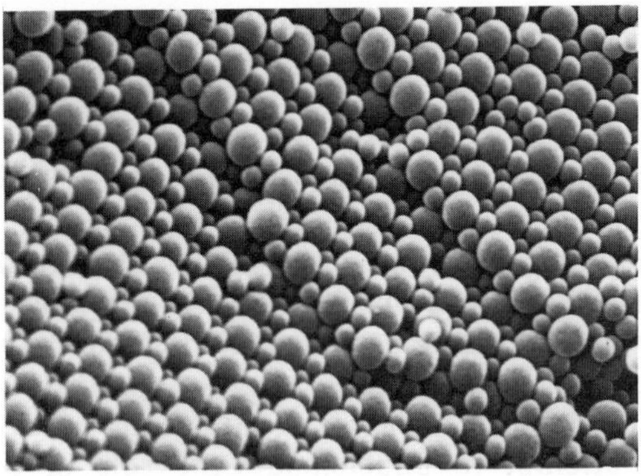

Figure 1: Scanning electron micrograph of a colloidal crystal formed from a bimodal mixture.

In the case of electrostatically stabilised dispersions mixtures of polystyrene particles (radius = 455 Å) and silica particles (radius = 270 Å) were examined by SANS both under equilibrium conditions and under shear [20,21]; the results were discussed in terms of conventional liquid-density theory. Very recently a detailed SANS examination has also been carried out on binary mixtures of polystyrene particles [22,23]. These results will be summarised below.

2. Binary Mixtures of Monodisperse Polystyrene Particles

The system used was composed of deuterated polystyrene particles (radius = 510 Å), designated the A particles, and hydrogenated polystyrene particles (radius = 168 Å) designated the B particles [21]. For a system of this type the differential scattering cross-section, I(Q), from the mixture can be written, for SANS, in the form [22],

$$\begin{aligned} I(Q) = {} & \phi_A V_A (\rho_A - \rho_m)^2 P(Q)_A S(Q)_{AA} \\ & + \phi_B V_B (\rho_B - \rho_m)^2 P(Q)_B S(Q)_{BB} \\ & + 2(\rho_A - \rho_m)(\rho_B - \rho_m) \\ & \times [\phi_A V_A \phi_B V_B P(Q)_A P(Q)_B]^{1/2} S(Q)_{AB} \end{aligned} \qquad (1)$$

with ϕ_A and ϕ_B the volume fractions of the A and B particles in the mixture, ρ_A and ρ_B their coherent scattering length densities and ρ_m that of the medium; V_A and V_B are the volumes of the single particles. $P(Q)_A$ and $P(Q)_B$ are the particle shape factors. In general, P(Q) is given by,

$$P(Q) = [3(\sin(QR) - QR\cos(QR))/(QR)^3]^2 \qquad (2)$$

with Q = the scattering vector for elastic scattering given by,

$$Q = 4\pi \sin(\theta/2)/\lambda \qquad (3)$$

with θ the scattering angle and λ the wavelength of the neutron beam. $S(Q)_{AA}$, $S(Q)_{BB}$ and $S(Q)_{AB}$ are the interparticle partial structure factors representing the interaction between the A particles in the mixture, the interaction between the B particles in the mixture and the interaction between the A and the B particles respectively under the same conditions.

The systems were examined at three contrasts using as the media H_2O, 75% D_2O and D_2O containing 6×10^{-5} mol dm^{-3} sodium chloride. The values of the coherent scattering lengths are listed in Table 1 and the compositions of the mixtures used in Table 2.

Table 1

Material	Designation	Radius/Å	$\rho/10^{10}$ cm^{-2}
d_8-polystyrene	A	510±10	6.35±0.1*
h_8-polystyrene	B	168± 5	1.41±0.05
H_2O	m	-	-0.56
75% D_2O	m	-	4.62
D_2O	m	-	6.35

* experimentally determined from contrast variation

Table 2

Composition of Binary Mixtures

$N_B/N_A = 15.0$; $R_B/R_A = 0.33$

Medium	ϕ_A	ϕ_B	ϕ_B/ϕ_A
H_2O	0.0194	0.0104	0.536
75% D_2O	0.0209	0.0112	0.536
D_2O	0.0214	0.0115	0.537

From a combination of experiments at three different contrasts it was possible to extract the three partial structure factors $S(Q)_{AA}$, $S(Q)_{BB}$ and $S(Q)_{AB}$. The results are shown in Figure 2.

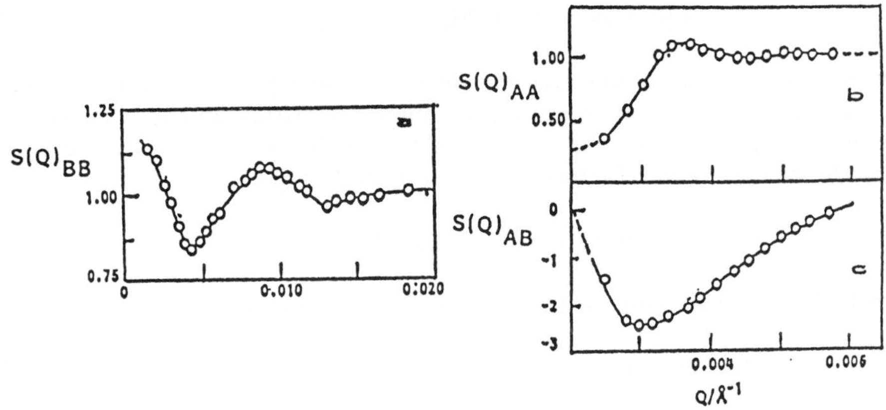

Figure 2: Partial structure factors against Q for $N_B/N_A = 15.0$ and $R_B/R_A = 0.33$

The present interpretation of these results is considered to be as follows:

2.1 $S(Q)_{AA}$ vs Q

The form of the curve indicates that the particles are colloidally stable and interacting by electrostatic repulsion as expected for charged particles in a low electrolyte concentration i.e. 6×10^{-5} mol dm^{-3} sodium chloride. The position of the main peak indicates an average correlation distance between the particles of ca. 2000 Å, indicative of short-range order. At the higher Q values the peaks are damped as expected for a "fluid-like" structure of the particles and as observed in earlier experiments with monodisperse latices [24]. In the binary system there was only a small displacement of the peak from its original position. The indication from these observations is that there is very little perturbation of the structural arrangements of the A particles at these volume fractions as a consequence of the presence of the B particles.

2.2 $S(Q)_{BB}$ vs Q

In the absence of the A particles the form of $S(Q)_B$ vs Q was very similar in shape to that shown for $S(Q)_{AA}$ but scaled according to particle size as expected for monodisperse electrostatically interacting spheres. In the mixture there is a profound difference in the behaviour of the B particles as indicated by the $S(Q)_{BB}$ vs Q curve. The upturn at low Q indicates that some clustering of the small particles has occurred in the "fluid-like" structure of the larger particles. This behaviour is a consequence of the much stronger electrostatic repulsion which occurs between the larger A particles. However, the peak at a Q value of ca 0.01 $Å^{-1}$ indicates that there is still some correlation between the B particles on a distance scale of the order of 700 Å.

2.3 $S(Q)_{AB}$ vs Q

This curve indicates a negative correlation between the A and the B particles suggesting some separation of the two species in the overall structure which are uncorrelated with the structures formed by the A-A and B-B interactions. In essence the presence of the B particles and their excluded volume as a consequence of their electric field means that the B particles are excluded from these regions. In addition, calculation of the potential energies of interaction for A-A, B-B and A-B pairs indicates that the force of repulsion has a nonlinear dependence on the size of the particles, in that, the force of the A-B repulsion is not the mean of the A-A and B-B repulsive forces; the A-A repulsive force is by far the strongest [22].

The form of the scattering curves and physical observation did not show any evidence of coagulation of the particles in the binary mixture.

3.1 Experiments with perfluorinated polymer particles

As a means of checking and confirming the results obtained with the polystyrene binary mixture further experiments were carried out in which the larger particles were contrast matched and the volume fraction of the smaller particles was incrementally increased. For these experiments perfluorinated particles, composed of poly-tetrafluorethylene and poly-fluoropropylene, were used - FEP latices. These particles were contrast matched for SANS experiments with 78.5% D_2O and thus this system had the advantage of moving away from the extreme ends of the contrast range; it also helped to reduce any effects of multiple scattering. In addition, since the particles had a refractive index of 1.347 it was possible to optically match the particles with a propanol-water mixture and hence to carry out light scattering experiments under similar conditions to SANS. The properties of the particles used are listed in Table 3. We follow the convention that A designates the large particles and B the small.

Table 3

Particles	Description	Radius/Å	$\rho/10^{10}$ cm^2
FEP/GJ1	A	760	$4.85 \pm 0.05^*$
h-polystyrene/FBS89	B	315	1.41
78.5% D_2O	m	-	4.85

* Experimentally determined from contrast variation

In terms of treatment both latices were ion-exchanged and then the medium adjusted to a composition of 78.5% D_2O-21.5% H_2O; samples were then dialysed against 78.5% D_2O.

Firstly, the polystyrene latex was examined by SANS in 78.5% D_2O at volume fractions of 0.0038, 0.0096, 0.019 and 0.038. The results of these experiments are shown in Figure 3.

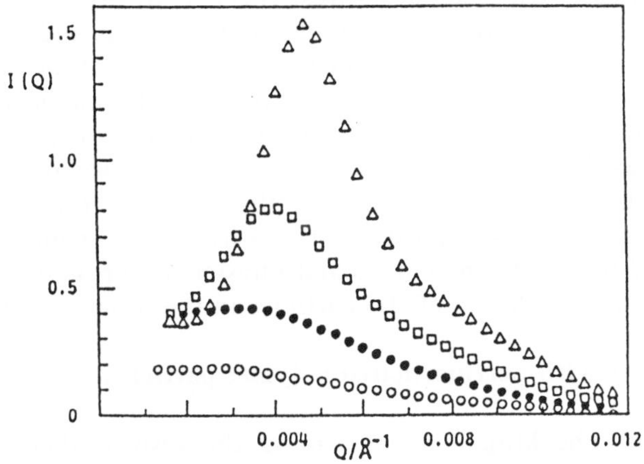

Figure 3: I(Q) vs Q for FBS89 latex with ϕ_B:- O , 0.0038; ● , 0.0096; □ , 0.019; Δ , 0.038.

In a further experiment binary mixtures were examined in which the FEP latex was kept at a volume fraction of 0.023 and the volume fraction of the polystyrene particles was increased from 0.002 to 0.0384. The ratios of ϕ_B/ϕ_A and N_B/N_A are given in Table 4. An experimental check confirmed that the scattering of the FEP particles under these conditions was zero within experimental error. The results are plotted in Figure 4. As can be seen there is a distinct increase in the scattered intensity from the small polystyrene particles at the low Q values which increases in magnitude as the volume fraction of the polystyrene particles increases.

Table 4

$R_B/R_A = 0.415$

ϕ_A	ϕ_B	ϕ_B/ϕ_A	N_B/N_A
0.023	0.002	0.09	1.2
0.023	0.0096	0.42	6.0
0.023	0.019	0.83	12
0.023	0.038	1.65	24

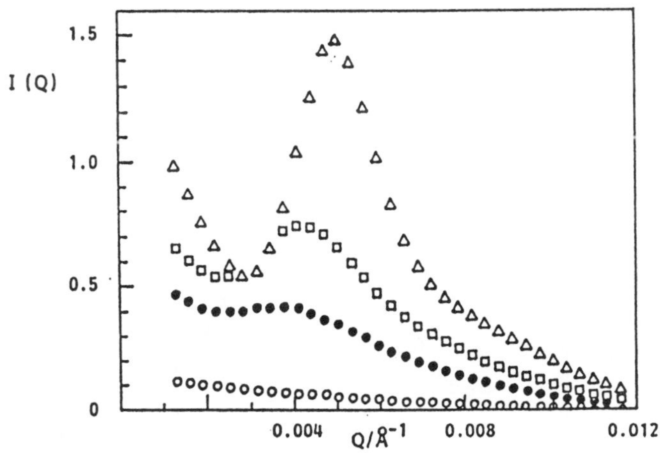

Figure 4:- I(Q) vs Q with ϕ_A = 0.023 (FEP latex) and ϕ_B :- O , 0.002; ● , 0.0096; □ , 0.019; △ , 0.038.

This effect shows up more clearly when a comparison is made of the spectra in the absence and the presence of the FEP particles. This is illustrated in Figure 5. In addition to the concentration dependent increase in intensity it can be seen that the position of the maximum in intensity moves to a higher Q value as the concentration of added polystyrene particles increases. This indicates that the correlation distance between the B particles is smaller when the A particles are present in the mixture suggesting that they are being squeezed. The minimum at low Q also moves to higher Q values as the volume fraction of polystyrene is increased.

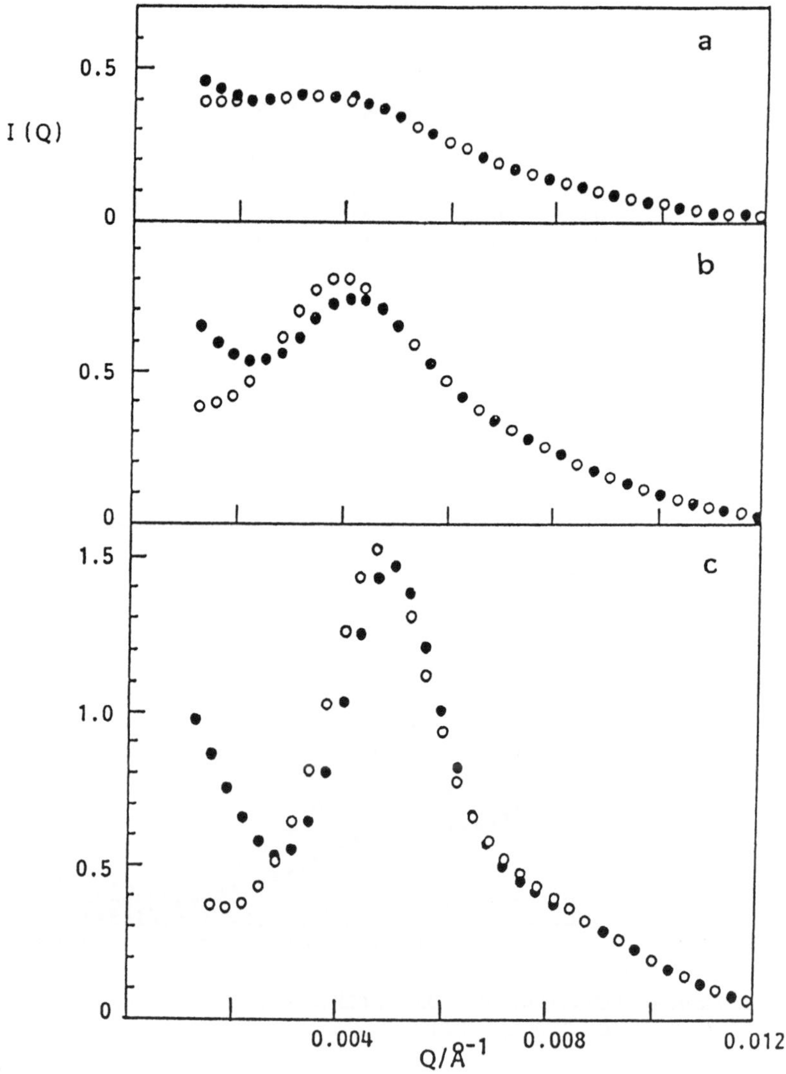

Figure 5: I(Q) vs Q. A comparison of scattering from FBS89 in the absence and presence of FEP:- O , FEP absent; ● , FEP present at ϕ_A = 0.023. FBS89 ϕ_B values, a) 0.0096; b) 0.019; c) 0.038.

The results shown in Figure 5c were also converted into curves of $S(Q)_B$ vs Q and $S(Q)_{BB}$ vs Q and these are shown in Figure 6; these confirm the comments made above on the basis of the intensity results. The extrapolation of $S(Q)_{BB}$ to zero Q is only tentative and it is possible that the curve could rise more steeply than shown by the dashed line in the figure.

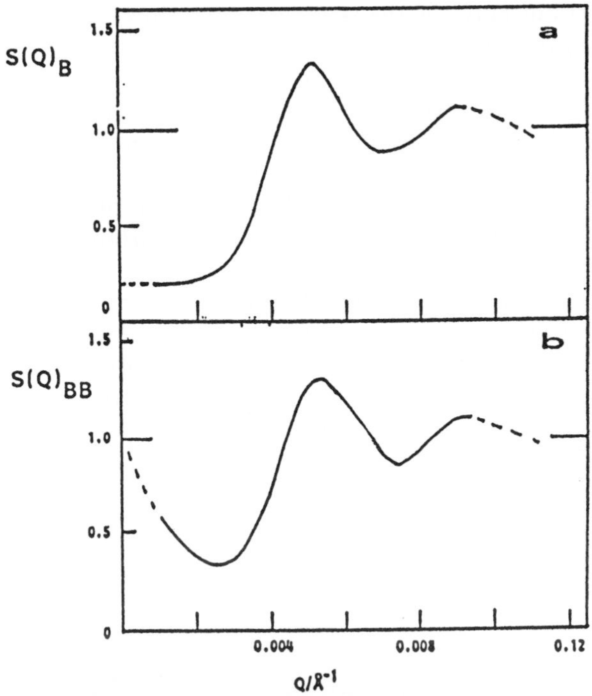

Figure 6: a) $S(Q)_B$ vs Q for FBS89 alone $\phi_B = 0.038$
b) $S(Q)_{BB}$ vs Q FBS89, $\phi_B = 0.038$ and FEP $\phi_A = 0.023$

4. Discussion

The results obtained with mixtures of FEP latices and polystyrene latices essentially confirm the results obtained in the earlier experiments [22,23]. They indicate that when small electrostatically charged particles are mixed with larger charged particles at 1:1 electrolyte concentrations of the order of 5×10^{-5} mol dm^{-3} there is a concentration dependent upturn at lower Q values in the curve of $S(Q)_{BB}$ which is dependent on the concentration of small particles.

The volume fractions recorded in the text are actual volume fractions. For charged particles the effective volume fractions will be larger than these. As an example, at 5×10^{-5} mol dm^{-3} electrolyte the Debye-Huckel reciprocal length of the double layer is 430 Å ($1/\kappa$). Thus taking the effective radius, as a first approximation, to be $R + 1/\kappa$ then the effective volume fractions for the FEP experiments become 0.09 and 0.51 instead of 0.023 and 0.038 thus suggesting very strong multiple electrostatic interactions.

An interpretation of these results is that the strong long-range electrostatic repulsion between the larger particles tends to force the smaller particles into interstitial positions between the larger particles thus forming at low volume fractions an organised but fluid lattice (superlattice). Thus in Figure 4 where $\phi_B = 0.002$ and the number ratio N_B/N_A is, within experimental error, 1.2 there is very little effect. However, the observed effect increases substantially as N_B/N_A rises from 1.2 to 6 and then to 12 and 24. In the latter situation there is a large excess of small particles over large; hence the number of smalls potentially exceeds the number of possible interstitial positions and hence the small particles cluster within the spatial volume available.

Some confirmation has been obtained for this hypothesis from crystallisation studies which were carried out by slow evaporation under time-average zero gravity conditions using polystyrene latices with large particles of radius 264 nm and small particles of radius 68 nm [25]. At low N_B/N_A ratios very well ordered binary crystals were obtained whereas when higher ratios were used clusters were formed. Scanning electron micrographs illustrating these effects are shown in Figure 7.

Figure 7: Scanning electron micrographs of binary crystals

5. Acknowledgements

Small-angle neutron scattering experiments were carried out at NIST, Gaithersburg, Maryland, U.S.A. and at the Institut Laue Langevin, Grenoble, France. We warmly thank both facilities for neutron beam time and their help and encouragement with the experiments. We also acknowledge with thanks support from the DTI Colloid Technology Project, including ICI, Schlumberger, Unilever and Zeneca and from EPSRC.

6. References

1. Birchall, J.D. (1989) *Chemistry and Industry*, 403-407.
2. Templeton-Knight, R. L. (1990) *J. Oil Colour Chem. Assoc.* **1**, 459-464.
3. Russel, W. B. (1987) *The Dynamics of Colloidal Systems*, University of Wisconsin Press Ltd., London, 89-115.
4. Bradford, E.B. and Vanderhoff, J.W. (1955) *J. Appl. Phys.* **26**, 864-
5. Ottewill, R. H. and Shaw, J. N. (1967) *Kolloid Z. u. Z. Polym.* **215**, 161-166.
6. Cairns, R.J.R., Ottewill, R.H., Osmond, D.J.W. and Wagstaff, I. (1976) *J. Colloid Int. Sci.*, **54**, 45-51.
7. Barclay, L., Harrington, A.H. and Ottewill, R.H. (1972) *Kolloid Z. u. Z. Polymere* **250**, 655-666.
8. Hachisu, S., Kobayashi, Y. and Kose, A. (1973) *J. Colloid Int. Sci.*, **42**, 342-348.
9. Brown, J.C., Pusey, P.N., Goodwin, J.W. and Ottewill, R.H. (1975) *J. Phys. A: Math. Gen.* **8**, 664-682.
10. Ottewill, R.H.(1980) *Prog. Colloid Polym Sci.*, **67**, 71-83.
11. Megen, W. van and Pusey, P. N. (1986) *Nature*, **320**, 340-342.
12. Bartlett, P., Pusey, P. N. and Ottewill, R.H. (1991) *Langmuir*, **7**, 213-215.
13. Chaiken, P. and Russel, W. B. (1997) in press.
14. Bartlett, P., Ottewill, R. H. and Pusey, P. N. (1990) *J. Chem. Phys.*, **93**, 1299-1312.
15. Bartlett, P. and Ottewill, R.H. (1992) *J. Chem. Phys.* **96**, 3306-3318.
16. Duits, M. H. G., May, R. P., Vrij, A. and de Kruif, C. G. (1991) *J. Chem. Phys.* **94**, 4521-4531.
17. Mendez-Alcarez, J.M., D'Aguanno and Klein, R. (1992) *Langmuir*, **8**, 2913-2920.
18. Bartlett, P., Ottewill, R. H., and Pusey, P.N. (1992) *Phys. Rev. Let.* **68**, 3801-3804.
19. Eldridge, M. D., Madden, P. A. and Frankel, D. (1993) *Nature* **365**,35-37.
20. Hanley, H. J. M., Pieper, J., Straty, G. C., Hjelm, R. and Seeger, P. A. (1990) *Faraday Discuss. Chem. Soc.* **90**, 91-106.
21. Hanley, H. J. M., Straty, G. C. and Lindner, P. (1994) *Langmuir*, **10**, 72-79.
22. Ottewill, R. H., Hanley, H. J. M., Rennie, A. R. and Straty, G. C. (1995) *Langmuir* **11**, 3757-3765.
23. Ottewill, R. H. and Rennie, A. R. (1996) *Prog. Colloid Polym. Sci.* **100**, 60-63.
24. Cebula, D. J., Goodwin, J. W., Jeffrey, G. C., Ottewill, R. H., Parentich, A. and Richardson, R. A. (1983) *Faraday Discuss. Chem. Soc.* **76**, 37-52.
25. Ottewill, R.H. and Stokes, D. (1997) to be published.

PROBING COLLOIDAL INTERACTIONS USING A MODIFIED ATOMIC FORCE MICROSCOPE

G. BRAITHWAITE, P. LUCKHAM, A. MEURK*, AND K. SMITH.
Department of Chemical Engineering, Imperial College of Science, Technology and Medicine, Prince Consort Road, London SW7 2BY, U.K..
** Current Address, Swedish Institute of Surface Chemistry, Stockholm, Sweden.*

The principles of atomic force microscopy (AFM) have been applied to study colloidal interactions. Rather than being used in the scanning mode an AFM tip, or a modified AFM tip is driven towards a surface and the interactions between the tip and the surface determined. Using this methodology van der Waals, electrical double layer and steric interactions have been determined and some of the data are presented here.

1. Introduction

It is the interactions between the particles in a concentrated colloidal suspension which determines the structure of the suspension and this structure in turn conveys the rheological properties of the suspension. Thus in order to control the rheology of a colloidal fluid information concerning the nature and range of the interactions between the particles is important. There are three long ranged interactions which dominate in colloidal suspensions, namely van der Waals, electrical double layer and steric interactions.

It was in the nineteen thirties that the theory for the interactions between macroscopic bodies due to van der Waals, or more specifically London dispersion forces, was proposed by Hamaker [1]. He postulated that when the surfaces were sufficiently close, so that the interactions are not retarded by molecules between the surfaces, the force of interaction varied as $-1/d^2$, where d is the separation between the surfaces.

The idea of the diffuse electrical double layer was first proposed by Gouy [2] at the turn of the century and the concept of a repulsive interaction occurring between two overlapping double layers was originally proposed by Derjaguin and Landau [3,4] and independently by Verwey and Overbeek [5]. These founding fathers of colloid science then combined their results with those of Hamaker to describe the interactions between two charged colloidal particles. The DLVO theory was born.

Steric interactions on the other hand were poorly understood until the early 1970's. Napper [6], Meier [7] and Hesselink et al [8] proposed independently that the steric interaction was comprised of two terms for overlapping terms. An osmotic pressure component due to the increase in polymer concentration and an entropic component due to the loss in configurational entropy. Clearly these two terms are not independent of

R. H. Ottewill and A.R. Rennie (eds.), Modern Aspects of Colloidal Dispersions, 101–112.

each other and Edwards [9,10] and particularly, de Gennes [11-13] have combined these terms. Even so, the steric interaction from a theoretical basis is not so well described as the van der Waals or electrical double layer interactions. These are quantitative, the steric interaction may only successfully be described at the qualitative level.

With regard to measurement, van der Waals forces drew the initial interest. The fifties and early sixties saw many attempts at measuring these interactions. However because of surface roughness problems the interactions were very weak and hence hard to determine [14-17]. The use of mica, which is molecularly smooth, circumvented this problem. The first direct evidence for the nature of the non-retarded van der Waals interaction, was produced by Tabor and Winterton [18-19] in their study of the van der Waals forces between two sheets of molecularly smooth mica in air. A $1/d^2$ dependence was observed. More accurate and extensive data were also obtained by Israelachvili and Tabor [20,21] confirming the theoretical predictions for van der Waals forces. A parallel study by Roberts and Tabor [22] managed to determine the nature of electrical double layer interactions between charged rubber surfaces. This technique however was superseded by a new mica surface forces apparatus, which as well as being able to determine van der Waals forces in air was also able to determine forces in liquids. Initially double layer interactions were studied [23] and then in the early eighties steric interactions [24,25] were probed. Until some five years ago this method proved to be the best for determining the interactions between surfaces and hence gave a handle to the interactions between colloidal particles. However the problem with the mica surfaces force method is that the substrate is limited to mica, although one or two other substrates have been used more recently (*e.g.* sapphire and mercury).

Recently, an alternative to the mica surface forces apparatus has been developed based on atomic force microscopy (AFM) [26]. In this method a small particle is attached to a very soft cantilever used in atomic force microscopy and the particle is driven towards the surface [27-29]. In this paper we report data concerning the three main types of colloidal interaction, van der Waals, electrical double layer and steric, in an attempt to understand how particulates interact.

2 Experimental

2.1 EQUIPMENT AND TECHNIQUES

In this study two machines have been utilised to study colloidal interactions. Firstly an apparatus has been designed and constructed with the specific aim of examining the surface forces at nanometer length scales and nanoNewton forces in a variety of media. It is built around the principles of the AFM, although we have not incorporated any scanning capability. This greatly simplifies the machine and considerably reduces the cost. This apparatus has been described fully elsewhere [30] but essentially consists of a glass particle stuck on the end of a commercial AFM lever. The lower surface is then raised and the deflection of the cantilever monitored using a laser/position sensitive photodiode set-up (see figure 1). The second machine which has been used was a commercial Topometrix Explorer AFM used in force spectroscopy mode. Both machines gave similar results for the data reported here. The Topometrix instrument was used to obtain data for van der Waals forces while the purpose built machine was used to determine electrical double layer and steric interactions. The purpose built machine proved to be more flexible than the commercial instrument.

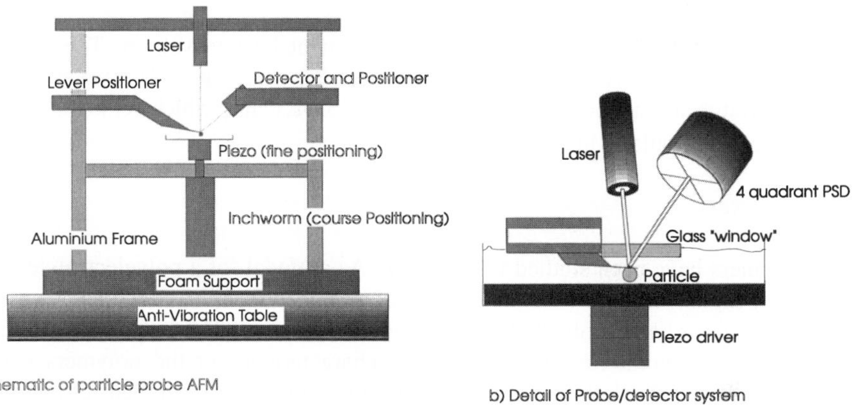

Figure 1 Schematic of modified AFM for force sensing. Figure a) displays the general construction of the apparatus without electrical connections. b) is a more detailed and pictorial representation of the sample section (not to scale)

The interactions were measured between an AFM tip, or a glass particle attached to an AFM tip, and a flat surface. When particles were used, the probe particles were typically 100 μm diameter glass spheres (BDH, UK). This particle was then mounted on a commercial silicon lever (Burleigh Instruments, UK) of quoted spring constant range 0.04-0.35 Nm^{-1}. The spring constant of the commercial levers was measured using the resonance shift of a loaded lever. The probe particle (the load) then became part of the calibration technique [31]. This proved an accurate way of determining the spring constant (to within ~5%). When an unmodified AFM tip was used however, the spring constant was determined directly from the resonance frequency of the lever. The inherent error in this measurement is greater, at around 20%.

The silicon nitride surface was a standard AFM tip, which had been treated with toluene to remove any surface contamination. The radius of curvature of the tip is typically 50 nm. Flat surfaces of quartz, glass or silicon nitride were studied. All were cleaned rigorously, the glass and quartz by sonication in a dilute RBS (a surfactant) solution, followed by rigorous washing with water, and the silicon nitride was treated with hydrofluoric acid for one hour to remove any surface film of silica which may have built up.

The results were obtained at room temperature (usually 25°C±2°C) and pressure. The scan speed was of the order of 0.02 Hz or 25 nm/s unless otherwise stated, which implies that a 100 nm layer is traversed in four seconds.

2.2 MATERIALS

2.2.1. Solvents

1-bromonaphthalene and diiodomethane were obtained from Aldrich. The electrolyte solutions were made from analytical grade materials and water. The water used was filtered through a Nanopure system (resistivity 18 MΩ cm), which removed any ions, organics and particulates in the water.

2.2.2. Polymers

Two polymers have been studied in this work. As a model for a polyelectrolyte, poly-l-lysine (ex. Sigma) of molecular weight 35,000 obtained from was used and as a model for a nonionic polymer polyethyleneoxide, PEO, was used. The polymer was obtained from Polymer Laboratories, UK. Molecular characteristics of the polymers used are given in table 1. All solutions were mixed at room temperature and then refrigerated for approximately 12 hours before use to allow the polymer to fully dissolve [32]. The working solution volume was 40 ml in all cases.

Table 1: The molecular characteristics of the PEO samples used

Polymer M_w	Batch Number	Polydispersity (M_w/M_n)	Calculated unperturbed radius of gyration
56,000	20833-6	1.05	7.4 nm
205,000	20837-3	1.10	14 nm

3. Results and Discussion

In the following section the results will be discussed and compared with respect to other work performed on similar systems. However all separation distances quoted here are relative to an assumed zero separation contact. This arises because calibration of the equipment is based on two assumptions: zero force is obtained when the lever is in equilibrium, well away from the surface and free from any interaction; zero separation is taken when the detector signal moves linearly with the drive voltage. This is known as constant compliance. Although there may have been material such as contamination or polymer sandwiched between the probe and the surface, the surface separation remained constant. There was no way to determine the absolute separation (and hence the layer thickness) or the contact area (unlike the SFA). There is therefore an offset error in any quoted distance value (of the order of 1-10 nm for two layers for any polymer system). These problems and assumptions are common to all AFM force sensing techniques. However, this does not alter the form of the interaction. It should also be noted that the surface energies (per unit area - E) reported here are in the form of $E=F/2\pi R$ (as required by the Derjaguin approximation [33,34].

3.1. VAN DER WAALS FORCES

Hamaker [1] postulated that when two macroscopic surfaces were sufficiently close, so that the interactions are not retarded by molecules between the surfaces, the force of interaction varied as $-1/d^2$, where d is the separation between the surfaces. Hamaker introduced the constant of proportionality, A, later known as the Hamaker constant, which is a material parameter and depends largely on the atomic polarisability of the material. For two like surfaces acting through any medium, this constant is positive and hence the interaction is attractive. However theory has suggested that between unlike surfaces immersed in the correct fluids, repulsive van der Waals forces should exist [35,36]. There has been some indirect evidence to support this based on particle rejection by solidification fronts [37,38], and there has been some indication of repulsive interaction between PTFE (Polytetrafluoroethylene) and gold surfaces immersed in various alkanes [39] although in this case there is the potential for repulsive interactions between the rough PTFE surfaces.

Theoretically repulsive van der Waals forces should be observed when the atomic polarisability of the material separating the two surfaces is intermediate between the atomic polarisability of the two surfaces themselves. The atomic polarisability is proportional to the refractive index, n, of the materials. Here we have measured the interactions between silicon nitride (n=1.98) and quartz (n=1.45) surfaces immersed in both 1-bromonaphthalene (n=1.66) and diiodomethane (n=1.76). As a control, the interactions between two silicon nitride surfaces immersed in the same fluids have also been determined. For all four systems, the Hamaker constants were calculated with Lifshitz theory [40], applying the procedure and parameters used by Bergström *et al.* [41]. Plotted on top of the data below is the theoretical van der Waals plot using equation 1.

$$W_{PLATE}(D) = \frac{F(D)}{2\pi R} = -\frac{1}{2\pi}\frac{A}{6D^2} \quad (1)$$

Figure 2a is the interaction between a silicon nitride AFM tip against a quartz surface immersed in 1-bromonaphthalene. The interaction is short range, commencing at surface separations of only some 2.5-3 nm and repulsive. Forces on compression and decompression of the surfaces are shown, the two sets of data are, within error the same. Shown as a solid line is the theoretical curve for the interaction between these surfaces in 1-bromonaphthalene. Good agreement is observed. In figure 2b is the van der Waals interaction between two silicon nitride surfaces immersed in 1-bromonaphthalene. Now only attractive interactions are observed which are again well fitted by the theoretical estimate. By increasing the refractive index of the fluid separating the surfaces such that it is closer to being intermediate between the two surfaces, silicon nitride (n=1.98) and quartz (n=1.45), stronger forces should be observed.

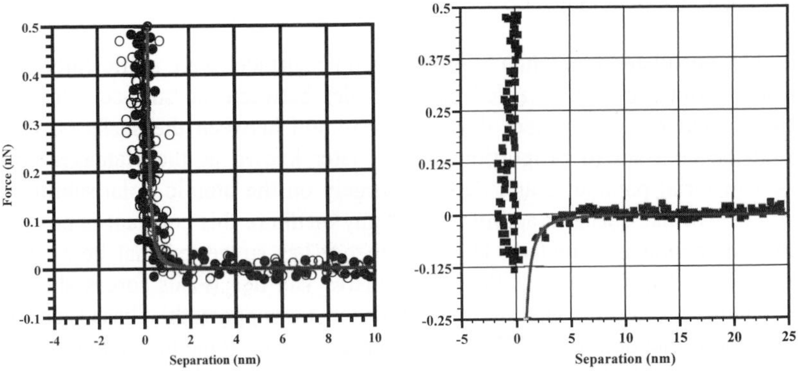

a b

Figure 2: a. Experimental and theoretical data for the interactions between a quartz plate and a silicon nitride AFM tip of radius 50 nm immersed in 1-bromonaphthalene. The theoretical curve is estimated by calculating the Hamaker constant to be $-1.5x10^{-21}$J.; b. Experimental and theoretical data for the interactions between a silicon nitride plate and a silicon nitride AFM tip of radius 50 nm immersed in 1-bromonaphthalene. The theoretical curve is estimated by calculating the Hamaker constant to be $2.6x10^{20}$J.

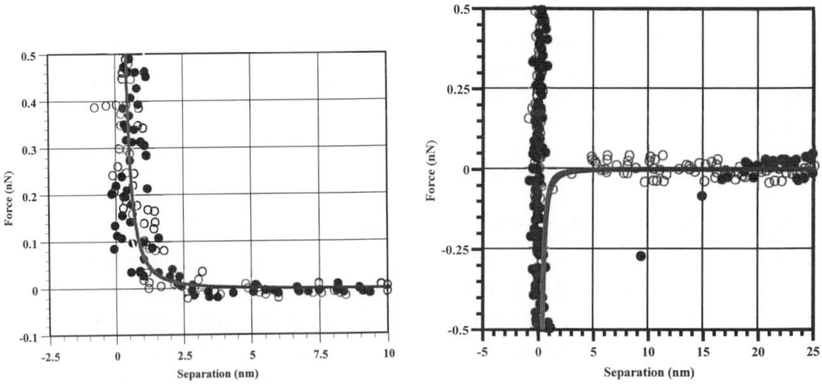

a b

Figure 3: a. Experimental and theoretical data for the interactions between a quartz plate and a silicon nitride AFM tip of radius 50 nm immersed in diioodomethane. The theoretical curve is estimated by calculating the Hamaker constant to be $-8.5x10^{-21}$J.; b. Experimental and theoretical data for the interactions between a silicon nitride plate and a silicon nitride AFM tip of radius 50 nm immersed in diiodomethane. The theoretical curve is estimated by calculating the Hamaker constant to be $8.3x10^{21}$J.

Figure 3a is the interaction between a silicon nitride AFM tip and a quartz surface immersed in diiodomethane (n=1.76). Repulsive interactions are observed commencing at surface separations of some 5.0 nm. The data are well fitted by the theoretical solid line. Again only attractive interactions are noted when the quartz surface is replaced by a second silicon nitride surface, which is well described by the theoretical estimate, as seen in figure 3b.

3.2 DOUBLE LAYER FORCES

There are many ways in which one can test the theories of the electrical double layer. One method is to vary the electrolyte concentration at a fixed surface potential, this may be achieved by simply adding electrolyte solution to water. An alternative method is to maintain the electrolyte constant and to vary the surface potential which can be achieved by adsorbing a polyelectrolyte to the surface.

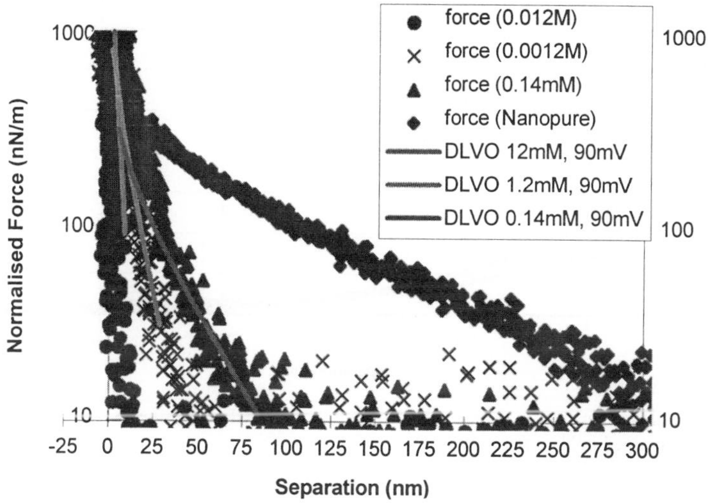

Figure 4. The interaction between two glass surfaces immersed in various aqueous electrolyte solutions.

Figure 4 is a plot of the interaction between two glass surfaces immersed in water and in various aqueous KNO_3 solutions [42]. In pure water the interaction is very long range, greater than 300 nm, but on addition of electrolyte the range of the interaction decreases. In 0.14 mM KNO_3 the interaction commences at a surface separation of some 85 nm, in 1.2 mM KNO_3 the interaction commences at a surface separation of 50 nm and increasing the electrolyte concentration by a further order of magnitude decreases the onset of interaction to around 10 nm. The solid lines on figure 4 correspond to the calculated electrical double layer interaction based on the Chan algorithm [42]. Agreement between theory and experiment is very good assuming a surface potential of -90 mV for the potential of the glass surfaces.

Figure 5 plots the interaction between two glass surfaces immersed in a dilute polylysine solution in the absence of any added electrolyte. Data are presented as a function of time of immersion in the solution. In all cases the interaction commences at surface separations of around 200 nm. The interaction is repulsive and remains repulsive until the surfaces are some 25 nm from contact whereupon they jump together to a separation of 10 nm where a steeper repulsion is noted. The long range part of this interaction is due to the overlap of the electrical double layers of the two surfaces. The attraction is due to van der Waals attraction. The short steep range repulsion is likely to be due to a small steric component due to some adsorbed polylysine, since these experiments were performed in polylysine solution. With time the magnitude of the long ranged repulsive interaction increases due to polylysine adsorption increasing the surface potential. The attractive interaction decreases for the same reason. Thus in these experiments we are essentially increasing the surface potential with time.

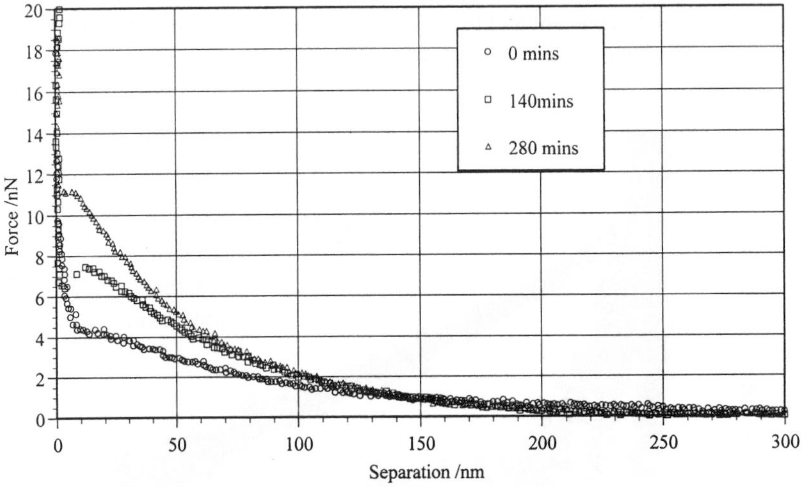

Figure 5 The interaction between two glass surfaces bearing adsorbed polylysine at various adsorption times. These data were obtained in Nanopure water.

3.3 STERIC FORCES

Data have been obtained as a function of time, and hence coverage for PEO adsorbed to glass surfaces. However due to limitations of space here we shall only present data at one partial coverage and at full coverage's.

Figure 6, for example, shows data gathered after 35 minutes incubation of the glass surface in the 56K polymer solution. The lever was present in the solution at all times and the probe was never removed more than 25μm away from the surface. Out at 200 nm the lever shows no deflection, and within experimental resolution it is subject to no interaction. This condition stays very stable until 45 nm is reached. At 45 nm there is a weak jump into some kind of contact at 35 nm (note that this corresponds to $\sim 5 \times R_g$). At 35 nm there is a negative interaction imposed on the probe of approximately -2μJm^{-2}. The energy then rises rapidly and monotonically up to hard contact at 70μJm^{-2}. Before this point is reached, the lever appears to be moving parallel to the surface from 50 μJm^{-2} to 60 μJm^{-2} at a separation of about 5 nm. At 65 μJm^{-2}, the separation finally collapses to hard contact and no further interaction is seen. Although the effect is not all that clear on this trace it is a common feature in other data not presented here and may well be due to the high surface pressures generated by the AFM

On reversal, the probe follows the surface without any apparent deviations down to the equilibrium position at 0 μJm^{-2}. Beyond this the lever parts slightly from the surface to 10 nm and an interaction energy of -20 μJm^{-2}, whereupon it is released. Note however that the probe takes a considerable time to release. Initially it reaches 30 nm over a period of about 0.1 seconds, but then beyond 40 nm (although it now appears to have recovered, it is still at a negative energy of -8 μJm^{-2}) it takes a further 3.5 seconds to recover gradually to its equilibrium value at 140 nm.

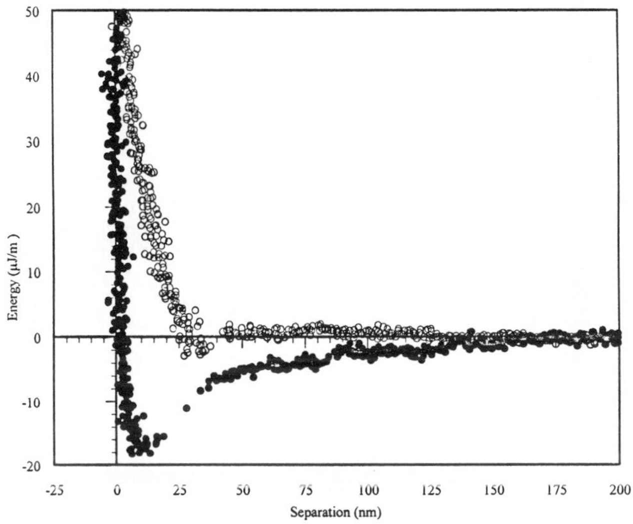

Figure 6 Interaction between a glass particle and a glass surface bearing adsorbed PEO of molar mass 56,000. The data were taken after 35 minutes adsorption of the polymer to both surfaces, a condition corresponding to partial coverages of the surfaces. Hard contact is assumed when the lever moves linearly with the surface. Inward (O) and outward (●) runs are plotted. Note the jump in at 40 nm and the non-zero jump out separation again due to probe roll.

Similar effects are seen for the higher molecular weight polymers except that the ranges of interaction are different. For PEO 200,000 on approach the interaction commences at a surface separation of 55 nm. Again the data are markedly different on separation of the surfaces with there being a pronounced long ranged attraction.

The data suggests that there is only partial coverage of the surface by the polymer. Although there is a layer of polymer on the two surfaces there is still also some free surface available for polymer to adsorb. For this reason, as the surfaces approach, there is a bridging effect where the "dangling" tails of the polymers on one surface can contact and adsorb to the other surface, thus exerting an attractive influence on it, however it was rare to see a jump-in in these experiments, although we did observe this for PEO 56K. The reason why we did not regularly see an attraction on approach of the surfaces as was observed by Luckham and Klein in their study of the interaction between PEO adsorbed to mica surfaces at low coverage is due to the rapid rate of approach in these experiments, where the entire compression of the surfaces takes some 10 s, whilst in the mica surface forces experiments the compression profile would typically take 10 minutes, or longer. We shall not discuss the details of the range of interaction here for the polymer at low surface coverage as the adsorption time and conditions, such as the separation of the tip from the surface, were somewhat variable.

However on separation of the surfaces a strong long ranged attraction was observed in all cases. The range of this attraction is noteworthy. It is very large, some 100-200 nm, thus it would appear that on separation the PEO is being pulled of one surface on separation rather like sticky tape.

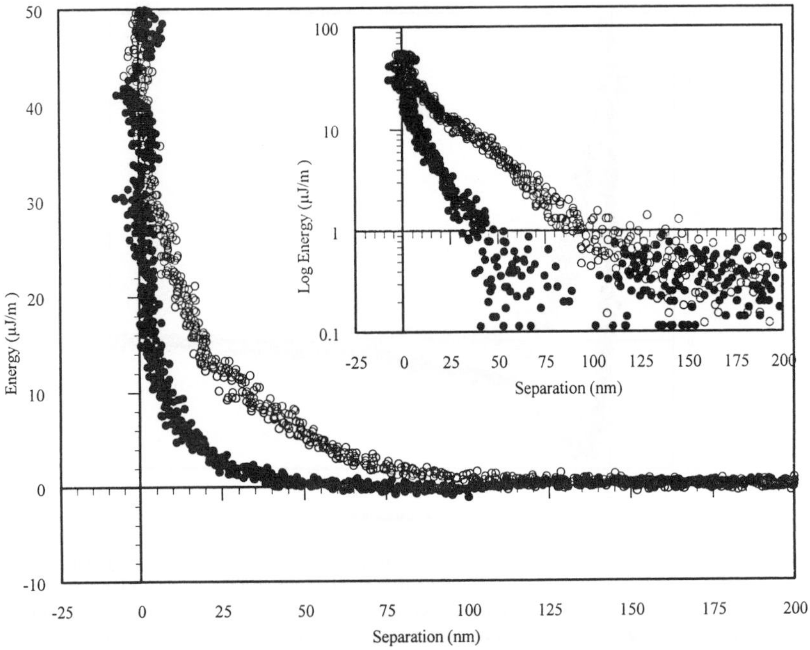

Figure 7 A plot of data taken after 24 hours for PEO 56,000. This is the equilibrium condition. A slight hysterisis is evident consistent with a rearrangement of the layer under pressure but no adhesion is seen. Inward (O) and outward (●) runs are plotted. A semi-log plot of the same data is inset.

Figure 7 represents the final equilibrium condition of PEO 56K after over 24 hours adsorption. Here there is no interaction out beyond 90 nm ($12 \times R_g$), however as the particle approaches the surface at 90 nm, it begins to feel an interaction. There is no attractive component, but instead the interaction approaches hard contact exponentially at an energy of 60 μJm^{-2}. When the motion is reversed, the decompression curve follows that of the compression curve down to 20 nm. From here, the decompression curve relaxes faster, returning to the zero interaction at 60 nm. Note that the ripple on the hard contact of the decompression curve is due to vibration. The semi-logarithmic plot of the data, shown inset to figure 7, has the same general form as the earlier data for PEO adsorbed to mica surfaces. A similar degree of hysteresis is also observed [19,20].

Figure 8 contains the data for PEO 200K at full coverage of adsorbed polymer. Here the interaction commences at a surface separation of 130 nm and increases monotonically with decreasing surface separation. On separation there is again a marked hysteresis although there is no adhesion of the surfaces together on separation.

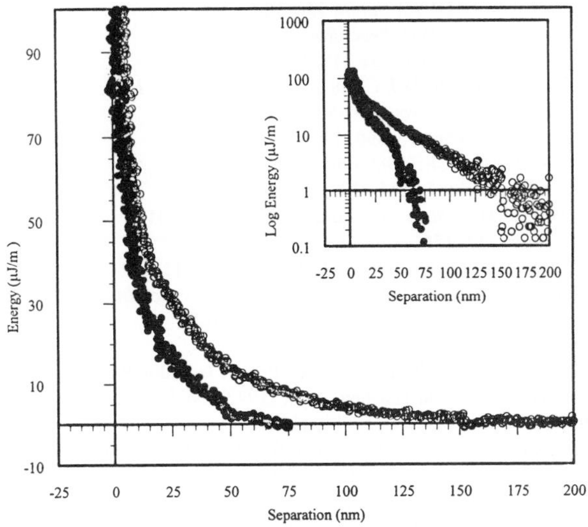

Figure 8 A plot of data taken after 24 hours for PEO 200,000. This is the equilibrium condition. A slight hysterisis is evident consistent with a rearrangement of the layer under pressure but not adhesion is evident. Inward (O) and outward (●) runs are plotted. A semi-log plot of the same data is inset

References

1. Hamaker, H. C. (1937) *Physica* **4**, 1058
2. Gouy, G. (1910) *J. Phys.*, **9**, 457
3. Derjaguin, B. V. (1939), *Trans. Faraday Soc.* **36**, 215
4. Derjaguin, B. V. and Landau, L., (1941) *Acta Physicochim. URSS*, **14**, 633.
5. Verwey, E. J. W. and Overbeek, J. Th. G. (1948) *Theory of the stability of lyophobic colloids,* Elsevier, Amsterdam
6. Napper, D. H. (1977), *J. Colloid Interface Sci.*, **58**, 390.
7. Meier, D. J., (1967) *J. Phys. Chem.*, **71**, 1861
8. Hesselink, F. Th., Vrij, A. and Overbeek, J. Th. G. (1971) *J. Phys. Chem.*, **75**, 2094.
9. Dolan, A. K. and Edwards, S. F., (1974) *Proc. Roy. Soc.Ser.A* **337**, 509
10. Dolan, A. K. and Edwards, S. F., (1975) *Proc. Roy. Soc.Ser.A* **343**, 427
11. De Gennes P. (1982)) *Macromolecules*, **14**, 1637.
12. De Gennes, P. (1982) *Macromolecules*, **15**, 492
13. De Gennes, P. (1987) *Adv. Colloid Interface Sci.*, **27**, 189
14. Black, W., de Jongh, J. G. V., Overbeek, J. Th. G. and Sparnaay, M. J. (1960) *Trans. Faraday Soc.* **56**, 1597.
15. Dejaguin, B. V., Abrikossova, I. I. and Lifshitz, E. M., (1956) *Quart. Rev. (London)* **10**, 292.
16. Kitchener, J. A. and Prosser, A.P . (1957) *Proc. Roy. Soc.* **242**, 403.
17. Rouweler G. C. J. and Overbeek, J. Th. G. (1971) *Trans. Faraday Soc.* **67**, 2117
18. Tabor, D. and Winterton R. H. W. S. (1969) *Nature* **219**, 1119.
19. Tabor, D. and Winterton R.H.W.S. (1969) *Proc. Roy. Soc.Sr A* **312**, 435

20. Israelachvili, J. N. and Tabor, D. (1972) *Nature* **236,** 106
21. Israelachvili, J.N . and Tabor, (1972) D. *Proc. Roy. Soc.*, **331,** 1.
22. Tabor, D, and Roberts, A. D., (1968) *Nature* **219**, 1121
23. Israelachvili, J.N.; Adams, G.E. (1978) *J. Chem Soc. Faraday Trans. I*, 74, 975-1025.
24. Klein, J.; Luckham, P. F. (1984) *Macromolecules*, **17**, 1041.
25. Klein, J.; Luckham, P. F. (1982) *Nature*, **300**, 429.
26. Binnig, G.; Quate, C. F.; Gerber, Ch. (1986)*Phys. Rev. Lett.*, **56**, 930.
27. Ducker, W.A.; Senden T.J.; Pashley, R.M. (1991)*Nature,* **353**, 239.
28 Biggs, S.; Healy, T. W. (1994)*J. Chem. Soc. Faraday Trans.*, **90**, 3415..
29. Biggs, S. (1995) *Langmuir,* , **11**, 156.
30. Braithwaite G. J. C.; Howe A.; Luckham P. F. (1996)*Langmuir*, **12**, 4224
31. Cleveland, J. P.; Manne, S.; Bocek, D.; Hansma, P.K. (1993)*Rev. Sci. Instrum.*, **64**, 403.
32. Porsch, B.; Sundelöf, L.-D.(1995) *Macromolecules*, 1995, 7165.
33 Derjaguin, B. V. (1934) *Kolloid Z.*, **69**, 155.
34. Hunter, R. J. (1991) *Foundations of Colloid Science Vol I*, Clarendon Press,.
35 Deryagin, B. V., Zheleznyi, B. V. and Tkachev, A. P., (1972).*Dokl. Akad. Nauk. SSSR* **206**, 1146.
36. Visser, J., *Adv. Colloid Interface Sci.* **15**, 157-169 (1981).
37. Neumann, A. W., Omenyi, S. N. & Van Oss, C. J., (1979) *Colloid and Polymer Sci.* **256**, 413.
38. Van Oss, C. J., Omenyi, S. N. & Neumann, A. W., (1979) *Colloid and Polymer Sci.* **257**, 737.
40. Lifshitz, E. M., (1956) *Sov. Phys. JETP* **2**, 73.
41. Bergström, L., Meurk, A., Arwin, H. & Rowcliffe, D. J, (1996) *J. Am. Ceram. Soc.*, **79**, 339.
42. Chan D. Y. C., Pashley, R. M. and White, L. R. (1980), *J. Colloid Interface Sci.,* **77**, 283

STRUCTURE OF DISPERSIONS UNDER SHEAR

S. M. CLARKE, A. R. RENNIE
Polymers & Colloids, Cavendish Laboratory, Madingley Road, Cambridge CB3 0HE, U.K.

R. H. OTTEWILL
School of Chemistry, University of Bristol, Cantock's Close, Bristol BS8 1TS, U.K.

Both neutron and light scattering has been used to study the structural arrangement and the orientation of particles in concentrated colloidal dispersions under shear. This article reviews the results of recent investigations on sterically stabilised and charge stabilised colloids: these highlight the importance of interparticle interactions and hydrodynamic forces in flow behaviour. Quantitative measurements of structural changes can be used to assess the validity of theoretical and computer models and provide understanding of flow under conditions of practice.

1. Introduction

The rheology of concentrated dispersions is central to many industrial processes and applications and, for example, is important in the preparation and use of detergents, drilling fluids, paints, paper coating and many other areas. There is a strong interplay between the structure and rheological properties of colloidal dispersions. To provide models of flow behaviour and to control the rheology it is necessary to understand this relationship. Further, the possibility of alignment, ordering or even flocculation under conditions of flow may be important in processing and properties of end products. In recent years considerable effort has been made to determine structure in complex fluids [1] under flow and to compare the results of these studies with computer simulations and theoretical models [2].

The rheology of concentrated colloidal dispersions has been studied extensively [3]. The usual behaviour is non-Newtonian with the viscosity varying extensively with shear rate. This is shown in Figure 1 with data for a sterically stabilised latex. Shear thinning occurs with a large drop in viscosity. Although not shown in this data, at the highest shear rates, shear thickening is often observed. This increase in viscosity is often associated with instabilities in the flow. In this article particular emphasis will be made of the delicate structural changes associated with shear thinning. The use of scattering

R. H. Ottewill and A.R. Rennie (eds.), Modern Aspects of Colloidal Dispersions, 113–122.

techniques with both light and neutrons will be described. A large amount of computer simulation data is available for 'hard-sphere' suspensions. These can be compared with experimental data for sterically stabilised dispersions. Brownian dynamics methods equilibrate particles under an imposed shear field with additional random forces to represent thermal motion.

It is convenient to consider the dimensionless quantity known as the Peclet number, Pe, defined as

$$Pe = 6\pi\ \eta\ \dot{\gamma}\ a^3 / 8k_BT \qquad (1)$$

where $\dot{\gamma}$ is the shear rate, η the viscosity, a the particle diameter, k_B Boltzmann's constant and T the absolute temperature. This can be thought of as the ratio of the time for a particle to be moved through one diameter by the shear field and the time for the particle to diffuse one diameter.

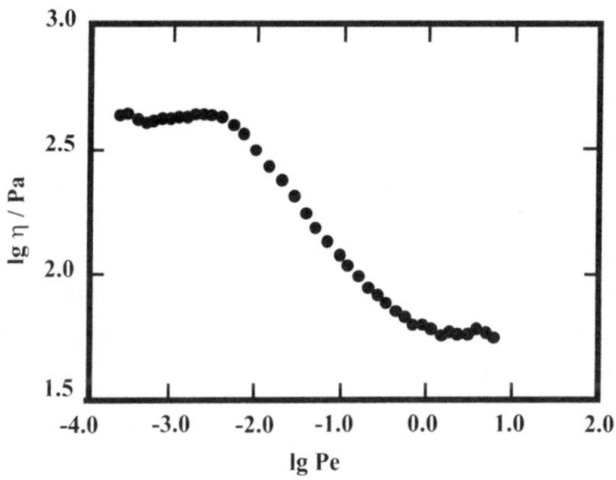

Fig. 1 Viscosity, η, as a function of shear strain rate $\dot{\gamma}$ for a polymethyl methacrylate (PMMA) dispersion in dodecane. The particles are 200 nm diameter and are sterically stabilised with poly 12-hydroxy stearic acid. For purposes of comparison, the shear rate is shown as a Peclet number.

Data for many materials can be mapped on to a single curve by using the quantities Pe and volume fraction. It is useful to report structural data as a function of this parameter so that comparison can be made with theory and computer models. In this way results from a few systems can be used to validate models of more general applicability. There are constraints on materials that are suitable for different experiments. Neutron scattering can provide results for many systems without problems of multiple scattering, but is restricted to particles of less than 0.5 μm. Light scattering can readily measure correlations between larger particles but care must be taken to reduce multiple scattering by choosing particles and dispersion media of similar refractive index. A further complication in studies of particles with a size appropriate to light scattering is that even small differences in density may cause significant sedimentation. In suitable model systems this can be avoided by making an appropriate choice of solvents with density equal to that of the dispersed phase.

2. Experimental Methods

Scattering techniques determine the spatial Fourier transform of a density distribution. The theory of light and neutron scattering is similar in the limit of weak scattering [4,5] with the difference that contrasts arise by means of nuclear interactions for neutrons and refractive index for light. The principle results are described in several texts [5] but can be simplified to the following form for intensity, I

$$I\,(Q) = k\ \ P\,(Q)\ \ S(Q) \qquad (2)$$

where Q is the scattering vector defined as $(4\pi/\lambda)\sin(\theta/2)$, λ is the wavelength, θ the scattering angle, P(Q) the form factor describing the scattering from an isolated particle and S(Q) describes the correlations between particles. The constant k depends on concentration of particles, scattering contrast and incident flux. For simple regular shapes such as spheres, discs or cylinders P(Q) can be calculated readily. In colloidal dispersions at high concentrations, it will often be S(Q) that is of primary interest. This can be related to the radial distribution function of the particles and thus to the structure and structural distortions. Further details on the use of scattering methods and the extension of Equation (2) to the more general case of mixtures of charged particles is presented in the paper by Johnson et al in this volume [6].

The three-dimensional nature of the structure requires that data is collected over as wide a range of the wave vector space **Q** as possible. The geometries that are available to small-angle neutron scattering (SANS) and to light scattering are shown in Figure 2. Details of the experimental arrangements for SANS [7,8] and light scattering [9] have been described elsewhere. Data is collected on a two-dimensional detector and converted to absolute scales. Scattering data can be compared with models of liquid and solid structure as well as with the results of computer simulations. Usually the effects of instrument resolution and the finite accessible range of **Q** cause Fourier transformation of experimental data to be difficult. In practice it is better to transform model and computer data making due allowance for instrument resolution by convolution [10]. Computer simulations often involve a fairly small number of particles and this can limit the resolution of the spatial information that is obtainable. Care must also be taken to average the data over a number of independent configurations [10].

3. Structural Changes under Flow

In this section we summarise briefly some main conclusions of recent work. The results are described in more detail in the original papers cited as references.

3.1 'HARD SPHERE' DISPERSION UNDER STEADY FLOW

Sterically stabilised colloidal dispersions have been considered as models for 'hard-sphere' fluids in several studies [11,12]. If the particles are large compared with the thickness of the layer giving rise to the repulsive interactions, the interparticle potential

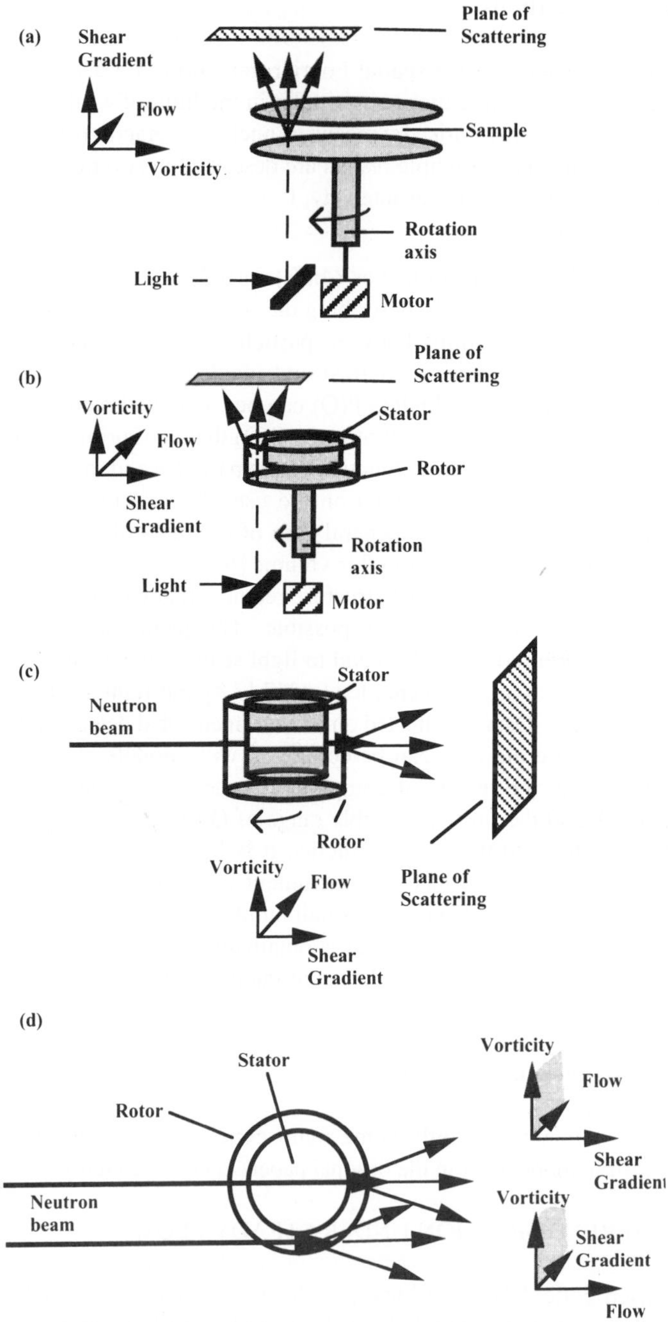

Figure 2. Schematic diagram of experimental geometries used to measure scattering from sheared dispersions. (a) disc-disc and (b) Couette cells used in light scattering experiments [9]. A Couette cell used in SANS experiments is shown in (c). The different positions of the incident beam with respect to the cell in attempts to observe all principal directions of flow are shown in (d).

may be reasonably approximated by a steep potential barrier. In practice this can be achieved to a reasonable extent although significant deviations from simple models of non-penetrating objects are found if structures are examined in detail. This section will describe work on large, 1.5 μm diameter, PMMA particles in organic media [13].

Qualitative descriptions of the increase in spacing of particles along the flow direction have been available for some time [8,9,14]. The extent of the increase in order can be determined from quantitative measurements. An example of this scattering data is shown in Figure 3 for a PMMA dispersion in a solvent which is density matched to the particles. Further confirmation of the structural changes under flow has come from measurement of scattering in the gradient-vorticity plane.

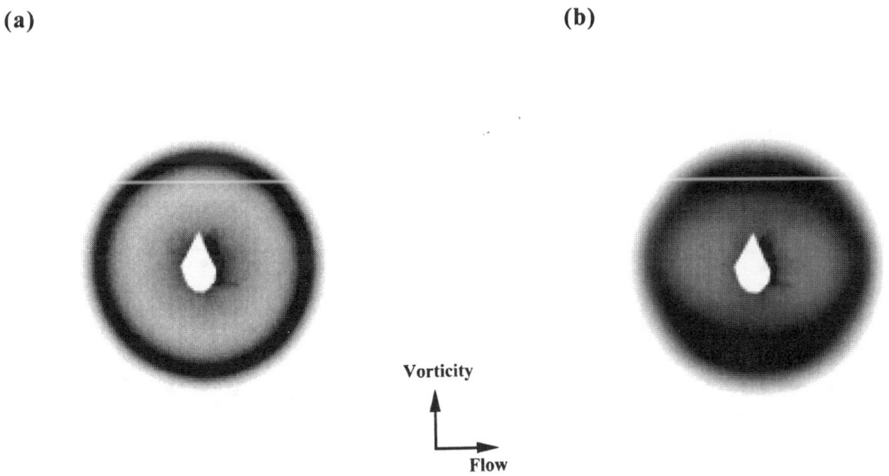

Figure 3. Scattering (a) at rest and (b) at Pe 70 for a sterically stabilised PMMA dispersion (1.5 μm diameter, volume fraction 0.4) in the flow-vorticity plane.

The data shown in Figure 3 can be analysed quantitatively. It is possible to fit models of 'hard-sphere' liquids to the data and quantify the structural distortions in terms of the density or volume fraction in different orientations [9]. Under conditions of flow the data is well described by the Ashcroft-Lekner model of a hard sphere fluid [15]; the results of some of these fits are shown in Table 1. The effective volume fraction is shown as ϕ and scale factor required to fit the data is a measure of the magnitude of the correlations. The changes during shear thinning are small but the density increase in the flow direction is in reasonable accord with that expected from the Krieger-Dougherty relation [16]. A further approach has been to compare the data with the scattering patterns calculated for structures observed in computer simulations [10,17]. The structure at high shear rates is clearly better modelled by simulations that include some interparticle hydrodynamic interactions [18] than those with just random Brownian forces. Quantitative comparison of the structural changes in computer models is still restricted by the finite size of the simulation data set and the number of independent configurations that are observed.

TABLE 1. Results of Fits of Ashcroft-Lekner Model to Light Scattering Data

	Flow		Vorticity	
Pe	ϕ	Scale	ϕ	Scale
0.0	0.46	64	0.46	72
0.7	0.42	65	0.45	73
1.4	0.41	70	0.45	77
7.0	0.38	71	0.45	74
14	0.38	75	0.45	77
28	0.38	75	0.45	75
70	0.37	69	0.45	69
140	0.37	77	0.45	77

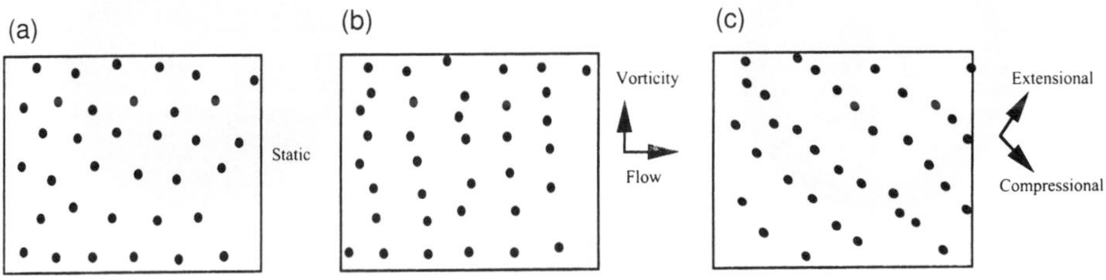

Figure 4. Schematic diagram of the structures of a sterically stabilised colloid (a) at rest, (b) under shear Flow-Vorticity plane, (c) under shear extensional-compressional plane.

The schematic drawing in Figure 4 shows the distortion in structures that have been observed by the light scattering experiments described and by earlier work with small-angle neutron scattering [9,14]. The separation of particles increases along the flow direction and on the extensional axis of the flow. There is also a significant increase in the correlation in the flow direction.

3.2 NON-UNIFORM SHEAR

The rheology of non-Newtonian fluids can show a wide range of behaviour. The examples of steady flow described in the previous section can be considered as a simple special example. Under many circumstances a process may involve oscillatory or discontinuous shear. The strain amplitude and frequency will then both play a role in determining the properties of the material. Another example of non-uniform flow fields is the behaviour of a dispersion that sediments while it undergoes shear. The regions of different particle density will have different viscosity and structures.

The dispersion described in the previous section can adopt very different structures under the influence of oscillatory shear. While the behaviour under continuous shear is readily interpreted as a distortion of a liquid structure, distinct diffraction peaks can be observed at moderate frequencies and amplitudes of oscillatory shear. Experiments have been made with a motor-drive system capable of high acceleration such that the shear profile is a square-wave. The strain can be characterised by a frequency or shear rate (given by the constant strain rate in each direction) and amplitude. Both the amplitude and shear rate were observed to have large effects. A typical scattering pattern is shown in Figure 5. This shows clear streaks or lines in the flow direction and is markedly different to the scattering from a continuously sheared sample which is shown in Figure 3.

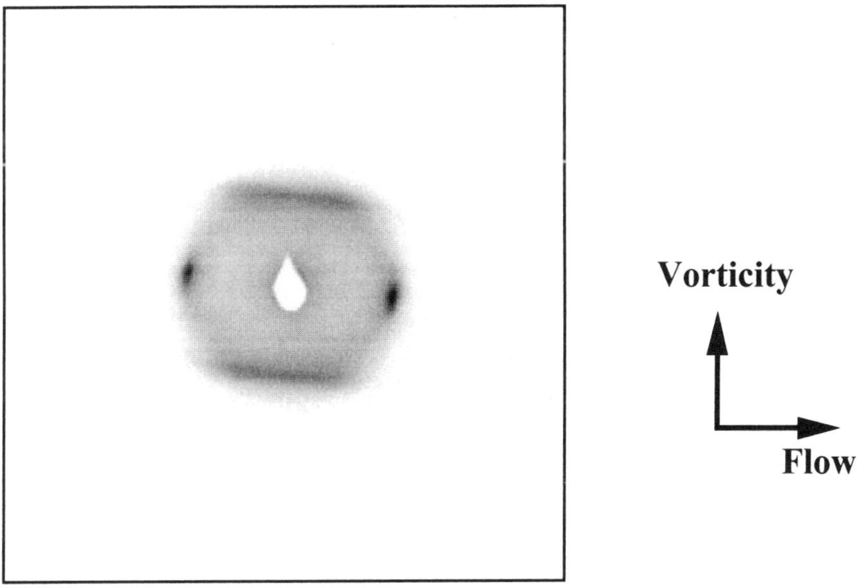

Figure 5. Scattering pattern for a dispersion under oscillatory shear at Pe 170, the lines of intensity in the flow direction are clearly visible.

Scattering in lines such as that shown in Figure 5 can be interpreted as a 'string-phase' [19]. Although such structures have been predicted in computer simulations of steady shear, they are observed in this material only under moderate amplitude oscillatory shear. The particles are thought to move in lines with no correlations between the positions of particles in adjacent strings. Another interesting feature of the scattering is the increase in intensity at small **Q** that correspond to distances larger than particle diameters. This is apparently some reversible association of the particles.

Sedimentation will often occur with particles of size about 1 μm and a density difference of 0.1 g cm^{-3} and this has been observed to cause highly ordered structures under the combined influences of gravity and shear [20]. In one system, PMMA particles in Decalin/Tetralin mixtures, a highly ordered crystalline sediment was observed to rotate with the moving lower plate of the disc-disc shear cell. All the shear was occuring in the dilute, upper liquid phase. This observation of a boundary within a colloidal dispersion and an effective solid layer may be important in understanding the flow in a variety of practical applications.

3.3 CHARGE STABILISED DISPERSIONS

It is interesting to contrast the flow behaviour of sterically stabilised colloidal dispersions with properties of charge stabilised particles that interact through a much softer, long-range potential. These materials can interact and form ordered structures at much lower volume fractions. The structure under shear has been investigated by small-angle neutron scattering [7]. This is illustrated in Figure 6 for a sample of polystyrene latex (diameter 205 nm) in deionized water at weight fraction 8.7 %. The structure at rest shows a highly ordered crystalline arrangement. Analysis of the intensities of the different Bragg diffraction spots has allowed this structure to be identified as hexagonal close packed [21]. On shearing at slow speeds, the regular structure is enhanced. At shear rates above 5000 s^{-1} the structure begins to 'melt' and the order decreases.

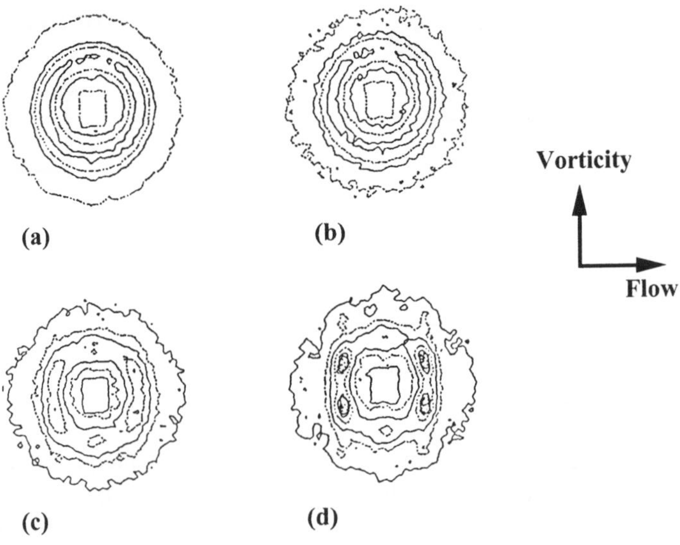

Figure 6 Small-angle neutron scattering from polystyrene latex under shear. (a) at rest, (b) 300 s^{-1}, (c) 5000 s^{-1} (initial), (d) 5000 s^{-1} after 15 min. This shows that the sample is aligned to a monocrystal under the influence of shear. The crystal remains stable after the shear stress is removed. Shear at higher rates can induce 'melting' of the crystal.

3.4 ANISOMETRIC PARTICLES, COMPLEX FLOW GEOMETRIES

Many practical dispersions are formed of either irregular or anisometric particles. The flow of these materials can be investigated to a limited extent in the manner described above. The requirements to optically match particles and dispersion media restrict but do not completely obviate the use of light scattering. A technique that has proved useful in determining orientational alignment is to measure the angular distribution of Bragg diffraction peak intensities. The intensity in a given position is directly proportional to the number of crystallites in that orientation. The method works well with neutrons for which many samples of a thickness up to 1 cm are adequately transparent. It has been applied to study the alignment of kaolinite dispersed in D_2O under pipe flow [22]. This material is charged at the high pH used in the measurements and the common edge to face aggregation was avoided with a polyacrylate stabiliser. The results show that several order parameters, not just the first, are important in describing the alignment.

Experiments can also be made to measure the alignment in complex flow geometries. The alignment and disorder arising from turbulent flow at different distances from a baffle in a pipe have been measured [23]. These methods can be applied to a variety of materials and a precipitated calcium carbonate (aragonite) which is formed as needles has also been studied [23].

4. Discussion and Conclusions

In this paper it has been possible only to describe some major features of recent rheo-optical studies of concentrated dispersions. The work has shown that such experiments can provide useful tests of computer models and theory. The requirement to predict the correct mechanical properties alone may not be sufficient to test a model. For example, this work has indicated that interparticle hydrodynamic interactions are important. In some circumstances, the structure can show that rheological measurements are not straightforward; the non-uniform flow gradients that arise in a sedimenting colloid may be found in many rheological measurement geometries as well as engineering practice.

The effects of boundaries and walls on colloidal structure may be important. Measurements on macroscopic, model hard dispersions have shown aligned structures at the cell surfaces. It has been reported that surface finish can significantly alter rheological measurements. The studies of structure at surface may lead to an understanding of these effects and improved control of flow. It is clear from the present work that shear strain amplitude and flow history are also important.

This area of research has much potential for further work. It is to be expected that in the future studies of mixtures will be extended over a wide range of systems such as particles dispersed in polymer solution and other mixtures of colloids under shear.

Acknowledgements

This work was supported by the DTI Programme in Colloid Technology with ICI, Schlumberger Cambridge Research, Unilever and Zeneca. Further support came from the EPSRC and ECC International for a project in collaboration with the University of Surrey. We are grateful for the provision of neutron facilities at the ILL, Grenoble, France and NIST, Gaithersburg Md. The facilities at NIST were supported by the National Science Foundation.

References

1. Clarke, S. M. and Rennie, A. R. (1996) *Curr. Opin. Coll. Interf. Sci.* **1**, 34-38.
2. Brady, J. F. (1996) *Curr. Opin. Coll. Interf. Sci.* **1**, 472-480.
3. Goodwin, J. W. (1990) pp 209-223 in *An Introduction to Polymer Colloids* Candau, F. and Ottewill, R. H. (Eds.) Kluwer Academic Publishers, Dordrecht.
4. Hukins, D. W. L. (1981) *X-ray Diffraction by Disordered and Ordered Systems,* Pergamon Press, Oxford.
5. Brumberger, H. (Ed.) (1995) *Modern Aspects of Small-Angle Scattering,* Kluwer Academic Publishers, Dordrecht.
6. Johnson, G. D. W., Ottewill, R. H. and Rennie, A. R. - Chapter 7 this volume
7. Ashdown, S., Markovic, I., Ottewill, R. H., Lindner, P., Oberthür, R. C. and Rennie, A. R. (1990) *Langmuir* **6**, 303-307.
8. Ottewill, R. H. and Rennie, A. R. (1990) *International Journal of Multiphase Flow* **16**, 681-690.
9. Clarke, S. M., Rennie, A. R. and Ottewill, R. H. (1995) *Adv. Coll. Interf. Sci.* **60**, 95-118. Dordrecht.
10. Clarke, S. M., Melrose, J. R., Rennie, A. R., Heyes, D. M. and Mitchell, P. J. *Physica* submitted.
11. Pusey, P. N., van Megen, W., Bartlett, P., Ackerson, B. J., Rarity, J. G. and Underwood, S. M. (1989) *Phys. Rev. Letts.* 63, 2753-2756.
12. Yan, Y. D., Dhont, J. K. G., Smits, C. and Lekkerkerker, H. N. W. (1994) *Physica A* **202,** 68-80.
13. Antl, L. , Goodwin, J. W., Hill, R. D., Ottewill, R. H., Owens, S. M. and Papworth, S. (1986) *Colloids and Surfaces* **17**, 67-78.
14. Lindner, P., Markovic, I., Oberthür, R. C., Ottewill, R. H. and Rennie, A. R. (1988) *Prog. Coll. Pol. Sci.* **76**, 47.
15. Ashcroft, N. W. and Lekner, J. (1966) *Phys. Rev.* **145**, 83-90.
16. Krieger, I. M. (1972) *Adv. Colloid Interface Sci.* **3**, 111.
17. Clarke, S. M., Melrose, J. R., Rennie, A. R., Heyes, D. M. and Mitchell, P. J., Ottewill, R. H., Hanley, H. J. M. and Straty, G. C. (1994) *Journal of Physics: Condensed Matter* **6**, A333-337.
18. Melrose, J. R. and Ball, R. C. (1995), Adv. Coll. Interf. Sci. **59**, 19-30.
19. Clarke, S. M. and Rennie, A. R. (1997) *Prog. Coll. Pol. Sci.* in Press.
20. Clarke, S. M. and Rennie, A. R. (1996) *Faraday Discussions*, **104** in Press.
21. Clarke, S. M., Rennie, A. R. and Ottewill, R. H. (1997) *Langmuir* **13**, 1964-1969.
22. Clarke, S. M., Rennie, A. R. and Convert, P. (1996) *Europhys. Letts.* **35**, 233-238.
23. Brown, A. B. D., Clarke, S. M. and Rennie, A. R. (1996) in *Dynamics of Complex Fluids* in Press.

SHEAR-THICKENING IN SUSPENSIONS OF DEFORMABLE PARTICLES

W. J. FRITH, A. LIPS
Unilever Research, Colworth House, Sharnbrook MK44 1LQ, UK

J. R. MELROSE, R. C. BALL
Cavendish Laboratory, University of Cambridge, Cambridge CB3 0HE, UK

Measurements of the shear thickening properties of various suspensions of deformable particles are discussed. The suspensions consist of small fraction wheat starch granules and agarose microgels in various hydrophilic media and show properties that are strongly dependent on the swelling behaviour of the particles. Results are compared with stokesian dynamics simulations of the flow of suspensions of spherical particles which include hydrodynamic interactions between particles and a model for a swollen polymer layer on the particle surface. This comparison confirms that the nature of the particle surface is crucial in determining the nature of the high shear behaviour of suspensions.

1. Introduction

Suspensions of deformable particles are of generic interest to the foods industry, being found in the form of foams, emulsions, and microgel suspensions. Such systems commonly display shear thinning behaviour arising from the deformability of the particles and, when the dispersed phase concentration is sufficiently high, viscoelastic behaviour is observed that is also closely linked to that of the particles themselves. Such behaviour has been studied, both experimentally and theoretically [1,2,3] for model suspensions relating to both food and surface coatings related systems.

Less frequently, systems such as these will display shear thickening behaviour whereby the steady shear viscosity is observed to increase rapidly with increasing shear-rate after some critical shear-rate has been exceeded. This phenomenon has long been studied and several theories have been proposed to explain the behaviour [4,5], however until recently, little systematic work on model systems had been conducted to test these models. One common feature of all of the proposed models was that the transition involved a structural change within the suspensions from an ordered free flowing arrangement of particles to one that is disordered and solid like. This hypothesis appears to be contradicted by the frequent observation of shear thickening in suspensions where ordering of the particles is extremely unlikely, i.e. where the particles are either irregularly shaped or highly polydisperse in nature, or both [6], examples of such systems include starch and pigment suspensions. In addition, recent studies on model suspensions [7,8,9] appear to indicate that most of the earlier models of shear thickening are inadequate and that a more appropriate mechanism involves the

R. H. Ottewill and A.R. Rennie (eds.), Modern Aspects of Colloidal Dispersions, 123–132.

formation of hydrodynamic clusters of particles [7]. Such behaviour has also been confirmed through the use of Stokesian dynamics simulations [10,11] which show that there is a strong tendency in concentrated suspensions toward the formation of such clusters.

One of the principle conclusions of the recent work in this area is that the properties of the particle surface are paramount in determining the shear-thickening behaviour [7,10,11]. A surprising result is that suspensions of Brownian hard spheres do not display strong shear thickening, because of the detailed nature of the forces between particles in such systems [11]. Particles that posses a swollen polymer layer on their surface form prime examples of suspensions of which will frequently display very strong shear thickening [7]. It is now believed that it is the detailed nature of this layer that determines the shape of the flow curve beyond the shear-thickening transition [11].

This paper discusses results from a collaboration between Unilever Research at Colworth House and the Cavendish Laboratory at Cambridge. At Colworth work has been carried out over the past 4 years to study the shear-thickening behaviour of suspensions of microgel particles of relevance to the food industry. Two classes of suspension have been considered: Starch suspensions have long been known to display shear thickening behaviour, and have been studied over a considerable period [6,12]. In view of recent advances in rheological techniques [6,7] and theory [10,11] it was felt worthwhile to revisit this system. Systematic studies have been made of the effects of granule size, overall concentration and the nature of the suspending medium on the behaviour of these systems and some work has been reported previously for maize starch [6], here we discuss some further results on a wheat starch with a smaller granule size. In conjunction with this work studies have also been carried out on the flow properties of model suspensions of agarose microgel particles, the work was primarily focused on the viscoelastic and flow properties of suspensions at concentrations above close packing [1]. However, at more moderate concentrations these suspensions also display shear thickening and some preliminary results in this area are reported here.

In parallel with the work at the Colworth Laboratory studies at the Cavendish have concentrated on the development of simulation and theoretical approaches to modelling the flow properties of very concentrated suspensions. It was this work that demonstrated that shear thickening arose through the formation of hydrodynamic clusters during shear [10,11], and that the particle surface properties are crucial in determining whether these clusters form. In particular the work has demonstrated that the pure hard sphere model is not appropriate as a model for suspension flow as the gaps between particles tend to collapse during shear leading to the suspension locking up. The fact that the gaps between particle surfaces are so small during flow in concentrated suspensions implies that in order to model such systems we must include a model for the particle surface that is accurate on the molecular scale, something which is as yet unfeasible.

However, the simulation techniques have developed to the extent where crude models for the nature of the particle surface can be included [13]. In particular a model for a swollen polymer layer has been included in the simulation which makes it possible to consider comparison with the observed behaviour of the starch suspensions, which are believed to posses such a layer. This is the subject of the present paper, along with some preliminary results on the shear-thickening properties of microgel suspensions.

In the following sections the materials and their microstructural properties are discussed, followed by a brief account of the simulation technique used, and finally a discussion of the results obtained and a comparison of the simulations with the experimental data.

2. Materials

2.1 WHEAT STARCH SUSPENSIONS

Wheat starch is usually comprised of a bimodal size distribution of granules, hence the material studied here was fractionated to remove the larger size portion, leaving a fairly narrow size distribution with a diameter of approximately 4μm. Suspensions were prepared in three different media, de-ionized water, pH 11 NaOH and ethane diol. The starch used was initially purified by repeated centrifugation and redispersion in de-ionized water to remove soluble material, and filtration to remove larger particulates and plant debris, the resulting suspension was used for all future studies. Particle size distributions were determined by static light scattering on a Malvern Mastersizer and are shown in figure 1.

Suspensions of the wheat starch granules in other media were prepared by repeated centrifugation and redispersion in the appropriate medium until sufficient purity was obtained, this was determined by the pH and viscosity of the supernatant for the NaOH and ethane diol respectively.

It is known that starch swells on contact with water usually taking up approximately its own volume in water. Previous work on maize starch [6] showed that for this case the swelling of the granules was also the same in ethane diol with a volumetric factor of

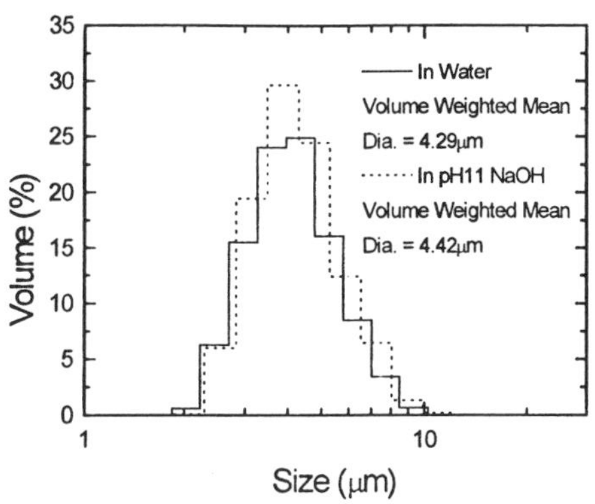

Figure 1 Particle size distribution for wheat starch suspensions.

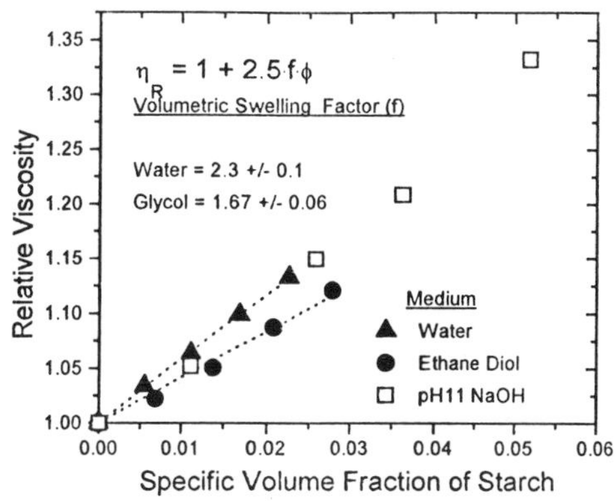

Figure 2 Intrinsic viscosities for wheat starch suspensions in water and ethane diol, demonstrating the different swelling factors in the two media.

1.76. Intrinsic viscosity measurements on the wheat starch suspensions were carried out in order to determine the degree of swelling and the results are shown in figure 2 as relative viscosity against starch specific volume fraction (0.62 times the w/v conc.). These results show clearly that the degree of swelling is significantly greater in the aqueous media than in the ethane diol. As a result, it would be expected that the granules suspended in aqueous media would be more deformable than those in ethane diol, and that there would be some consequent effect on rheology.

2.2 MICROGEL SUSPENSIONS

Spherical microgel particles were produced by emulsifying a 5% w/w solution of agarose (low viscosity type XII ex Sigma) in oil (sunflower oil) at 90°C using Hypermer B246 (ICI Surfactants) as an emulsifier. This w/o emulsion was then cooled to 20°C to gel the polysaccharide. The resulting microgel was then transferred to water by centrifugation and purified by further dilution and centrifugation in water. These particles are spherical and non-aggregated and have an average diameter in the region of 10μm. It is expected that the concentration of the polymer within the particles will be approximately the same as that of the original solution, since it is known that an agarose gel will not exhibit additional swelling on contact with excess water.

Particle size distributions were again determined using static light scattering on the Malvern Mastersizer and are shown in figure 3. This result must be treated with some caution, however, since the refractive index difference between the particles and the medium is slight and as a result measurements had to be made at rather high concentrations in order to get sufficient scattering intensity. In particular it is believed that the second peak at smaller sizes is an artefact. This is primarily because the rate of centrifugation was adjusted so as to remove particles smaller than 3μm and further fractionation had no effect on the measured particle size distribution. Also visualisation of the sample under a microscope did not reveal the presence of small particles.

As mentioned above, it is expected that the concentration of polymer within the

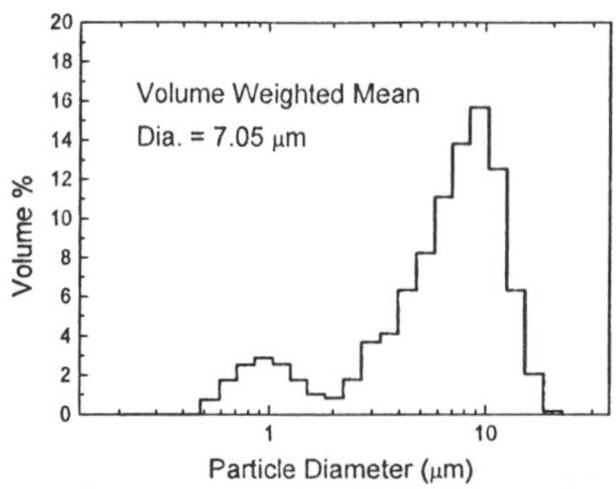

Figure 3 Particle size distribution for agarose microgel particles.

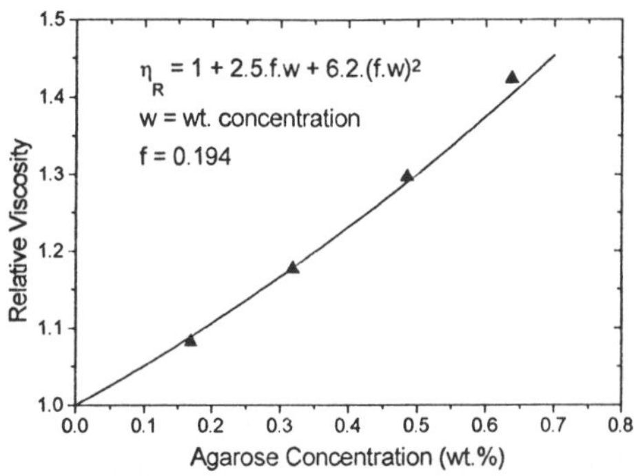

Figure 4 Intrinsic viscosities for agarose microgel suspensions.

microgel particles will be the same as that of the original polymer solution, i.e. the particles undergo no additional swelling upon being transferred into water. Intrinsic viscosity measurements on these samples, shown in figure 4, confirm this hypothesis. The results give an effective swelling ratio for the microgel particles of 20 v/w, i.e. the particle volume fraction is 20 times that of the weight fraction of the polymer in the suspension.

3. Rheological Methods.

The flow properties of the suspensions were studied using a Rheometrics DSR Controlled Stress Rheometer. Parallel plate geometries were used for all measurements, the surfaces of which were coated with a layer of P600 Emery Paper to provide a rough surface in order to avoid errors due to wall slip. A measurement strategy involving shearing the sample at high stresses in between each measurement was adopted in order to prevent sedimentation of the sample during the measurement, the method is described in detail in [6]. Finally in order to prevent evaporation, or the uptake of water in the case of ethane diol, the sample was covered with dodecane during measurement.

4. Simulation Technique for Concentrated Particles in a Fluid

Modelling the flow of particles in a concentrated suspension in a fluid medium is a significant challenge. However, in the limit of concentrated systems, the relative motion of the surfaces of close particles must dominate the viscous interactions. Indeed, for purely hard spheres, the fluid within narrow gaps determines a viscous force which diverges to infinity for decreasing gaps between surfaces in relative motion.

Only these 'near field' viscous terms are retained in the simulation algorithm. The motion of the particles is assumed to be quasi-static, that is it occurs under a balance between:

- Dissipative (or viscous) forces and torques which depend on the relative motion of the particles,
- The more usual conservative forces which are defined by a potential between particles.

$$-\mathbf{R} \bullet \mathbf{V} + \mathbf{F}_C = 0 \qquad (1)$$

The equation of motion is that of Newton with the accelerations set to zero. The viscous force $-\boldsymbol{R}\cdot\boldsymbol{V}$, is assumed to be linear in the 6N velocity/angular velocity vector $\boldsymbol{V}$ and the resistance matrix $\boldsymbol{R}$ has coefficients which depend only on the separation of the surfaces of nearest-neighbour particles. The conservative forces $\boldsymbol{F_C}$, are given by the direct interactions between the granules. Flow is driven by what are known as Lees-Edwards boundary conditions [14] on systems of spheres within a box surrounded by and interacting with periodic images of themselves. Nearest neighbours are identified by the construction of Delaunay tetrahedra [15]. This guarantees that all particles with gaps below $\sqrt{2}-1$ diameters are identified as neighbours.

Note that this algorithm is quite different from traditional molecular dynamics (MD) where masses and accelerations are included. To move the particles the code needs to compute the forces F_C and find velocities by inverting the sparse matrix R. This leads to a highly correlated motion of the particles; for example, if a force were applied to a line of particles, in one computer time step the whole line would move. In traditional MD only the nearest neighbours would feel the force in the next time step.

The simulation algorithm has now been developed to the extent that it is possible to model the behaviour of particles surrounded by a swollen polymer layer. Flow of the suspension medium through this layer is hindered, giving a combination of conservative and dissipative forces between particles that interact during flow [11, 13].

5. Results and Discussion

5.1 WHEAT STARCH SUSPENSIONS

Flow curves for the wheat starch in de-ionized water at a range of concentrations are shown in figure 5, whilst figures 6 and 7 show similar data for the wheat starch suspensions in ethane diol and pH 11 NaOH respectively. Data are shown as relative viscosity against Peclet Number $(\eta_M \dot{\gamma} a^3 / kT)$, where η_M is the medium viscosity, $\dot{\gamma}$ is the shear-rate, and a is the particle radius. Strong shear thickening is displayed by all the suspensions, which gradually decreases as concentration is reduced. It is interesting to compare the detailed nature of the shear-thickening behaviour displayed in figures 5 and 6. For the suspensions in water the increase in viscosity with shear-rate is large but

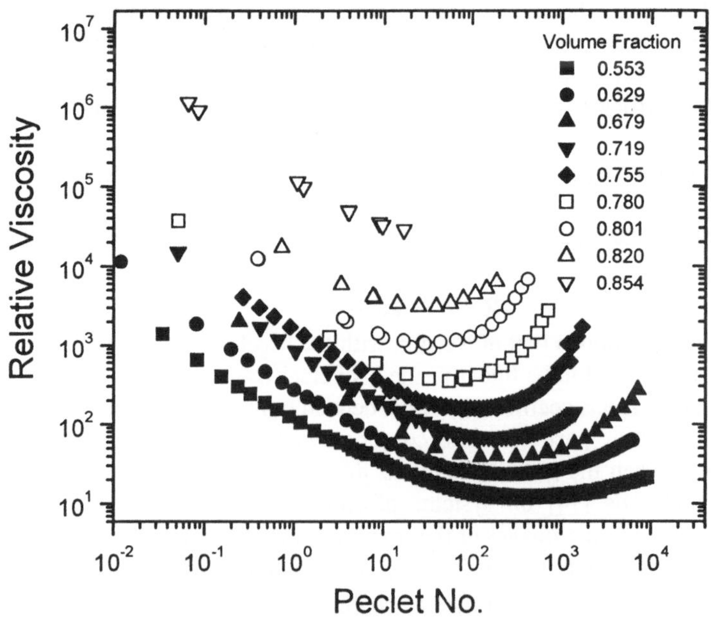

Figure 5 Flow curves for wheat starch suspensions in de-ionized water, effect of concentration.

always gradual. This behaviour contrasts with that of the suspensions in ethane diol where the transition is very abrupt above a certain concentration, and the viscosity increase is much larger. It is thought that this difference in behaviour arises from the particles in the two media having different deformabilities, which in turn is due to the different degrees of swelling in the two systems. It is worth noting that the behaviour of the suspensions in ethane diol more closely reflects that of data obtained for model hard-sphere suspensions [7] lending support to this hypothesis.

An additional feature displayed by the suspensions in de-ionized water is the strong shear-thinning behaviour at lower shear-rates, a phenomenon not observed in the suspensions in ethane diol. This behaviour is accompanied by thixotropic time dependent behaviour which implies that the starch granules are weakly aggregated. This behaviour was not observed in other starch suspensions [6] and it was also noted that the pH of the wheat starch suspensions was rather lower at pH 4 than that displayed by other starch suspensions. An investigation of the pH dependency of the suspensions showed that a minimum viscosity was displayed at a pH of 11. It is thought that this behaviour arises from the presence of charged groups on the starch granules which only dissociate above a certain pH leading to a reduction in the degree of aggregation and of the viscosity at low shear-rates. A minimum in viscosity is observed because, as the pH is increased past 11 the starch granules begin to swell significantly.

As can be seen in figure 7 the flow curves for suspensions of wheat starch in pH 11 NaOH display a far smaller degree of shear-thinning than do those in de-ionized water.

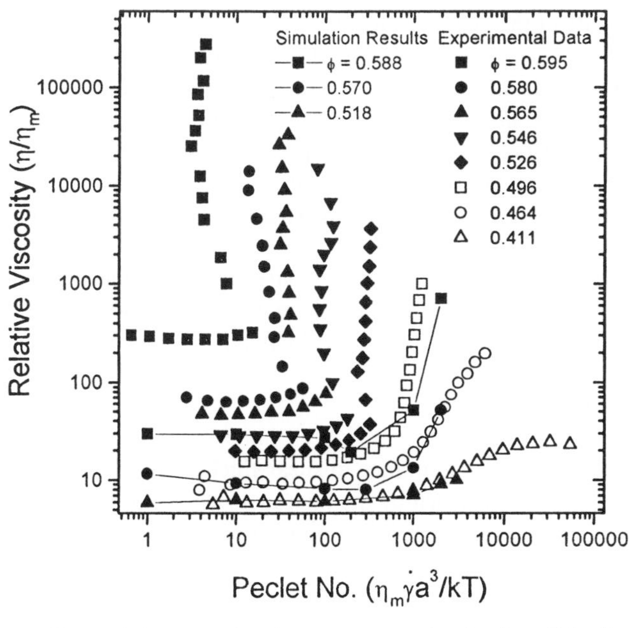

Figure 6 Flow curves for wheat starch suspensions in ethane diol, effect of concentration. Points with lines are simulation results.

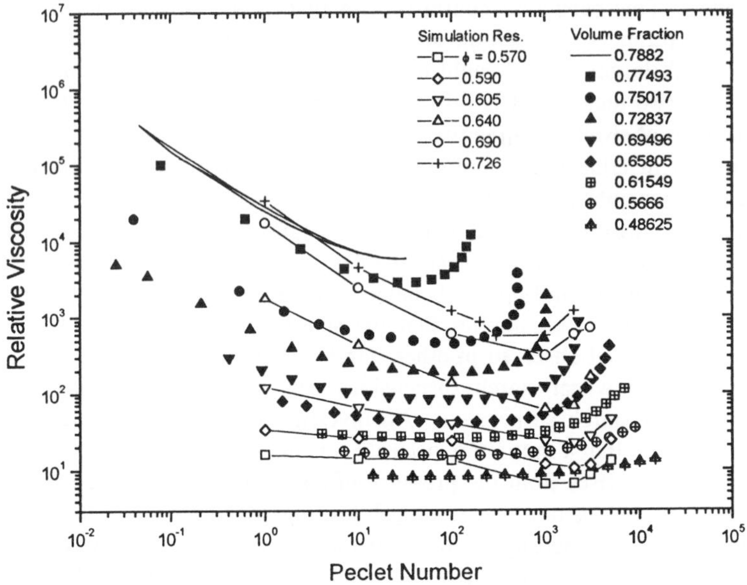

Figure 7 Flow curves for wheat starch suspensions in pH 11 NaOH, effect of concentration. Points with lines are simulation results.

However, the shape of the shear thickening portion of the curve is not significantly changed, leading to the conclusion that this part of the flow curve is primarily governed by the deformability of the starch granules.

Also shown in figures 6 and 7 are the results of simulations that have been conducted using reasonable estimates for the thickness and the deformability of the swollen layer on the surface of the starch granules. We find that the different rheological characteristics of the suspensions in the two media can be accounted for simply by the changing porosity of the coat, taking a higher value for the aqueous suspensions than for those in ethane diol as would be expected based on their molecular size.

The simulation results obtained show remarkable agreement with the wheat starch data despite the rather arbitrary values for the polymer layer properties. A more detailed account of the work and the model used for the polymer coat can be found in [11,13].

5.2 AGAROSE SUSPENSIONS

A further illustration of the importance to shear-thickening of the surface properties of particles in suspension is provided by the microgel suspensions. These systems are both more deformable than the starch granules and better characterised in that their internal structure is believed to be uniform and basically the same as that of a bulk gel of the same concentration. As a result the particles have well defined mechanical properties and a deformability which is well described by the Hertz model [1,3]. In addition, these particles carry a small charge arising from sulphate groups, this enables study of the effects of varying electrostatic interactions to be made independently of the mechanical properties of the particles.

Figure 8 shows flow curves of viscosity as a function of shear-rate for agarose suspensions suspended in media with different salt concentrations. It is clear that there is a strong effect of the screening of the charge on the shear-thickening properties of the suspensions. The suspensions show a minimum viscosity in de-ionized water, indicating that there is a very small amount of charge on the particles, which is only just sufficient to counteract the attractive interactions between them. Figure 9 shows the flow curves as a function of concentration for these systems in de-ionized water, these materials show pronounced shear-thickening, not at all dissimilar to that observed in the starch suspensions, despite these particles being far more deformable.

These are clearly preliminary results and there is tremendous scope for a systematic study of the flow properties of the suspensions and how they are determined by the effects of charge and particle deformability. Interestingly agar, in contrast to agarose, carries quite a significant charge, this would allow the electrostatic interaction between the particles to be manipulated through both screening by salt concentration and manipulation of the number of charged groups per particle.

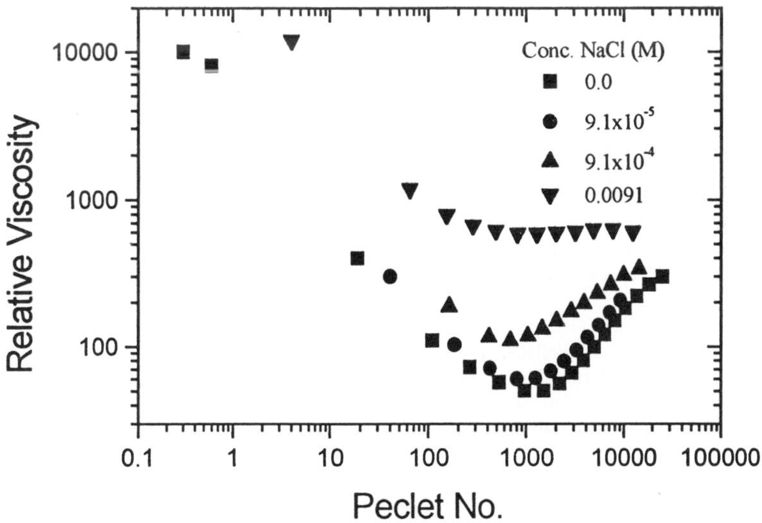

Figure 8 Flow curves for agarose microgel suspensions, effect of ionic strength.

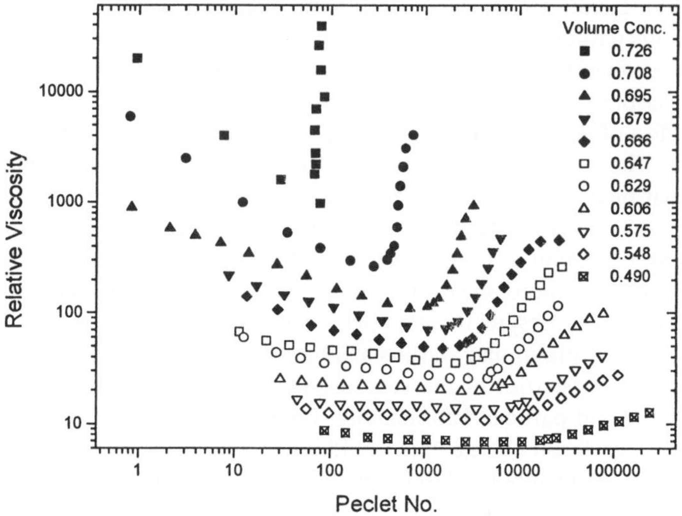

Figure 9 Flow curves for agarose suspensions in de-ionized water, effect of concentration.

6. References

1. Frith W.J., Lips A., (1996) *'Proceedings of the XIIth International Congress on Rheology'*, Quebec City, 558.
2. Buscall R., (1991) *J. Chem. Soc., Faraday Trans.*, **87**(9), 1365.
3. Evans I.D., Lips A., (1990) *J. Chem. Soc. Faraday Trans.*, **86**, 3413.
4. Hoffman R.L., (1974) *J. Colloid & Interface Sci.*, **46**(3), 491.
5. Boersma W.H., Laven J., Stein H.N., (1990) *AIChE J.*, **36**(3), 321.
6. Frith W.J., Lips A., (1995) *Adv. Colloid Interface Sci.*, **61**, 161.
7. Frith W.J., D'Haene P., Buscall R., Mewis J., (1996) *J. Rheol.*, **40** (4), 531.
8. Laun H.M., Bung R., Hess S., Loose W., Hess O., Hahn K., Hadicke E., Hingmann R., Schmidt F., Lindner P., (1992) *J. Rheol.*, **36**(4), 743.
9. D'Haene P., Mewis J., Fuller G.C., (1993) *J. Colloid & Interface Sci.*, **156**, 350.
10. Melrose J. R., van Vliet J. H., Ball R. C., (1996) *Phys. Rev. Lett.*, **77**, 4660.
11. Melrose J. R., van Vliet J. H., Silbert L., Ball R. C. and Farr R., *This Volume - Chapter 11.*
12. Hoffman R.L., (1982) *Adv. Colloid Interface Sci.*, **17**, 161.
13. Melrose J.R., Van Vliet J.H., Ball R.C., Frith W.J., Lips A., (1996) *Proc. R.S.C, Starch: Structure and Function.*
14. Lees A.W., Edwards S.F., (1972) *J. Phys.*, **C5**,1921.
15. Ball R.C., Melrose J.R., (1995) *Adv. Colloid & Interface Sci.*, **59**, 19.

THE SHEAR FLOW OF CONCENTRATED COLLOIDS

J. R . MELROSE†, J. H van VLIET† , L. E. SILBERT†
R. C. BALL*, R. FARR*
*Polymers & Colloids† Theory of Condensed Matter**
Cavendish Laboratory, Madingly Road
Cambridge CB3 0HE U. K.

We report our current understanding of the shear flow of concentrated colloids based on insight from simulations of an approximate model. We find that strong ordering of suspensions of rigid spheres under flow only occurs at volume fractions above 50%. We show how surface effects such as those of polymer coated particles and Brownian forces can determine shear thickening effects and the second Newtonian plateau. We present preliminary data on the distribution of stress within a flowing suspension and examine the structure factor in the flow-gradient plane.

1. Introduction

Controlling colloid rheology is key to many technological issues. Computer modelling offers the ability to gain much insight into such systems, but fully accurate models are not yet feasible. We therefore study a model that includes a reduced system of hydrodynamic interactions appropriate for the limiting case of concentrated systems. In concentrated suspensions of particles under steady shear, effects such as ordering under flow and the origins of shear thickening and shear thinning are poorly understood. In particular, a number of micro-structural origins of shear thickening have been suggested. Some have seen it as a general feature of non-equilibrium phase diagrams [1], others as a mechanical instability of layered flow [2]. Some theories [3,4] emphasise the role of conservative repulsive forces. Some have speculated on the role of Brownian forces [5]. Here we follow [6] and find that hydrodynamic clustering is key to this effect. We will introduce a simple model for polymer coated particles and show how the thickening is sensitive to the coat. Ordering of colloids under flow has been investigated by scattering techniques [7-9] and for 'near hard spheres' [7,9] found only to be present above volume fractions close to 50% - this is in contrast to simulations without hydrodynamics which have shown strong ordering at much lower volume fractions. The trends of the simulations here agree well with these experiments, however we point out that large system studies are required in regions of co-existing

R. H. Ottewill and A.R. Rennie (eds.), Modern Aspects of Colloidal Dispersions, 133–147.

phases. We still lack a true many-body theory of the flow of concentrated systems, but insights from the simulations should help. To this end we present new data on stress distributions and report on the correlation of particles in the shear-gradient plane.

2. The simulation model

2.1 TECHNICAL REVIEW AND THE NEED FOR A REDUCED MODEL.

Modelling the flow of particles concentrated in a hydrodynamic medium is a significant challenge. Schemes capable of high accuracy, either through higher order moment expansion [10], boundary collocation [11] or explicit numerical solution of discretised Stokes equations [12], are still not numerically efficient enough for a study of concentrated systems in shear flow. Either the periodic moving boundary conditions cannot be implemented or the need to handle very narrow gaps and high moment interactions between close surfaces demands too high a refinement of the descritisation or too high a truncation of a moment expansion. Less accurate methods have been proposed [6,13-15]. The most studied algorithm [6] involves $O(N^3)$ inversions of matrices which limit studies to small systems ($N < 100$) in 2D - although results for Brownian spheres in 3D have been reported recently [15]. It appears that dissipative particle dynamic [14] methods for representing the fluid phase are not viable in concentrated systems (again due to difficulty in handling the lubrication regime).

In the limit of concentrated systems the divergent hydrodynamic lubrication interactions due to the relative motion of the surfaces of close particles must dominate the resistance to particle motion. The leading squeeze mode diverges as the inverse gap between particle surfaces. Indeed Brady and co-workers [6,13,15] have already found that hydrodynamic clustering is determined by pair lubrication. Our algorithm identifies nearest neighbours and forms a resistance tensor pair-wise out of terms including the divergent parts. This truncation of the resistance tensor is at worst a truncation of $(1/r^3)$ terms : each row in the resistance matrix relates to Brinkman-like problem of one sphere moving in a fixed array, this has a velocity field which decays as $1/r^3$ [16]. However, a near neighbour truncation of the resistance tensor does, on inversion, generate a many-body mobility tensor which has a long range $1/r$ dependence and leading terms determined by the 1/gap divergence of the pair resistance matrix. However, it is likely an inaccurate approximation to the full hydrodynamic matrices as it ignores local N-body effects in which the flows within the narrow gaps are coupled to the flows in the local pores space around the particle. We suspect that the errors are of the order of terms in the logarithm of the inverse gap, which suggest errors at the 20-50 % level in low shear rate hydrodynamic contribution to the viscosity. Numerically, the algorithm requires $O(N^2)$ inversion of a sparse matrix, but in practice sufficient accuracy is gained from $O(N)$ operations. Others have made a similar approximation [17] but the scope of our results is unprecedented. Furthermore and contrary to others [17], we set up the matrix **R** solely out of Galilean invariant terms (those that depend just on relative velocities of particles) . In the absence of an explicit fluid velocity field in the simulation, we argue that it is physically incorrect to break Galilean invariance in the particle velocities by coupling them to a background flow through a set of drag forces. We drive shear flow through the boundary conditions . We refer to our model as a *frame invariant pair drag* model.

2.2 THE HYDRODYNAMIC MODEL

Between close surfaces, pair hydrodynamic interactions can be divided into squeeze modes along the line of centres (the leading singular interaction diverging as the inverse gap between particle surfaces) and modes arising from the transverse motion of neighbours. The code includes terms for all these modes. The hydrodynamic, colloidal and Brownian force and torque balance on each particle to give the equation of motion :

$$-\mathbf{R} \cdot \mathbf{V} + \mathbf{F}_C + \mathbf{F}_B = \mathbf{0} \qquad (1)$$

The hydrodynamic force and torque's, $\mathbf{R}\cdot\mathbf{V}$, are related linearly to the 6N velocity/angular vector $\mathbf{V}$ by the resistance matrix $\mathbf{R}$ formed from a sparse sum of pair terms between nearest neighbours. Flow is driven by Lees-Edwards [18] periodic boundary conditions on the particle velocity field. At each time step Equation (1) was solved for $\mathbf{V}$ by conjugate gradient techniques. Fixed time step computations can lead to overlap, we adopt a variable time step such that particles never overlap but can approach arbitrarily close : given a set of velocities from solving Equation (1), the time step is chosen as the smaller of a default value and the value such that the gap between the first pair of colliding particles is reduced to a pre-set factor of its current value. For Brownian simulations we generate random force and torque, $\mathbf{F}_B$, on each particle through a sum of weighted pair terms on each bond such that obey the fluctuation-dissipation theorem: $< \mathbf{F}_B \mathbf{F}_B > = 2\, k_B T\, \mathbf{R}$. Terms in the gradient of the diffusion tensor enter the first integral of the Langevin equation, and we include these by use of a time symmetric order dt^2 predictor-corrector algorithm for the particle trajectory. The Peclet number, Pe, sets the dimensionless shear rate - the ratio of the Brownian relaxation time to that of the shear rate, here we define

$$Pe = d^3 \mu \gamma / (k_B T) \qquad (2)$$

where μ is the solvent viscosity , d the particle diameter and γ the applied shear rate.

3. Structure and rheology under flow.

We have presented a number of results elsewhere [19-24]. Here we first summarise those results and then report new data.

3.1 PREVIOUS WORK BY THE AUTHORS

The 'bare' model of sheared hard spheres with only hydrodynamic interactions is a logical place to start the study - in the traditional view this is thought to be an ultimate high shear rate steady state, it is often associated with the second Newtonian plateau - a high shear regime of approximately constant viscosity. However we found that this view is incorrect. Rather the bare model is pathological under shear flow, it does not have a steady state solution [19] and shows extreme strain thickening. Clusters of particles form under shear down the compression axis and the divergent viscous interaction determines a strong coupling between the closeness of particle surfaces in a given cluster and the time scale for relaxation of that cluster under an applied force. For bare hard spheres, however, there are no mechanisms for relaxation and inter-particle gaps collapse to ever smaller values by low applied strains We have developed [20] a kinetic clustering theory of this effect, defining jamming in shear flow as the volume fraction (we find 0.51-0.53) above which clusters form and gel before they tumble.

The bare model is unphysical. In any real colloid Brownian forces and conservative forces provide mechanisms which can control the close approach of particles, introduce additional time scales into the problem and determine a variety of steady states. We have studied a number of models. Brownian spheres [21] in which just Brownian forces are included. Hookean spheres with short range repulsive springs on particle surfaces [22]. One problem we have made good progress on is the shear thickening effect [21]. We have extended the theory that this is due to hydrodynamic clustering [6], differentiating between discontinuous (related to the jamming described above) and continuous thickening and quantitatively simulating experiment [5] by adopting models of polymer coated particles [21]. We find a role for both Brownian and repulsive forces [3-5] but find that although thickening can couple to an order-disorder effect [2], thickening is not in principle an order-disorder transition. Core-corona models [23] of particles with porous surface layers were able to fit the rheological data and shear thickening effect in very high concentration native starch granules in suspension - this work is reported elsewhere in the proceedings [24]. The hydrodynamic clustering theory has been confirmed through experiment by others [25]. We studied systems with aggregation forces [26] due to depletion interactions and induced dipoles and found a universal shear thinning for these systems at high concentration and shear banding at moderate concentrations.

3.2 ORDERING UNDER FLOW OF BROWNIAN SPHERES

Considerable effort has been extended [7-9,15,27] to study the relationship between order, shear flow and volume fraction, Φ. In particular, if spheres with only short range surface forces at volume fractions below the equilibrium transition are sheared [7,9], can they order induced by the steady shear ? The experimental data for polymer coated spheres [7] first (on going up in Φ) suggests a weak string ordering at Φ = 0.48-0.5, then above 0.5 stronger ordering is reported. Only small changes in scattering intensity are seeen at lower Φ [9]. As a function of Pe number strings are found at the onset of ordering but sliding layers are the dominant structure at higher Pe. Simulations of Brownian dynamics and other models [27] without hydrodynamic interactions grossly fail in that they show shear induced ordering at much lower Φ. Simulations [15] with the hydrodynamics approximated as in ref. [13] first show ordering at Φ = 0.45 (for systems of N=27 and 108 spheres) at volume fractions for Pe ~ 10 and then order over a broadening range of Pe at higher volume fractions.

We simulated systems of N=200 Brownian hard spheres at Φ =0.42,0.45,0.47,0.48, 0.51 and 0.53 at Pe=10,80. Although the ordering is best seen looking down the flow direction, in Figure 1 we show structure factors in the vorticity-flow plane from Φ = 0.45, 0.51, 0.53 at Pe=10 - we choose this plane as it is the most accessible plane for light scattering experiments. [7,9]

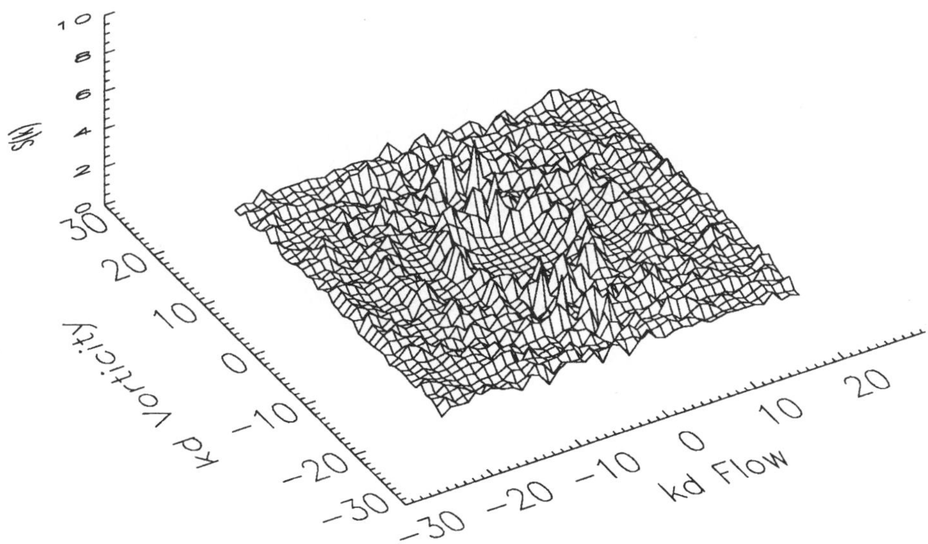

Figure 1a (see over page for caption)

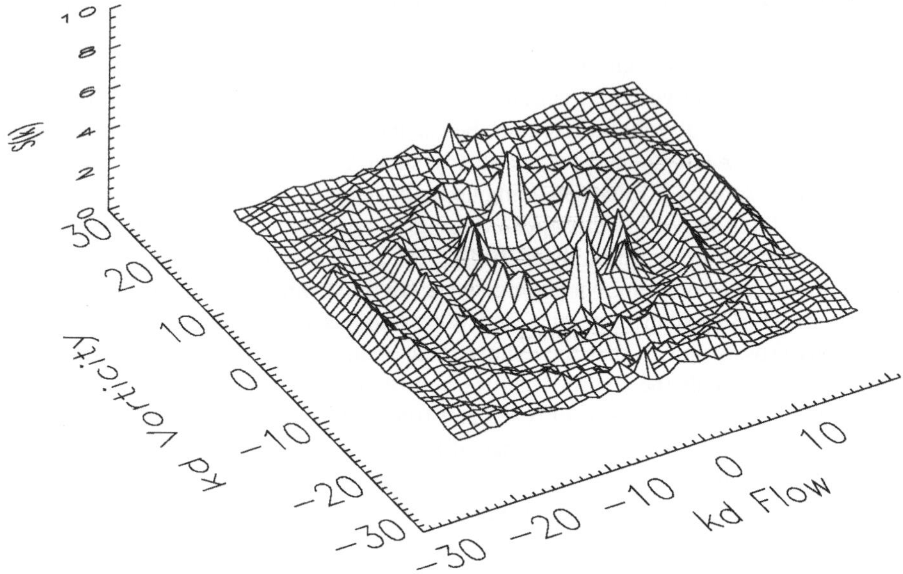

Figure 1b

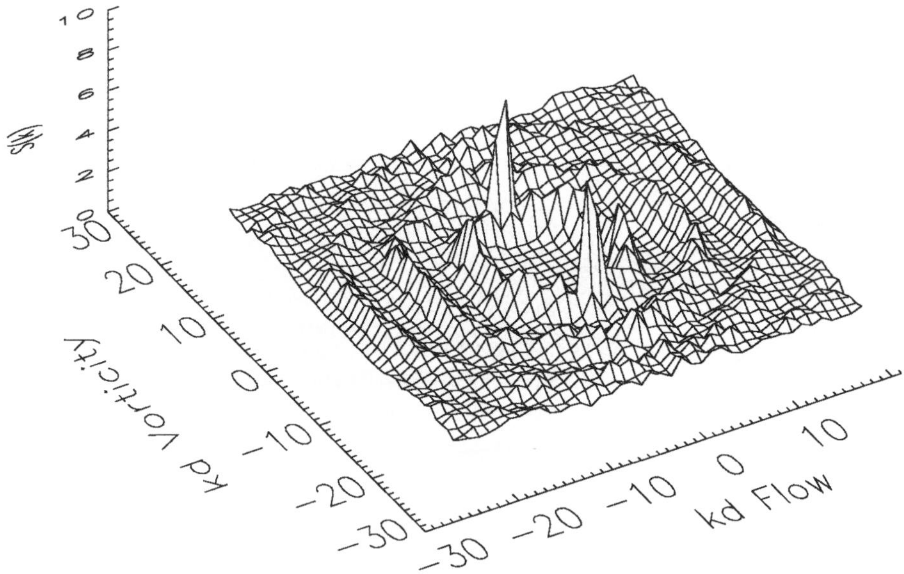

Figure 1c

Figure 1 Structure factors in the flow-vorticity plane from simulations of N=200 Brownian spheres at Pe=10 at $\Phi = 0.45$ (a),0.51(b),0.53(c).

The data at Φ = 0.51 is characteristic of string ordering. The bands of intensity along the vorticity direction, normal to the flow direction indicate strings of particles aligned parallel to the flow direction (see this proceedings for a string scattering pattern found under oscillatory shear [9]). The increasing intensity with increasing Φ of the higher orders of these bands indicate the increasing correlation of inter-particle spacing along the individual strings. However, with increasing Φ there is an increasing variation of intensity within the bands themselves (a double peaked structure emerges by Φ =0.53), this indicates increasing near neighbour correlation between strings and a transition from strings to sliding layers. The peaks on the vorticity axis indicate spacing of strings in the vorticity direction. With increasing Pe the order decreases and systems become disordered by hydrodynamic clustering. We find no order at any Pe at Φ =0.45, but in some configurations at Pe=10 Φ = 0.48 few strings appear within the bulk. Figure 2 shows a snapshot of a configuration with such weak ordering; strings also appear at Pe=80 Φ = 0.48 but are not evident by Pe=200 Φ = 0.48.

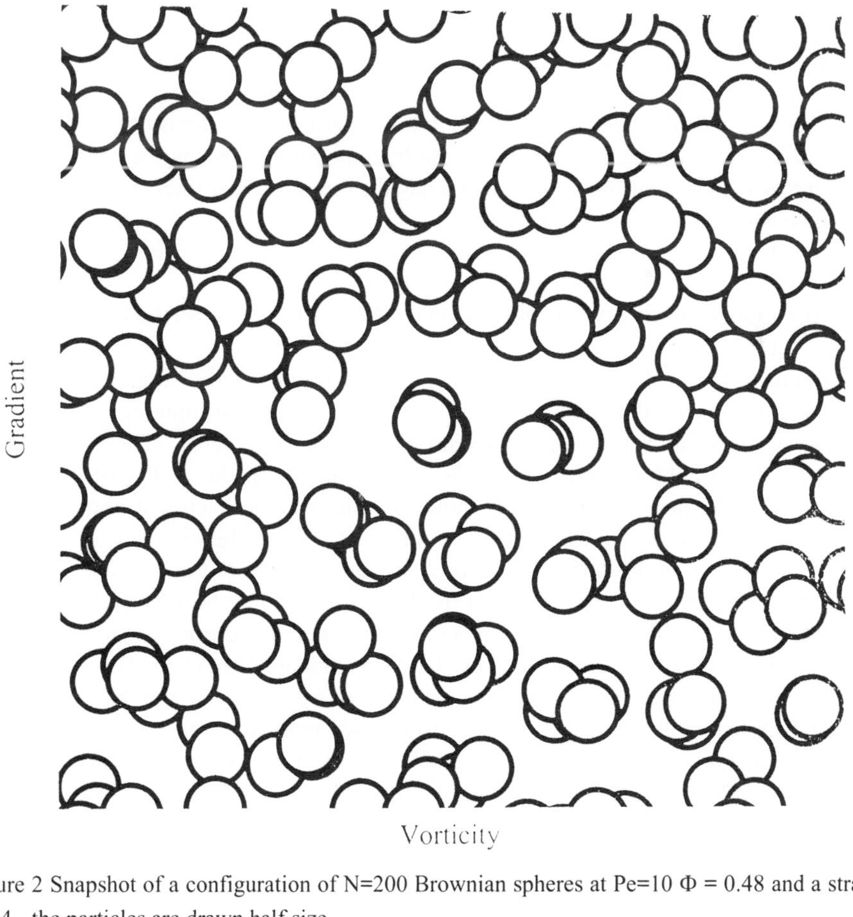

Figure 2 Snapshot of a configuration of N=200 Brownian spheres at Pe=10 Φ = 0.48 and a strain of 54 - the particles are drawn half size.

Although the model here is more consistent with experiment than that of ref. [15] (which showed relatively strong order beginning at Φ = 0.45) it is hard to identify why. The models are different in that ref. [13,15] includes some low moment contributions to the mobility matrix, computes the Brownian terms differently and systems sizes were also smaller. We checked the latter with N=50 simulations and found no difference as to the onset of order. What is clear is that both simulations suggest that any region of shear induced ordering below the Φ of the equilibrium transition is small. There are slight differences between the structures for simulations with squeeze-only hydrodynamics and for the model in which shear interactions were also included (as in all data shown here). At a given Pe in the ordering regime there is slightly more inter-string correlation in the presence of shear interactions. There are other issues which complicate the comparison of Brownian spheres with experiment. Firstly, ordering under flow in simulations can be affected by system size - in particular in regions of phase co-existence, ordered regions can percolate through the periodic boundaries. In some case we have found it necessary to extend the simulations to large boxes elongated along the flow direction. In these large simulations [19] we have found co-existing strings, layers and disorder and the regions are descrete both in time and space. Secondly, at high volume fractions, the presence of repulsive surface forces of range even 1% the diameter of the particle can influence the structures - polymer coats provide such an interaction in the experimental systems of ref. [7,9]. Part of the effect is simply the increase in volume fraction but the coats will extend the range of Pe across which ordering is found and delay the onset of shear thickening leading the second Newtonian plateau of experiment.

3.3 RHEOLOGY OF BROWNIAN AND POLYMER COATED SPHERES

We examined the rheology of systems of N=50 Brownian spheres. Figure 3 shows the simulation data at Φ= 0.45 and 0.5. The curves show the shear thinning characteristic of these systems but also a shear thickening at higher shear rates. These results are within 10-20% of the those of ref. [15]. Note, the sharp dip at Pe=10 Φ = 0.5 , corresponding to the onset of shear induced ordering, as discussed above. It is somewhat surprising to find that at the extreme of the dip the viscosity at Φ=0.5 is below the viscosity at Φ=0.45. This is probably an anomaly of the box size. Similar but stronger effects are seen in the data of [15] corresponding to the higher degree of ordering seen in those simulations. Note also that there is no true second Newtonian plateau in the data. We believe the simulation results are correct for Brownian hard spheres. Brownian forces act like a repulsion between surfaces, but the thickening with increased applied stress corresponds to an ever closer approach of particles as they pass each other in flow. We have shown [21] that the typical gap on close approach scales as the inverse of the reduced stress, $Pe\eta_r$, and that -contrary to textbook statements- for Brownian hard spheres, Brownian forces are relevant at all Pe. However experiments often show with increasing Pe both a second Newtonian plateau followed by a thickening response much stronger than that of Figure 3. We believe this is due to other interactions, in particular in suspensions of polymer coated particles. The interaction due to polymer coats has both conservative and dissipative parts. The conservative part arises from mutual compression of their polymer layers. The porous coat enhances the dissipation of fluid forced to flow through the coat under relative motion of the particles, leading for gaps at which coats contact to a higher dissipation than that of hard spheres at the same gap. The latter is a complex problem depending on the solvent polymer interactions, the pore space and concentration profile of the polymer coat and its dynamics. In particular the lowering of the quality of the solvent for the coat polymer gives an enhanced thickening response [5]. A crude model for the squeeze mode interaction has been developed [28]. This assumes a step function concentration profile for a coat of semi-dilute polymers and assumes the relative motions are such that the coats remain in their regime of linear response. We have added this model to our simulations and shown how it gives quantitatively similar results to experiment [21]. On the one hand in there is a region beyond shear thinning where the applied stress is insufficient to lead to compression of contacting coats and the simulations show flow in a relatively well ordered state and have an approximate plateau in viscosity. On the other hand at even higher stress, the coats do compress leading to thickening and disordering stronger than that of Figure 3 (see Figure 4 below).

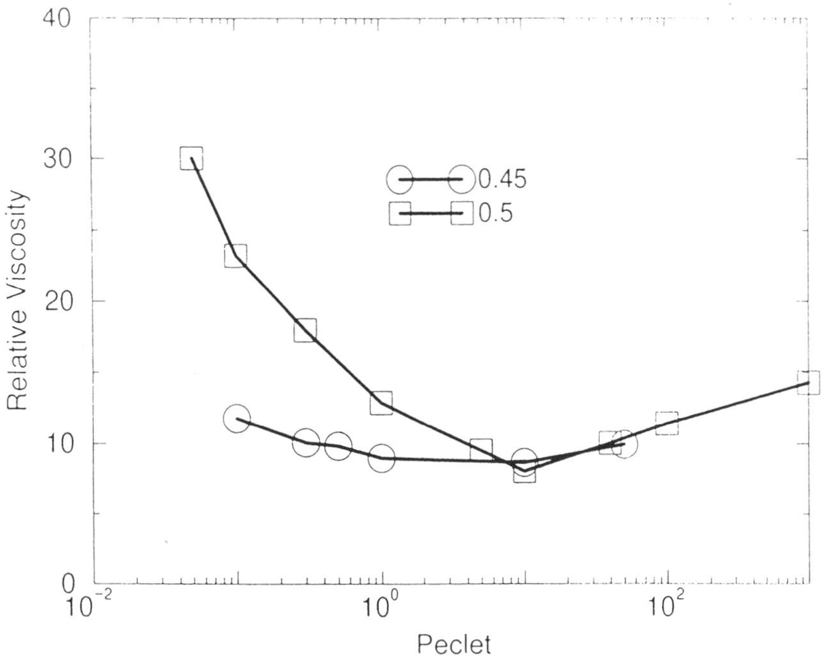

Figure 3 Rheology of simulations of N=50 Brownian spheres at $\Phi = 0.45,0.50$

Although the mechanism for thickening is hydrodynamic clustering, there can be a coupling to order-disorder effects in that such clustering disrupts order flow and often (but not exclusively) occurs from out of ordered regions. The thickening response is highly sensitive to the dissipative part as shown in Figure 4.

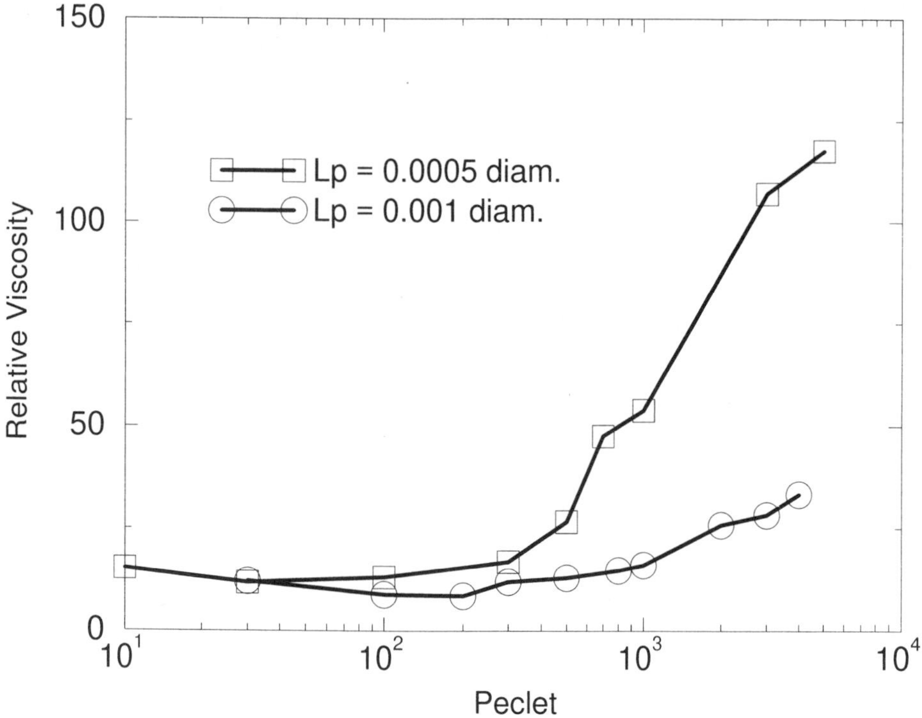

Figure 4 Simulations of N=50 spheres with polymer coat interactions at $\Phi = 0.54$, coat thickenss 0.01 diam. , data for coats of two different nominal pore size are given.

3.4 Nature of Bulk Flow

Here we give some preliminary data on our current effort to understand bulk flow of suspensions. Firstly we can separate out the different contributions to the stress. Figure 5 shows the rheology data of figure 3 but with the hydrodynamic and Brownian contributions identified.

The shear thinning is seen to be due to the Brownian contribution whilst the shear thickening is seen to be due to the hydrodynamic contribution. This feature was first identified in 2D simulations of ref. [6,13] and it has recently become possible to resolve the different contributions in experiment [25] confirming the dominance of the hydrodynamic part in the shear thickening regime.

The simulations can provide information. it is hard to obtain otherwise. Figure 6 shows the distribution of the shear stress per inter-particle bond between nearest neighbours. We show data for two systems Hookean spheres which just a short range repulsive spring at their surface and the same but with additional aggregation forces given by the depletion interaction law. We compare the Hookean system at low and high shear rates in the shear thinning region and at the beginning of the thickening regime respectively. Most bonds fall into a central peak which dominates the average. However at high shear rates in the Hookean model and at low shear rates in the aggregating systems there are a set of bonds which carry far more than the average stress. We are currently studying these distributions in more detail [29].

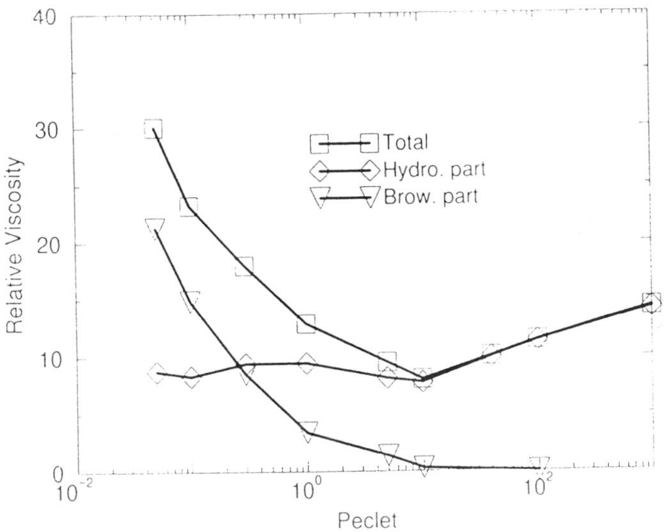

Figure 5 Brownian and hydrodynamic components of the viscosity for N=50 at $\Phi = 0.50$.

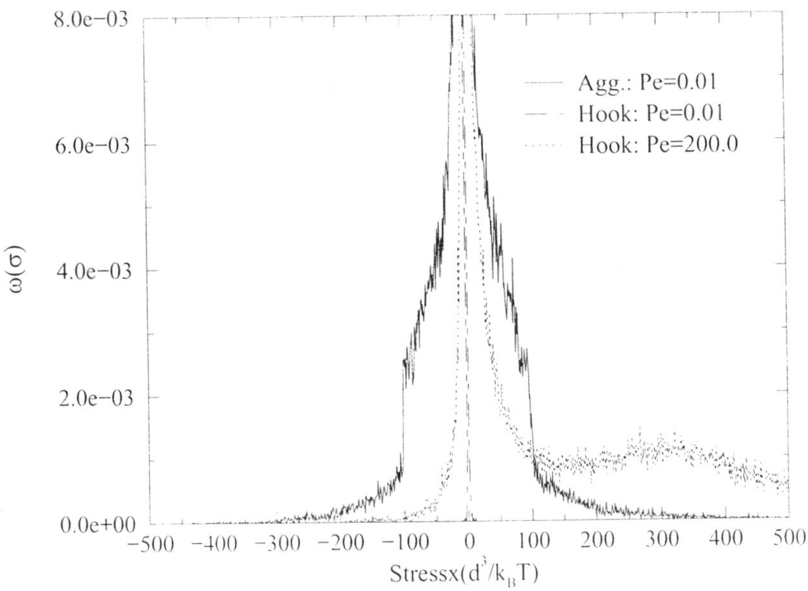

Figure 6a (caption see over)

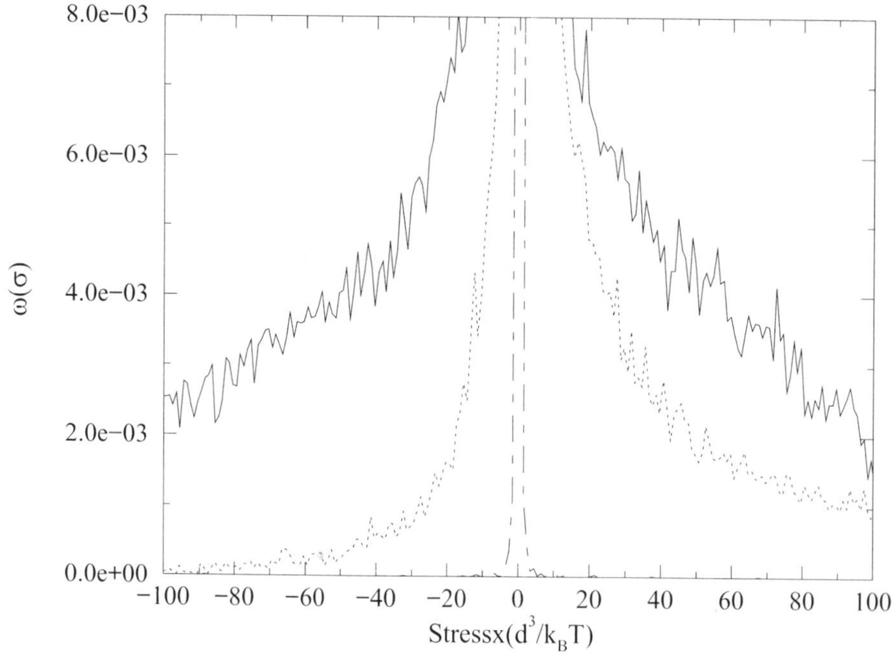

Figure 6 Distribution of shear stress per bond for simulations at $\Phi = 0.50$. of N=700 Hookian spheres (at Pe=0.01 mid shear thinning abd Pe=200 deep on the second Newtonian plateau) and Hookian spheres with and aggregation potential of -8k_BT at Pe=0.01 mid range of shear thinning. The average values are Hook(low Pe=0.01) = 0.1466, Hook(high Pe=200) = 109.95 and Agg(low Pe=0.01) = 2.199 . Figure 6b a close up of the central region of figure 6a

Figure 7 shows a structure factor in the shear gradient-flow plane. In the total intensity there is a distinct dip along the compression axis, there is also some indication of this in the structure factor. This feature suggests a dense packed and disordered structure along this direction, we find such dips even in regions of ordered flow. Intuitively we must expect such clustering along the compression axis direction, however, the temporal behaviour of such clustering is unknown. Note also the peaks on the bands which lie in the sector between the flow direction and the compression axis, suggesting correlation between neighbouring particles from different strings within this sector. We are examining this issue and extending our clustering theory of jamming to the steady state regime [30].

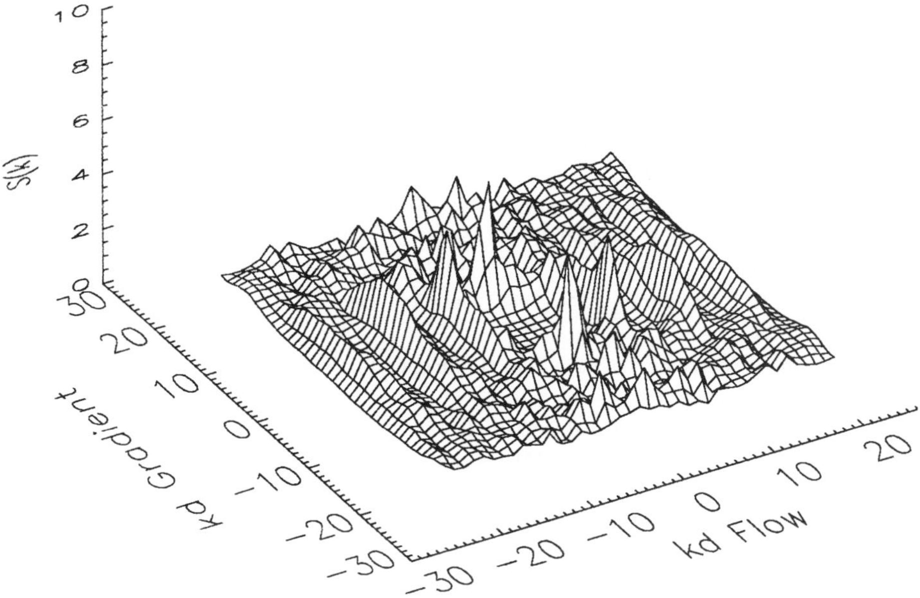

Figure 7 Structure factor in the shear gradient-flow plane from simulation of N=200 Brownian spheres at Pe=10 $\Phi = 0.53$.

4. Conclusions

The model system has taught us much about the flow of suspensions : the jamming of hard spheres, the true nature of high shear rate states, the origins of shear thickening and the important role of short range inter-particle interactions. The model is only an approximate representation of just the leading parts of the hydrodynamic interactions dominating in the close packed limit. Although, we suggest that it is sufficiently close to real systems but flexible enough to be a useful tool to gain insight into the physics of flow in such systems, it is hoped that more accurate techniques will be feasible with advance of computing power.

5. Acknowledgements

We thank S. F. Edwards, A Lips, R. Buscall , A Rennie, S. Clarke, and in particular W. Frith for many useful discussions. We thank the DTI/colloid technology project co-funded by the DTI, Unilever, Schlumberger Zenaca and ICI for funding.

6. References

1. Woodcock, L. V. (1984) *Chem. Phys. Lett..* **111,** 455.
2. Hoffman, R. L. (1974)*Col. and Intf. Sci.* **46,** 491.
3. Boersma, W. H., Laven, J. and Stein, H. H. (1990) *AIChE Jou.* **36**, 321; (1992) *J. Colloid and Interface Sci.* **149**, 10.
4. Chow, M. K. and Zukoski, C. F. (1995) *J. Rheol.* **39,** 15,33.
5. Frith, W. J., d'Haene, P., Buscall, R. and Mewis, J. (1996) *J. Rheol.* **40,** 531.
6. Bossis, G. and Brady, J. F. (1984) *J. Chem. Phys.,* **80**, 5141.
7. Ackerson, B. J., and Pusey, P. N. (1988) *Phy.s Rev. Lett.* **61**, 1033; Ackerson, B. J., (1990) . *J. Rheol.* **34**, 553.
8. Ashdown, S., Markovic, I., Ottewill, R. H., Lindner, P., Oberthur, R. C. and Rennie, A. R. (1990) *Langmuir* **6**, 303.
9. Clarke, S. M., Rennie, A. R. and Ottewill, R. H. (1997) *see this proceedings, chapter 9.*
10. Ladd A. J. (1994) *J. Fluid Mech.* **271**, 285 and references therein; Cichocki, B., Felderhorf, B. U., Hinsen, K., Wajnryb, E. and Blawzdziewicz, J. (1994) *J. Chem. Phys.* **100**, 3780 and references therein.
11. Kim, S. and Karrila, S. J. (1992) *Microhydrodynamics* Butterworths, Stoneham ; Hassonjee Q., Pfeffer, P. and Grantos, P. (1992) *Int. J. Multiphase Flow* **18**, 353.
12. Ladd, A. J. (1994) *J. Fluid. Mech.* **271**, 285 ; 331 ; Ladd, A.J. (1996) *Phys. Rev. Lett.* **76,** 1392.
13. Durlofsky, L., Brady, J. F. and Bossis, G. (1987) *J. Fluid. Mech.,* **180**, 21.
14. Koelman, J. M. V. A. and Hoogerbrugge, P. J. (1992) *Europhys. Lett.* **19**, 155; (1993) *ibid* **21**, 363.
15. Phung, T. H., Brady, J. F. and Bossis, G. (1996) *J. Fluid. Mech.* **313**, 181. Phung, T. H. (1995) *Ph D. Thesis* California Inst of Tech.
16. Treloar, R. L. and Masters, A. J. (1989) *Mol. Phys.* **67**, 1273.
17. Doi, M., Chen, D. and Saco, K., (1987) *Ordering and Organising in Ionic Solutions* pg 482 World Scientific, Singapore.; Toivakka, M., Eklund, D. and Bousfield, D. W., (1995) J. Non. *Newtonian Fluid Mech.* **56** 49; Yanamoto. S. and Matsuka T. (1996) *J. Chem. Phys.* **102**, 254; F. Torres (1996) private communication.
18. Lees, A. W. and Edwards, S. F. (1972) *J. Phys.*, **C5** 921.
19. Melrose, J. R. and Ball , R. C., (1995) *Europhys. Lett.* **32**, 535; (1995) *Adv. in Coll & Intf. Sci* **59**, 19.
20. Farr R., Melrose, J. R. and Ball, R. C., (1996) unpublished.
22. Melrose, J. R., van Vliet, J. H. and Ball, R. C. (1996) *Phys. Rev. Lett.* **77**, 4660.
22. Melrose, J. R., van Vliet, J. H. and Ball, R. C. (1996) unpublished.
23. Melrose, J. R., van Vliet, J. H., Ball, R. C., Frith, W. and Lips. A. (1996) in *Proc. Int. Conf. on Starch Structure and function* Royal Society of Chemistry, London.
24. Frith, W., Lips, A. Melrose, J. R. and Ball, R. C., (1997) *see this proceedings.*
25. Bender, J. and Wagner, N. J. (1996) *J. Rheol.* **40**, 899; Butera, R. J., Wolfe, M. S., Bender, J. and Wagner, N. J. (1996) *Phys. Rev. Lett.* **77**, 2117.
26. Silbert, L. E., Melrose, J. R., and Ball, R. C. (1997) unpublished; Melrose, J. R., Itoh, S. and

Ball, R. C. (1996) in Bullough, W. *Proc. 5th Int. Conf. on Electrorheological fluids* World Scientific, Singapore. pp 404.

27. Heyes, D. M. and Melrose, J. R., (1993) *Jol. Non. Newt. Fluid Mech.* **46** 1; Heyes, D. M. (1995) *J. Phys.: Condens. Matter* **7** 8857. Xue, W. and Grest, G. S. (1990) *Phys. Rev. Lett.* **64**, 1409.
28. Fredrickson, G. H. and Pincus, P. (1991) *Langmuir* **7** 786; Potanin, A. A and Russel, W. B. (1995) *Phys. Rev. E* **52** 730.
29. Silbert, L. S., Melrose, J. R., and Ball, R. C. (1997) unpublished.
30. Farr, R., Melrose, J. R. and Ball, R. C. (1996) unpublished

POLYMER DYNAMICS IN THIN FILMS

J.L. KEDDIE[1], R.A. CORY AND R.A.L. JONES
Polymer and Colloid Group,
Cavendish Laboratory,
Madingley Road,
Cambridge CB3 0HE

Spectroscopic ellipsometry has been used to measure the glass transition temperature of thin polymer films. Polystyrene films on a silicon substrate show a substantial molecular weight independent depression of glass transition temperature for films of decreasing thickness. Poly (methyl methacrylate) films show an increase when prepared on a silicon native oxide surface, but a decrease on gold substrates. The competing roles of the free surface, interactions with the substrate and finite size effects in determining the dynamics of polymers in confined geometries are discussed.

1. Introduction

Much is known about the dynamics of polymers in bulk melts[1]. In many applications polymers are used in a form involving colloidal dimensions; examples include polymer latices, polymer thin films and coatings, as well as polymer blends and alloys containing dispersed phases of colloidal dimensions. Extreme examples of such material are provided by so-called nano-composites, in which polymer is intercalated within the galleries in clay minerals [2] - here the polymers are confined within layers whose thickness is of order 10 Å. In such systems one might wonder if the confinement of the polymer chains on a sub-micron scale and the proximity to surfaces and interfaces in such systems might not have an effect on the dynamics of the polymer chain, which in turn will be responsible for properties such as adhesion and the coalescence of latices. Unfortunately there are few, if any, direct surface-sensitive probes of polymer dynamics available. We have, however, been able to detect substantial changes in dynamics in systems of colloidal dimensions by studying the glass transition temperature of very thin films.

The glass transition is a non-equilibrium transition at which second derivative thermodynamic quantities such as heat capacity and expansivity are discontinuous [3]. Essentially it occurs when a characteristic relaxation time associated with motion of the polymer chain at the level of a number of segments becomes larger than the experimental time on which the system is being probed. At this point translational degrees of freedom of the liquid no longer can come to equilibrium and the system

---

[1] Present address: Department of Physics, University of Surrey, Guildford, Surrey GU2 5XH

R. H. Ottewill and A.R. Rennie (eds.), Modern Aspects of Colloidal Dispersions, 149–157.

becomes a glass. Thus if the relevant relaxation time in a system of reduced dimensions changes this will be reflected in a size dependence of the glass transition temperature.

2. The glass transition for small systems

Potential causes of such a size-dependent change in relaxation time fall into two categories. Firstly, we would expect a pure finite size effect. The origins of this effect can be seen in the following simple argument in the spirit of Adam and Gibbs[4]. The starting point of their picture of a glass forming liquid is the assumption that segmental motion becomes increasingly cooperative as the glass transition is approached. They introduced the idea of a "cooperatively rearranging region" (CRR); in order for one segment to move then all the other segments in this region have to move also. If the CRR contains n* segments, then the relaxation time τ can be written as

$$\tau = \tau_0 \, exp\left(\frac{n^* \Delta\mu}{kT}\right) \quad (1)$$

where $\Delta\mu$ is the free energy barrier for rearrangement per segment. If the size of our system is now reduced below the size of the CRR, assuming all the segments in our small system to be dynamically equivalent, we find a reduction in the relaxation time[5]. Supposing the number of segments is a factor z smaller than the n*, we find $\tau(z) \sim \tau^z$. Experimentally it is known that relaxation times in glass forming systems obey the Vogel-Fulcher equation

$$\tau = \tau_0 \, exp\left(\frac{B}{T - T_0}\right) \quad (2)$$

where the Vogel-Fulcher temperature T_0 is about 50 K below the experimental glass transition temperature T_g. Since T_g is defined by the condition that at this temperature the relaxation time τ becomes comparable to a time-scale characteristic of the experiment τ_{exp}, we conclude that in a small system the glass transition temperature $T_g(z)$ should obey

$$T_g(z) = T_0 + z(T_g - T_0) \quad (3)$$

where T_g is the bulk glass transition temperature. We cannot use this expression to give a detailed prediction of the size dependence of the glass transition temperature, because we do not know how the size of the cooperatively rearranging region varies with temperature. However, from this argument we can draw two important conclusions:

1. If one reduces the system-size of a glass-forming liquid *in a way which maintains the dynamic equivalence of all the segments within the system* the glass transition temperature must be *reduced over the bulk value.*

2. The Vogel-Fulcher temperature T_0 sets a lower limit on the glass transition temperature as reduced by a finite size effect. For polymers this implies that the maximum reduction in glass transition temperature thus obtainable is about 50 K.

We stress the fact that these conclusions rest on the assumption that the effective free energy barrier for cooperative rearrangement is simply proportional to the number of segments involved. This clearly relies on the assumption that all the segments are dynamically equivalent. This is easily realised in computer simulations, where such an enhancement in mobility has been seen in a 2-d model of a glass-forming liquid in a finite system with periodic boundary conditions[6]. However this condition will certainly not be met if segments near the boundaries of the confined system have intrinsically different dynamics, for example by virtue of being physically attached to solid walls.

We should, therefore, consider the effect of confining walls on the dynamics of nearby segments quite separately to the pure finite size effect discussed above. There is essentially little or no theory relating to the way in which the interactions between segments and a surface or interface affect the glass transition behaviour. Some computer simulation work has been carried out; for example a molecular dynamics study of the surface of glassy polypropylene[7] revealed that the segmental mobility near a free surface was enhanced over the bulk value, and that the enhancement persisted over a longer range than the perturbation of the density near the surface. Recent measurements of the mechanical properties of polystyrene using surface force microscopy[8] have revealed that for low molecular weight polystyrene the surface has a rubber-like mechanical response at room temperature, reflecting greater mobility near the surface.

While it is intuitively reasonable to suppose that a segment close to a free surface would have a greater degree of mobility than a segment in the bulk, and conversely that a segment physically attached to a wall might have a lower degree of mobility, what is not obvious is over what length scale this dynamic perturbation would persist. A recent computer simulation[9] suggests that as a glass-forming liquid becomes supercooled, the distance into the bulk over which the dynamics are perturbed increases, ultimately to a value in excess of the chain radius of gyration. This is a significant result, because it suggests that cooperativity of motion in glass forming liquids can cause the influence of a surface or wall to extend beyond its immediate vicinity.

3. The glass transition in thin polymer films: experimental method.

The glass transition is signaled by a discontinuity in quantities that are second derivatives of free energies, such as specific heat capacity and expansivity. This suggests that one could measure the glass transition of thin films simply by measuring their thickness as a function of temperature. We have done this using ellipsometry. The method is described in detail elsewhere[10,11,12]; briefly polymer films are spun cast onto silicon substrates following various surface treatments. After annealing, the samples are heated while the ellipsometric angles ψ and δ are continuously monitored using a photoelastically modulated spectroscopic ellipsometer (Jobin-Yvon Uvisel). In order to attain maximum sensitivity to the very small changes in thickness that take place during heating the experiments are performed at the pseudo-Brewster angle of the substrate, at a wavelength which is selected to maximise the difference in ellipsometric angles between spectra taken at room temperature and a temperature well above the glass transition temperature.

Figure 1 shows typical data, in this case for polystyrene on a substrate of silicon from which the native oxide has been stripped using buffered hydrofluoric acid. For the small changes in thickness involved, the ellipsometric angle ψ is essentially a linear function of the thickness, which is shown on the right hand axis. Thus this curve indicates two regions of constant expansivity separated by a relatively sharp transition, which we can identify with the glass transition. For this 18 nm film the transition is already at a substantially lower temperature than the bulk glass transition temperature of polystyrene.

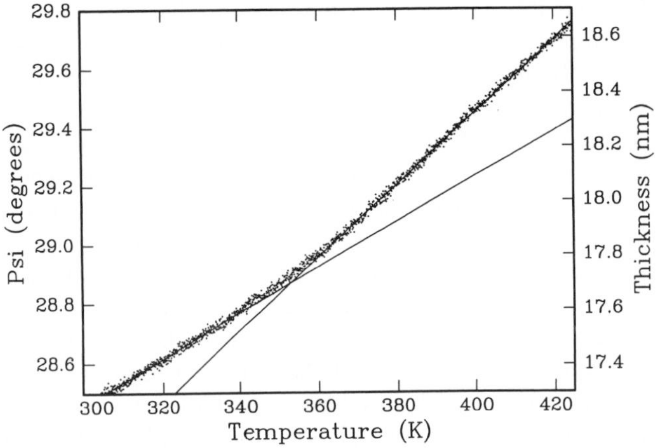

Figure 1. The ellipsometric angle ψ as a function of temperature for a thin polystyrene film, initial thickness 17.3 nm and molecular weight 2,900,000. An approximate conversion to film thickness is shown on the right hand axis. The scan was carried out at an angle of incidence of 80° and with light of wavelength 387.5 nm. The glass transition is marked by the intersection of the two straight lines, and in this case occurs at 353 K.

4. The glass transition in thin polymer films: results and discussion

We expect to find a marked contrast between the situation in which a strong interaction exists between the polymer and the substrate, and the opposite case, when the interaction is much weaker. This expectation is born out, as shown in figure 2, which shows glass transition temperatures for poly(methyl methacrylate) films. Films were made from identical polymer, cast in one case onto native oxide coated silicon wafers, and in the other case onto gold coated wafers. The films on gold show a substantial *reduction* in glass transition temperature with decreasing film thickness, while the films on the native oxide show a much weaker thickness dependence, with a slight *increase* in the glass transition temperature with decreasing thickness. So it seems that the presence of a substrate with which the polymer can strongly interact - in this case, presumably, via hydrogen bonds between the silanol groups at the native oxide surface - the local mobility is reduced and the glass transition slightly increases. Other workers have also found increases in glass transition temperatures for PMMA on silicon native oxide[13] and for poly (2- vinyl pyridine) on the same substrate[14], while picosecond acoustic measurements[15] reveal a substantial increase in the elastic modulus of PMMA in ultra-thin films on an aluminium substrate. This evidence, then, all supports a picture in which strong interactions between the polymer and a substrate lead to a reduced mobility and an increased glass transition temperature.

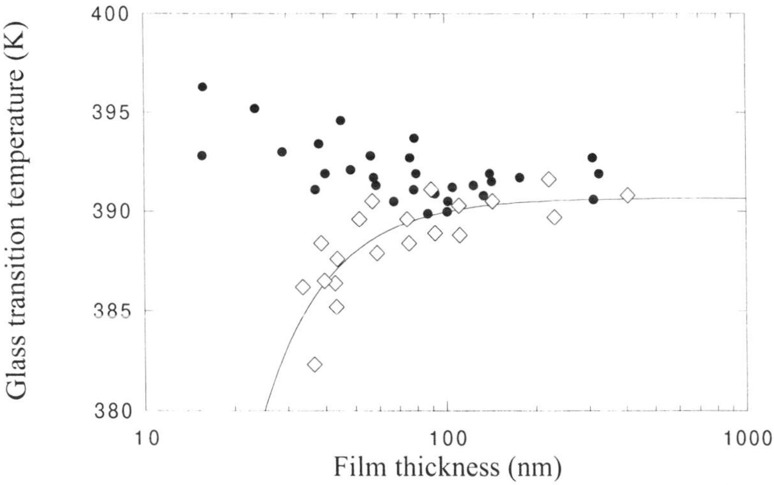

Figure 2. The glass transition temperature of thin PMMA films on gold coated substrates (diamonds) and native silicon oxide (filled circles). The molecular weight of the PMMA was 100 250. The solid line is a fit to the empirical function of equation 4.

The reduction in glass transition temperature was represented by an empirical function, with the form

$$T_g(d) = T_g(\infty)\left[1-\left(\frac{A}{d}\right)^{\delta}\right], \qquad (4)$$

where $T_g(d)$ is the glass transition temperature for a film of thickness d, and A and δ are constants. For the case of PMMA on gold, the best fit values of the constants were A=7.2 ± 3.6 nm, δ = 1.9 ± 0.6, and the bulk glass transition temperature $T_g(\infty)$ = 390.7 ± 0.5 K.

If an increase in glass transition temperature for thin films can be attributed to strong interactions with the substrate, what is the cause of the decrease in glass transition temperature for films on the more weakly interacting substrates? As we argued above, one might argue that a decrease in glass transition temperature could arise either from a pure finite size effect or from the effect of increased mobility at the free surface. In addition, other suggestions have been made; that an increase in mobility in thin films is connected with confinement of the chains below their natural radius of gyration, or that it is connected with the segregation of chain ends to the surface. Further insight into the origins of the depression of Tg for polymer films on weakly interacting substrates can be obtained by studying the influence of molecular weight on the effect.

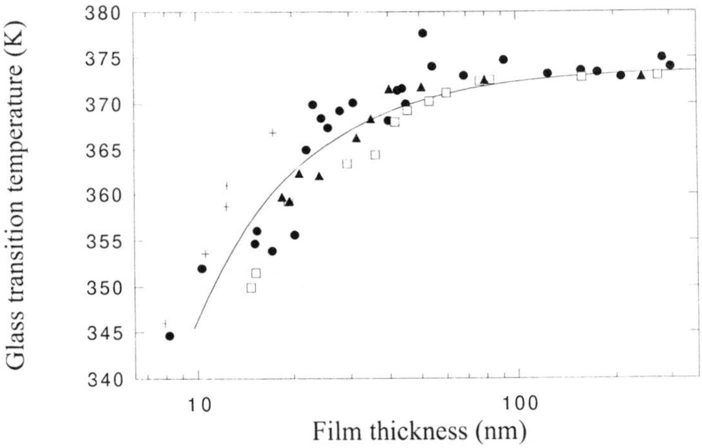

Figure 3. Glass transition temperatures of thin polystyrene films on silicon substrates. The molecular weights were (circles) 2,900,000; (squares) 500,800; (diamonds) 120,000; (crosses) 225,000 (carboxy terminated); the solid line is equation 4 with values of the constants A=1.3 nm, δ = 1.28, and $T_g(\infty)$ = 373.8 K.

Figure 3 shows results obtained for glass transition temperatures for polystyrene films of differing molecular weights (this includes some data for carboxy terminated polystyrene which has not been grafted to the interface; we discuss results for end-grafted carboxy-terminated polystyrene below). Qualitatively, the results have similar form to those for the PMMA films, and we may fit the data to the same empirical function given in equation 4; we find best fit values of the constants A=1.3 ± 0.1 nm, δ = 1.28 ± 0.2, and $T_g(\infty)$ = 373.8 ± 0.7 K. But of special significance is the fact that the results for three different molecular weights seem remarkably similar. If the effect were purely a result of the confinement of polymer chains to films thinner than the chain dimension, then we might expect the thickness at which the reduction of glass transition temperature commenced to vary with molecular weight in the same way as the radius of gyration. However the observed dependence on molecular weight, if any, is much weaker than this.

The absence of molecular weight dependence thus leads us to the conclusion that the relevant length scale controlling the T_g reduction is not the overall chain dimension. Can we conclude, then, that this is a finite size effect as discussed above, in which the relaxation time is reduced because the film is confined to a length scale smaller than the cooperatively rearranging region? We recall that we argued that a pure finite size effect, as distinguished from a surface or interface effect, could not cause the T_g to be reduced below the Vogel-Fulcher temperature T_0. This is about 320 K for polystyrene, so from our data we cannot rule this out. However, Brillouin scattering has recently been used to measure the glass transition temperature of free-standing polystyrene films [16]. These experiments showed even greater depressions in glass transition temperatures than our experiments on weakly interacting substrates, and in particular these workers found that for a film of thickness around 30 nm, the T_g was around 300 K, well below the Vogel-Fulcher temperature.

What is suggested by these results, then, is that the observed changes in glass transition temperature with thickness are predominantly the result of the competition between the effect of the substrate, which given relatively strong interactions with the substrate causes a tendency for the glass transition temperature to *increase* with decreasing thickness, and the effect of the free surface, which leads to a tendency for the glass transition temperature to *decrease* with decreasing thickness. Thus a free standing film, with two free surfaces and no substrate, exhibits a very pronounced depression in T_g, while a film on a weakly interacting substrate, with only one free surface exhibits a rather smaller decrease in T_g. As the interaction with the substrate becomes stronger, the effect of the substrate apparently outweighs the effect of the interaction at the substrate and leads to an overall increase in T_g. This picture is supported by experiments using x-ray reflectivity to probe T_g in thin films of polystyrene supported on silicon wafers in which the native oxide was stripped and the experiments carried out in vacuum[17]. In these experiments the T_g increased with decreasing thickness, presumably because in this case the substrate presented a truly hydrogen-passivated silicon surface with which the polystyrene interacts rather strongly, whereas in our experiments, carried out in air, some regrowth of oxide takes place during the annealing stage leading to a less strongly interacting substrate.

To probe the effect of strong interactions with the substrate further, we studied a situation in which these interactions could be rather precisely controlled. These systems comprised ultra-thin polystyrene films that were end-grafted onto a native oxide coated silicon wafer. We know that homopolymer polystyrene interacts rather weakly with native oxide, because such films dewet on being heated. But when end-functionalised polystyrene, in which every polystyrene chain was terminated by a carboxy group, is spun onto silicon coated with native oxide, and heated, a physical bond is formed between a chain and silanol groups on the oxide surface. Unreacted chains may be removed using a solvent, leaving an ultrathin polymer film stabilised against dewetting by the grafting points.

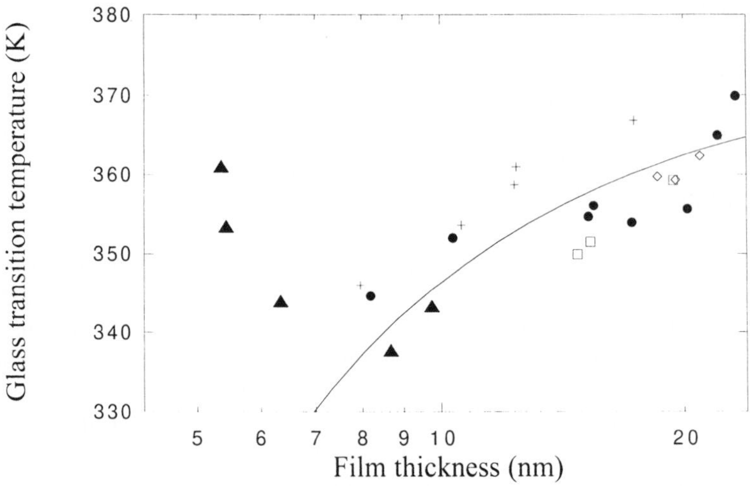

Figure 4. Glass transition temperatures of ultra-thin, end-grafted polystyrene layers, of molecular weight 225,000 (triangles). For comparison are shown glass transition temperatures for thin, non-grafted polystyrene films - other symbols and solid line are as for figure 3.

Glass transition temperatures for such ultrathin grafted polymer layers are shown in figure 4, together with points from the thinnest ungrafted films. The glass transition temperatures are still greatly reduced over the bulk values. There does, however, seem to be a tendency for the very thinnest films to slightly increase in T_g, though without recovering their bulk values. It is possible that this increase in T_g for the very thinnest films reflects the increasing influence of the tethering point at the substrate in restricting segment mobility in these films.

5. Conclusions

The glass transition behaviour of polymers in thin films can be very different to their bulk behaviour. The influence of substrates, free surfaces and finite size effects can combine to produce either increases or decreases in glass transition temperature according to the relative importance of each influence. This emphasises that the dynamics of polymers confined to colloidal dimensions is likely to be strongly perturbed with respect to bulk dynamics.

Acknowledgements

This work was supported by the DTI programme in Colloid Technology with ICI plc, Zeneca plc, Schlumberger Cambridge Research and Unilever.

References

1. Doi, M. and S.F. Edwards (1986) *The theory of polymer dynamics*. Oxford University Press. Oxford
2. Vaia, R.A., *et al.* (1997) *Journal of Polymer Science: Part B: Polymer Physics*, **35,** 59.
3. Donth, E.-J. (1992) *Relaxation and thermodynamics in polymers: glass transition*. Akademie Verlag. Berlin
4. Adam, G. and J.H. Gibbs (1965) *Journal of Chemical Physics*, **43,** 139.
5. Jérôme, B. and J. Commandeur (1997) *Nature (London)*, **386,** 589.
6. Ray, P. and K. Binder (1994) *Europhysics Letters*, **27,** 53.
7. Mansfield, K.F. and D.N. Theodorou (1991) *Macromolecules*, **24,** 6283.
8. Kajiyama, T., K. Tanaka, and A. Takahara (1997) *Macromolecules*, **30,** 280.
9. Baschnagel, J. and K. Binder (1995) *Macromolecules*, **28,** 6808.
10. Keddie, J.L., R.A.L. Jones, and R.A. Cory (1994) *Europhysics Letters*, **27,** 59.
11. Keddie, J.L., R.A.L. Jones, and R.A. Cory (1995) *Faraday Discussions*, **98,** 219.
12. Keddie, J.L. and R.A.L. Jones (1995) *Israel Journal of Chemistry*, **35,** 21.
13. Wu, W., J.H.v. Zanten, and W.J. Orts (1995) *Macromolecules*, **28,** 771.
14. Zanten, J.H.v., W.E. Wallace, and W.-L. Wu (1996) *Physical Review E*, **53,** R2053.
15. Lee, Y.-C., *et al.* (1996) *Applied Physics Letters*, **69,** 1692.
16. Forrest, J.A., *et al.* (1996) *Physical Review Letters*, **77,** 2002.
17. Wallace, W.E., J.H.v. Zanten, and W.L. Wu (1995) *Physical Review E*, **52,** R3329.

THE STRUCTURE OF BLOCK COPOLYMERS OF ETHYLENE OXIDE AND BUTYLENE OXIDE AT FLUID/FLUID INTERFACES.

S. CALPIN-DAVIES, B. J. CLIFTON, T. COSGROVE,
R. M. RICHARDSON AND A. ZARBAKHSH
School of Chemistry, University of Bristol, Cantock's Close, Bristol BS8 1TS. UK
and
C. BOOTH AND G. YU.
Department of Chemistry, University of Manchester, Manchester M13 9PL UK

Neutron reflection and small-angle scattering have been used to investigate the structure of block copolyethers at fluid-fluid interfaces. Contrast variation methods have been used in both techniques to give volume fraction profiles and adsorbed amounts. The block copolymers are found to straddle the oil-water interface with the more hydrophobic butylene oxide group being attracted into the oil phase. The extent of this effect depends on the ratio of the block sizes. The ethylene oxide layer is stretched significantly into the aqueous phase. The results are compared with self-consistent mean-field predictions of the detailed interfacial structure.

1. Introduction

Block copolymers based on ether monomers have been widely used in formulations to impart steric stability [1]. The most common of these polymers are tri-blocks based on ethylene oxide [EO] and propylene oxide [PO]. However, many of the commercial polymers based on these monomers are polydisperse and in the worst case contaminated with di-blocks. The PO block is less soluble in water than EO and accounts for the success of these polymers in stabilising formulations containing emulsions. However, the hydrophilic-hydrophobic balance [HLB] of these polymers is quite low and an improvement in their ability to partition across oil-water interfaces can be made by substituting the PO block by butylene oxide [BO]. Recently [2], a range of these copolymers based on EO and BO have been synthesised by anionic polymerisation and have narrow molecular weight distributions [typically less than 1.1].

Existing data on the adsorption properties of EO-BO polymers include studies at the liquid/air[3] and liquid/solid interfaces [4]. In the latter paper it was found that the polymers blocks strongly segregate at the polystyrene-latex interface, where the BO block is virtually completely adsorbed as trains at the latex surface. The greater HLB

R. H. Ottewill and A.R. Rennie (eds.), Modern Aspects of Colloidal Dispersions, 159–171.

factors for the BO/EO copolymers suggest that they will form more stable interfacial layers than the equivalent PO based polymers .

In this paper we present both neutron reflection and small angle scattering studies of EO-BO polymers at macroscopic hexane-water and water-air interfaces and in low polydispersity emulsions comprising of toluene and water. The results are compared with self-consistent mean-field calculations based on the Scheutjens-Fleer model[5]

2. Experimental

Two experimental techniques were used in this study, neutron reflection and small-angle neutron scattering and these are discussed below together with details of the sample preparation and the simulation parameters.

2.1 NEUTRON REFLECTION

The neutron reflection data were obtained using the SURF multi-wavelength reflectometer at ISIS and the fixed wavelength reflectometer at NIST. For the liquid-liquid interface, the method requires the formation of a stable (0.2 - 4μm) oil layer on top of an aqueous sub-phase. As the neutron beam enters the sample from above, the top layer must remain extremely thin because of the loss of intensity due to absorption. This is particularly problematic with highly protonaceous material. The maintenance of a thin oil layer of constant thickness throughout a reflectivity experiment lasting several hours, is not straightforward [6]. The current apparatus consists of a PTFE trough filled with an aqueous polymer solution sub-phase (approximate volume 70 cm^3). This is enclosed in a sealed double walled temperature controlled compartment. The aluminium body of the compartment is set at 26°C, while the sub-phase is cooled to 18°C by the use of Peltier devices attached to the trough's underside. This temperature differential is maintained by using a Marlow Instruments temperature controller. To form the double layer, partially deuterated hexane (~3 cm^{-3}) is syringed on top of the aqueous phase and into its surrounding gutter through a small re-sealable inlet. Previous work with hexane, has shown that such deposition techniques lead to the formation oil islands [7] and to remove these we raise the temperature of the trough to 23 °C, which vaporises most of the deposited hexane. The trough is then slowly cooled back to 18 °C, which condenses the hexane vapour on to the aqueous sub-phase. This results in the hexane wetting the surface evenly, forming a uniform film [6]. The temperature differential is maintained throughout the experiment to within ±0.1°C. In this way, the hexane evaporation, and its slow drainage over the water meniscus, is balanced by vapour condensation thus dynamically controlling the film thickness. The PTFE troughs were cleaned with Decon 90 and rinsed in ultrapure water prior to use.

The SURF instrument was calibrated using D_2O, and four angles of incidence were employed throughout (0.35, 0.5, 0.8, 1.5°) giving an effective Q range [the modulus of the scattering vector] of 0.01 to 0.6 $Å^{-1}$. Over this range, the complete reflectivity profile, including a second critical edge from a partially deuterated oil layer, can be observed. Two contrasts for the deuterated polymer were used and the initial bulk concentration of polymer was ≈ 900 ppm. Liquid-air experiments were also carried out using the fixed wavelength source at NIST using the same cell, but maintained at 25°C. By varying the incident angle a Q range of 0.01 and 0.2 Å^{-1} was achieved . For these experiments null-reflecting water [8% D_2O, 92% H_2O] was used as solvent

2.1.1 Reflectivity data analysis

The results of the reflection experiment have been analysed using the matrix method combined with a thick film attenuation function [6]

2.2 SMALL-ANGLE SCATTERING

SANS measurements were also performed on the LOQ diffractometer at ISIS. This is a fixed-geometry, time-of-flight (TOF), instrument which utilises a spread of neutron wavelengths between 2 and 10 Å. These give an effective Q-range of approximately 0.008 to 0.22 Å^{-1}. Typical wavelength resolution is ≈ 4 %. The samples were run in 2mm pathlength, UV-spectrophotometer grade, quartz cuvettes (Hellma) and mounted in aluminium holders on top of an enclosed, computer-controlled, sample changer. The effective sample volumes were approximately 0.4cm^{-3}. Temperature control of 25 ± 0.5°C. was achieved during the experiments. Measurement times were between 120 and 180 minutes.

The scattering data were (a) normalised for the sample transmission and incident wavelength distribution, (b) background corrected using an empty quartz cell (this also removes any significant instrumental background) and (c) corrected for the linearity and efficiency of the detector response. The data was then converted to absolute units by scaling it to the scattering from a well-characterised partially deuterated polystyrene-blend standard sample.

2.2.1 Scattering data analysis

When the disperse phase substrate (p) is close to contrast match with the solvent (s), the total small-angle scattering arises from a number of contributions [8]

$$I(Q) = \left(\rho_p - \rho_s\right)^2 I_{pp} + 2\left(\rho_\ell - \rho_s\right)^2 \bar{I}_{p\ell} + \tilde{I}_{p\ell} + B_{inc} \qquad (1)$$

where ρ corresponds to the scattering length density of the particles (subscript p), polymer layer (subscript ℓ) or solvent (subscript s). I_{pp} is any residual particle scattering, $I_{p\ell}$ is the scattering from the average layer structure, $\tilde{I}_{p\ell}$ is the scattering arising from concentration fluctuations within the polymer layer structure and B_{inc} is the incoherent background scattering mainly due to protons. Under these contrast conditions, the evaluation of the magnitude of the $\tilde{I}_{p\ell}$ term, which represents spatial inhomogeneities in the average structure of the polymer layer, is essential. For (relatively) low molecular weight polymers adsorbed onto large particles this term is insignificant, but this is not always the case and when present this extra contribution to the scattering must be properly accounted for. In these particular experiments no substantial contribution from fluctuations was observed.

The layer term $I_{p\ell}$ is given by

$$I_{p\ell}(Q) = -\frac{2\pi(S/V)(\rho_p - \rho_\ell)^2}{Q^2}\left|\int_0^\infty \phi(z)e^{iQz}dz\right|^2 \qquad (2)$$

In emulsion systems further problems arise from the way in which the samples are prepared. In particular, neither the size nor the volume fraction [ϕ_p] of the droplets are known accurately. However, by contrast variation it is possible to 'match out' the scattering from the adsorbed layer and obtain the surface to volume ratio (S/V) of the samples explicitly, if the various scattering length densities are known. Under these conditions the particle scattering can be written as

$$I(Q) = \frac{2\pi(S/V)}{Q^4}(\rho_p - \rho_s)^2 \qquad (3)$$

By combining equations (2) and (3) and using appropriate values for the scattering length densities of the continuous phase [see Table 1a], the unknown (S/V) term can be eliminated. For spheres $(S/V)=\phi_p/2r_o$ where r_o is the radius of the droplet assuming minimal scattering from the adsorbed layer when $\rho_p - \rho_s >> \rho_p - \rho_\ell$.

2.3 PREPARATION OF THE AEROSOL EMULSION

The emulsions were formed by using an adapted Weinstock and Rapport aerosol generator [9]. The procedure involves the evaporation of the oil to give a coarse aerosol, which is then passed atomised in a furnace held at 74°C. By subsequent cooling to -79 °C in an insulated chimney nucleation and growth occur. The carrier gas was dry nitrogen and the final emulsification was achieved by bubbling the aerosol through a 0.1% w/w polymer solution. A flow rate of 0.1 m^3/hour was employed. The emulsions whilst not completely monodisperse are very much superior to those prepared by conventional methods. The physical properties of the emulsions are given in Table 1b. The emulsion droplet size was determined by Photon Correlation Spectroscopy [PCS], the samples having been diluted to 0.01 volume fraction by the addition of polymer solution of the same concentration as that with which the emulsion was formed. From the SANS data it was found that the emulsion could be contrast matched with 30% H_2O/ 70% D_2O mixture. This corresponds to a mixture of 95%D/5%H toluene. The off-contrast samples required to measure the S/V ratio were made up at 1% H_2O/ 99% D_2O.

2.4 MATERIALS

The general methods of polymerisation of the block copolymers have been described previously [2] and the triblock copolymer synthesis has been described in a recent paper [4]. The copolymer was characterised by GPC and NMR using methods similar to those described previously, the latter being used to check the composition, the overall chain length and the purity.

The molecular characteristics and physical properties of the copolymers used are given in Table 1a. The water used was obtained from a Millipore water purifier. D_2O was supplied by MSD Isotopes, n-hexane (Analar), d_{14} n-hexane (99%+), d-toluene and toluene were obtained from Aldrich without further purification.

Table 1a Properties of samples used

Sample		RMM / g mole^{-1}	M_w/M_n (GPC)	Density / g cm^{-3}	Scattering length density* / x10^6 Å^{-2}
d-$E_{98}B_{15}$d-E_{98}	[DE98]	10,488	1.09	1.1	6.42$^+$
$E_{36}B_{27}E_{36}$	[E36]	5,110	1.09	1.1	0.50
$E_{18}B_{27}E_{18}$	[E18]	3,530	-	1.1	0.44
Hexane		86	-	0.66	-0.58
D_{14}-hexane		100	-	0.77	6.16
D_8 Toluene		100	-		5.5
H_2O		18	-	1.0	-0.56
D_2O		20	-	1.11	6.40

*calculated from the molar composition of the two blocks.

Table 1b Emulsion samples

Sample	Size/PCS/Å	Polydispersity Index	Volume fraction
E18B27E18	2630	0.321	0.036
E36B27E36	2840	0.38	0.038

3. Self-consistent mean-field calculations

The theoretical calculations were performed using the mean-field lattice model for polymer adsorption developed by Scheutjens, Fleer and Evers [5] *via* the Goliad program. The adsorbed amount for each sample was obtained from the experimental data assuming that 1 mg m^{-2} is equivalent to of an adsorbed monolayer. The Flory solution parameters for the polymers were taken as $\chi_{EB} = 2.0$; $\chi_{EW} = 0.43$; $\chi_{BW} = 2.0$; χ_{ET} , $\chi_{EH} = 2.0$. and χ_{BT} , $\chi_{BH} = 0.5$. where E is ethylene oxide, B butylene oxide, W water, T toluene and H hexane. The parameters where chosen to reflect that the B and E groups are incompatible and B is virtually insoluble in water. The χ parameters for the pure solvents were calculated from their solubility parameters[10] .This gave a value of 5.2 for toluene/water and 7.8 for hexane water and for air-water a value of 8. Furthermore, both monomers and all solvents were taken to occupy one lattice site. The above approximations are not ideal but despite these limitations the results of the calculations are still qualitatively informative.

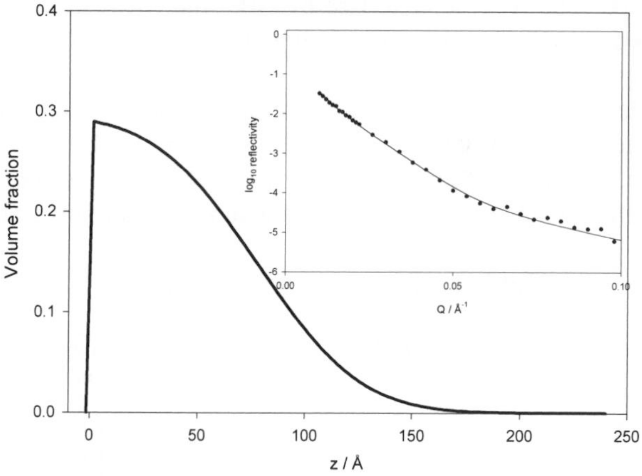

Figure 1 Volume fraction profile from DE98 at the null water/air interface. The insert is the reflectivity data and the resulting fit.

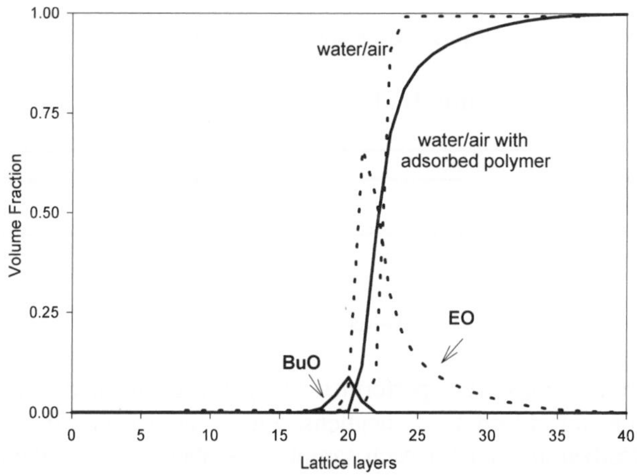

Figure 2 Volume fraction profile for E100B15E100 block copolymer at the air/water interface using the Scheutjens Fleer model for block copolymers. The fine dotted line corresponds to a naked water/air interface.

4 Results and Discussion

The most straightforward interfacial experiment to undertake is to determine the structure of the block copolymers at the water/air interface. The results shown in Figure 1 represent the reflectivity data from the DE98 copolymer at the null-water/air interface and the fitted scattering length density profile. The data show a highly extended polymer layer which stretches ≈ 6 R_G ,but does not extend significantly into air. The adsorbed amount calculated assuming that the scattering length density is defined by the deuterated EO segments is 2.5 mg m^{-2}.

For comparison in Figure 2 we present calculations based on the Scheutjens-Fleer theory [5]. In the simulation the BO block is virtually constrained to the air phase whereas the EO block partitions across the interface. For the cubic lattice the radius of gyration for the EO chains [100 segments] is 4 lattice layers. The lattice chains therefore stretch about $8R_G$ compared to ≈ 6 for the real data. Given a typical monomer length ≈ 4Å then the lattice layer stretches to about 120Å which is comparable to the experimental data. The block copolymer is behaving very much like a terminally attached chain with the more hydrophobic BO groups anchoring the molecules to the interface.

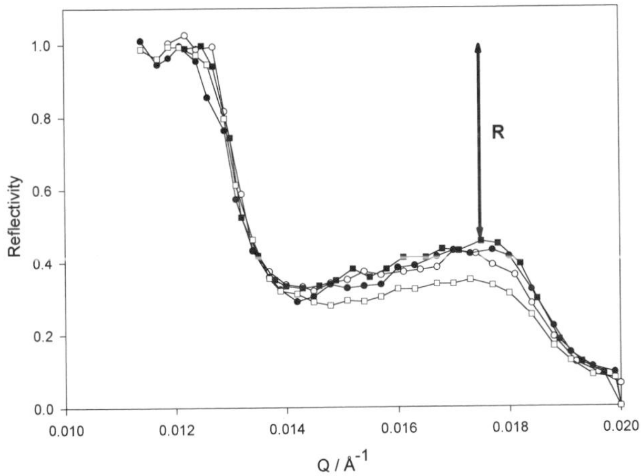

Figure 3 Reflectivity data from a 60% d-hexane+40% hexane/D_2O interface as a function of time. B 0 min. d= 3.26μm, G 30 min. d = 4.17 μm , J 90 mins d = 3.47 μm, E 120mins d = 3.53μm . R is the change in reflectivity due to the attenuation by the hexane layer.

Obtaining results from the liquid-liquid interface is rather more difficult. The main problem is trying to keep a consistent top film [hexane] at the water interface. Based on our previous work we have used the absorption coefficient of H-hexane to monitor the change in this layer [6]. Figure 3 shows the raw reflectivity data as a function of time. From the drop in intensity between the two critical edges, which correspond to total external reflection at the hexane/air and hexane/water interfaces respectively, the layer thickness can be obtained. As can be seen over two hours the layer once stabilised does not vary much from a nominal 4μm. Given a moderately stable layer we have obtained data for the DE98 polymer at the water hexane interface. Two different contrasts were used and data collected at up to four different angles. This resulted in 5 separate experimental runs. Each of the runs has a different scaling factor as it is impossible to reproduce the same thickness hexane layer. For the data in D_2O and below $Q \approx 0.02$ Å^{-1}, the appearance of the critical edge makes scaling of the reflectivity profiles straightforward. At the higher Q ranges, when the edges are invisible or for the null-water/hexane sample, the absolute reflectivity must be fitted with a scaling parameter. However, unambiguous values of the layer second moment can be found from a surface Guinier approximation. In order to optimise the value of the data all 3 sets were fitted simultaneously and these fits are shown in Figure 4. It should be emphasised that the data are not particularly sensitive to the adsorbed amount not only because of the scaling factors but also because of the poor contrast of the deuterated polymer in D_2O.

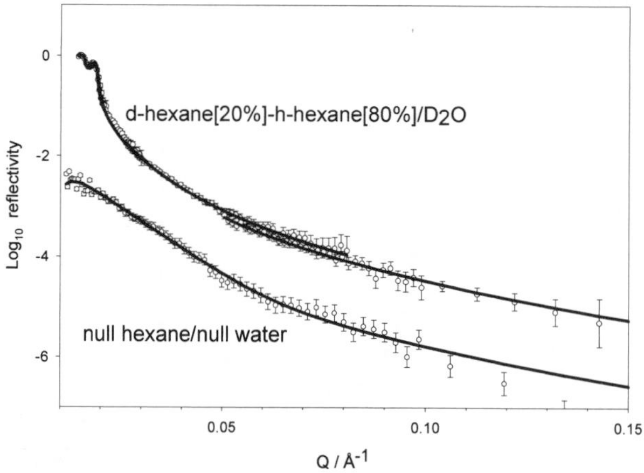

Figure 4 Reflectivity data for 80% d-hexane-20% h-hexane: on D_2O and null reflecting hexane on null reflecting water for the DE98 copolymer. Solid lines are the best overall fit to the data leading to the volume fraction profile shown in Figure 5.

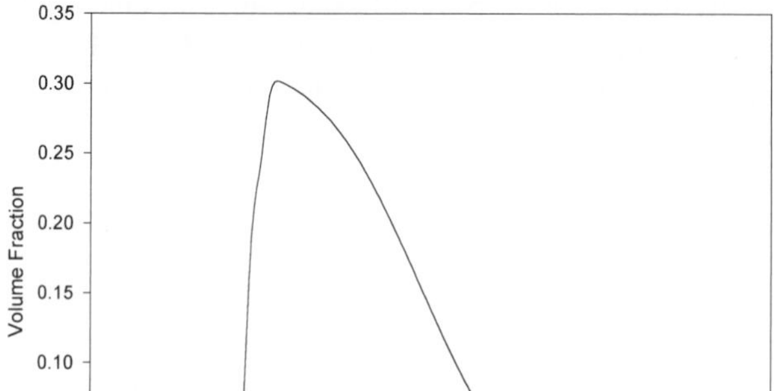

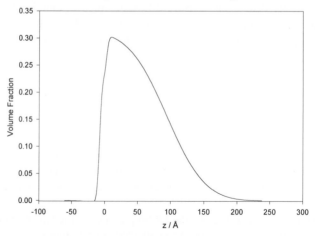

Figure 5 Overall best fit volume fraction profile for the multiple data sets shown in Figure 4 for DE98 at the hexane/water interface.

The combined data set was fitted using a two-block model and assuming that only DEO and water contribute to the scattering length density. The two blocks represent incursions of the polymer into the hexane and water interfaces respectively. The resultant fit is shown in Figure 5.

In comparing this data with the air/water interface it can be seen that a very similar layer extent is found. This is not surprising given that the EO block is insoluble in hexane. The adsorbed amount resulting from this data is 3.3 mg m^{-2} which compares favourably with the previous data [2.5mg m^{-2}]. The simulation results for the hexane/water interface are very similar to those shown in Figure 2, though the layer extent is slightly greater in this case because of the higher adsorbed amount. The DE98 polymer is highly asymmetric and it will behave more like a terminally attached chain. The adsorbed amount and the layer thickness however, will be determined by a balance of the stretching of the EO chains in the water and the solvation of the BO block in the hexane. The final layer structure is a balance of these competing effects

The other polymers used in this study are very much more symmetric than DE98 and their details are found in Table 1a. Figure 6 shows the reflectivity data obtained from the E36 polymer at the air/D_2O interface. The inset shows the fitted single block profile. As can be seen the layer is quite diffuse with a second moment of 47±5Å and adsorbed amount of 2.9± 0.5 mg m^{-2} . The incursion into the air layer for this polymer can be seen to be fairly substantial compared to DE98 and this is directly related to the BO block which is almost twice the size. The structure at the interface is also no longer limited by the crowding of longer EO arms. The overall molecular weight is also much smaller and neither the buoy (EO) or anchor (BO) blocks dominate the interfacial behaviour. We shall discuss this data more fully in the following sections.

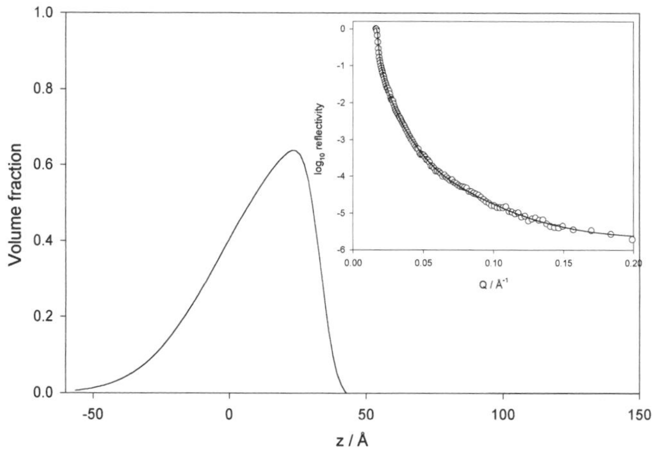

Figure 6 Scattering length density plot for the E36 polymer at the air/water interface. Inset is the reflectivity data with the derived fit.

Both the E36 and E18 polymers have been studied using small-angle scattering. Because of way in which the emulsions are formed we do not have accurate measurements of the particle volume fraction and hence the surface area of the dispersed phase. However, by using multiple contrasts it is possible to determine the (S/V) ratio using the Porod analysis described above (Equation 2)

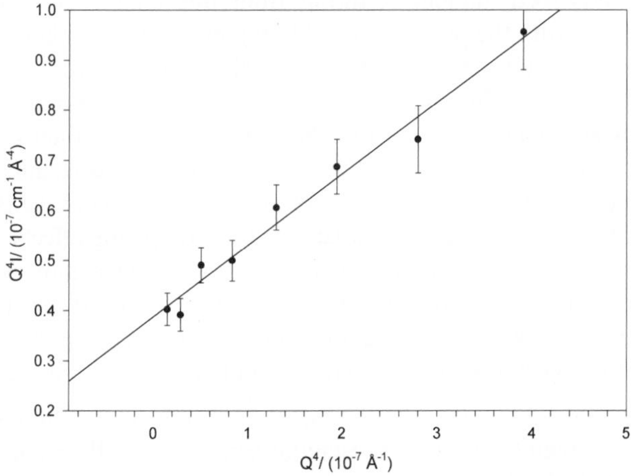

Figure 7 Porod plot for an 95%d +5%d toluene emulsion in D_2O. The line is a linear regression to estimate the intercept.

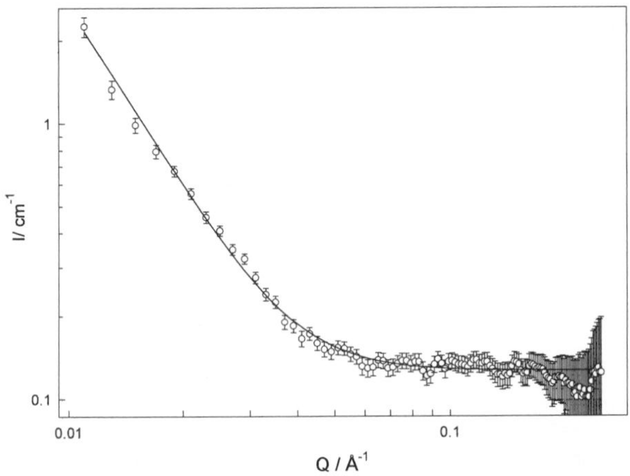

Figure 8 Small-angle scattering data for the E36 polymer adsorbed on a 95%d+5%h toluene emulsion in contrast matched water [30%h+70%d]. The solid line is the best fit to an exponential volume fraction profile shown in Figure 9.

The data in Figure 7 represent the low Q region of the scattering of the emulsion in virtually pure D_2O. The data are plotted as Q^4I vs. Q^4 and the intercept [using Equation 3] gives the required (S/V) value. Using the PCS data from Table 1b we can estimate the actual volume fractions and they are quite close to the nominal 0.04 values aimed at in the preparation.

Figure 8 shows the raw scattering data from the E36 polymer when the droplets of toluene are matched to water [30%D_2O/70%H_2O] .The data have been fitted with an exponential volume fraction profile. A similar set of data and fit has been obtained for the E18 polymer. The adsorbed amount and second moments of the profiles obtained by integration are given in Table 2. These data are similar to those that were obtained from a surface Guinier analysis [8].

Table 2 Charteristics of the emulsion adsorbed layers

Polymer	second moment σ /Å	Γ /mg m^{-2}
E36	34 ± 3	3.0 ± 0.2
E18	27 ± 3	1.9 ± 0.2

Figure 9 compares the experimental volume fraction profiles for the two polymers E18 and E36. Firstly, comparing the data for the E36 polymer at the toluene/water interface and the air/water interface [Figure 8] shows that the profiles both extend to ≈ 150 Å and have a relative large volume fraction at the interface [0.8 and 0.7 respectively]. These data again show very similar behaviour at both interfaces despite the solvency differences between toluene and air. Both sets of data give very similar adsorbed amounts though slightly different second moments..

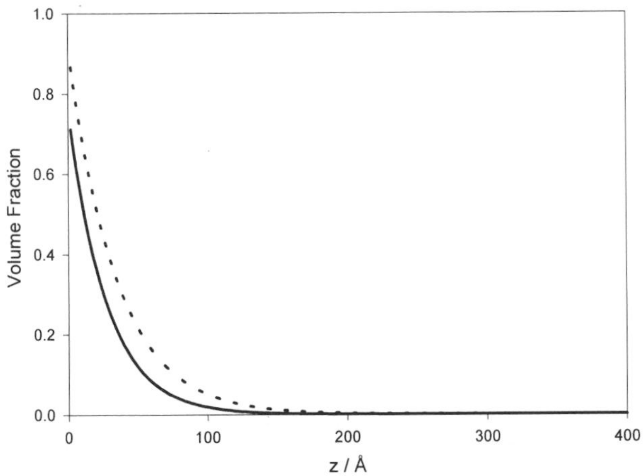

Figure 9 Volume fraction profiles for the two polymer E36 [- - -] and E18[—] at the toluene/water interface.

Comparing the data for the two different polymers [Figure 9] shows that despite both polymers having the same number of potential buoy segments the volume fraction at the surface was greater for the polymer with the larger number of EO groups. This would suggest that both EO and BO groups are found at/in the interface. On examining the extent of the layers it can be seen that both polymers form highly extended layers ≈ 3-4 times their R_G.

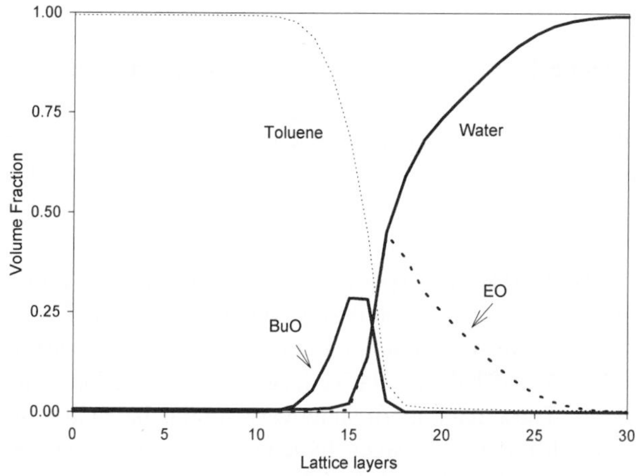

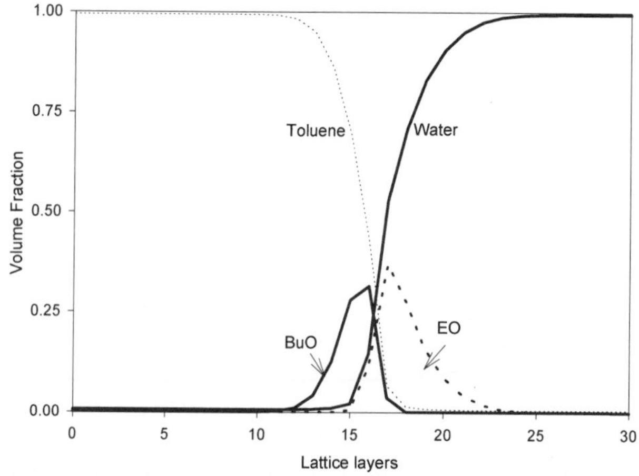

Figure 10 Theoretical volume fraction profiles for the two polymer [Top] E36B27E36 and [Bottom] E18B27E18 at the toluene/water interface using the Scheutjens-Fleer block copolymer theory.

Although the profile fits do not account for the two different segment types [the average scattering length density has been used in the calculations] this breakdown is easily achieved in the simulation. Figures 10a,b compare the SF predictions for the two polymers taking into account the actual experimentally observed adsorbed amounts. For both polymers the EO and BO segments are found in both phases. The volume fraction of the two polymers follows almost exactly the volume fractions of the respective solvents.

The SF volume fraction at/in the interface for the theoretical profiles mirror the experimental values in terms of their extent and concentration at the interface, though the absolute values in the latter case are somewhat larger.

Conclusion

Two neutron techniques have been used to identify the structure of block copolymers at fluid/fluid interlaces. The results indicate that the adsorbed amounts depend on both the relative and absolute block sizes. The larger block is dominant in determining the interfacial structure. For symmetric block polymers there is a much greater invasion of segments into the 'opposite' phase. There are not great differences in the structure of the layers at the air/water and oil/water interfaces at least for the systems studied in this paper.

Acknowledgements

BC would like to acknowledge the DTI/ICI link scheme for support for a PDRA grant. SCD would like to acknowledge the University of Bristol and ICI [Dr. T. Blease] for the provision of a joint award. TC and RMR would like to acknowledge DRAL [Dr. J.R.P. Webster] and NIST [Dr. S. Satija] for their help and for the provision of neutron beam time

References

1 Alexandridis, P.; Hatton, T. A. *Colloids and Surfaces Physicochemical and Engineering Aspects* **1995**, *96*, 1-46.
2 Yang, Y. W.; Tanodekaew, S.; Mai, S. M.; Booth, C.; Ryan, A. J.; Bras, W.; Viras, K. *Macromolecules* **1995**, *28*, 6029-60412
3 Schillen, K.; Claesson, P. M.; Malmsten, M.; Linse, P.; Booth, C. *submitted* **1997**.
4 Griffiths , P.C.; Cosgrove , T.;Shah, J.; King , S. M.; Yu , G.; Booth, C. *Langmuir* submitted **1997**
5 Evers, O.A,; Scheutjens, J.M.H.M.,; Fleer, G.J., *Macromolecules* **1991,24**, 5558
6 Phipps, J. S.; Richardson, R. M.; Cosgrove, T.; Eaglesham, A. *Langmuir* **1993**, *9*, 3530-3537.
7 Hauxwell, F.; Otterwill, R. H. *J. Coll. Int. Sci.* **1970**, *34*, 473.
8 Cosgrove, T.; Heath, T. G.; Ryan, K.; Crowley, T. L. *Macromolecules* **1987**, *20*, 2879-2882.
9 Mallagh, L. The Adsorption of Block Copolymers at the Liquid-Liquid Interface. Ph.D., Bristol, 1989.
10 Brandrup, J. and. Immergut. E. H.. *Polymer Handbook*; J. Wiley & Son. Ltd.: 1989.

THEORIES OF POLYMER-MEDIATED FORCES

M. E. CATES

Department of Physics and Astronomy
University of Edinburgh
JCMB, Kings Buildings
Mayfield Road, Edinburgh EH9 3JZ, UK

This paper reviews recent theoretical contributions on the role of polymers (either as stabilizers or as depletants) in colloidal stability. These include (i) a multiple scattering theory for the dynamics of adsorbed homopolymer layers; (ii) a scaling theory of the deformation and desorption of end-adsorbed polymer layers in strong flows (permeation and shear); (iii) a study of the depletion force between colloidal particles mediated by small spherical or rodlike polymers.

1. Introduction

Polymer mediated forces come in two broad categories: those between adsorbed polymer layers (usually repulsive, unless things go wrong), and the depletion force, which arises from the exclusion of polymers from the space between larger colloidal particles (generally attractive). In the case of adsorbed layers, one distinguishes homopolymer adsorption (where all monomers of a chain are equally attracted to a surface), in which the density profile falls rapidly as one moves away from the surface, from the formation of a polymer brush (of more uniform density) in the case where only the ends of the chains are attached to the wall. For an introduction to these concepts, see Ref.[1]. The study of such forces is central to colloid physics, particularly through the issue of colloidal stability. In many cases one is interested in stability under transient conditions [2], for example the role of polymer coats in determining the rheology of rapidly sheared colloids [3], or the role of depletant in the phase separation kinetics (ageing and sedimentation) of colloids [4].

R. H. Ottewill and A.R. Rennie (eds.), Modern Aspects of Colloidal Dispersions, 173–184.

2. Adsorbed Homopolymer Layers

The understanding of homopolymer layers under flow conditions is relevant to lubrication and filtration problems, as well as the transport and rheology of polymer-stabilized colloidal suspensions. In the last case, it is very often the stability of a colloid under flow conditions (for example in a mixing tank) that matters, rather than the purely thermodynamic stability which would be appropriate in the absence of flow. Crudely one can estimate the flow stability by comparing the magnitude of thermodynamic and hydrodynamic forces [5]; however, this neglects the fact that the fluid and the polymer interact directly with one another, so that additivity cannot be assumed. Also, such arguments do not allow for the finite dynamical response times of polymer chains [6].

Previous models of the local fluid motion in a homopolymer layer rely on a Darcy flow approximation, where the permeability is an (ad-hoc) function of the local segment density [7, 8]. The polymer is effectively regarded as a static porous medium with the same density profile as the original polymer layer. Models based on this type of description neglect the non-local response due to hydrodynamic coupling of the polymer to the fluid. If a section of polymer is subjected to a drag force, this propagates along the chain as a change in shape which exerts forces elsewhere on the fluid. Non-local hydrodynamic effects are known to be dominant in the dynamics of isolated polymer coils [6], and we expect them to play an important role also in the adsorbed layer problem.

Our approach to the problem of adsorbed layer rheology sets out to calculate the hydrodynamic response of the polymers to the flow, and conversely, of the flow profile to the polymer distribution. Some preliminary results were presented in [9, 10]; more accurate and fuller ones can be found in [11]. The method involves a multiple-scattering formalism based on the work of Freed and Edwards [12] for polymer solutions. To make progress, we have had to simplify and restrict the calculations in several ways. Firstly, the adsorbed homopolymers are modelled as a collection of loops, each attached to the adsorbing surface at a single point. By choosing a power law distribution of loop sizes [13], it is possible to mimic rather accurately the structure expected in a real layer. (Tails can then be added if required; this is discussed in [11].) Secondly, for most of our calculations, we treated the adsorbing wall as penetrable both to the polymers and to fluid flow. This assumption is not important, however, since the high polymer density at the adsorbing wall itself leads to the formation of an impenetrable screen on which fluid motion (and hence, within our model, polymer motion) is frozen [11].

Our third assumption was to use gaussian loops to model an adsorbed layer, made of chains at the Theta temperature, in which three body interactions dominate. This model, in three dimensions, gives a self-similar density profile [7] in which polymer loops of any given size are close to the overlap threshold with each other. This geometrical feature is shared by adsorbed layers of swollen chains in a good solvent; we believe that our results yield insight into this more important case. Fourthly, based on this geometrical picture, we allow for the Zimm dynamics of individual loops, but ignore certain "screening corrections" [6]. Such corrections should have only marginal effects in the layer since loops on a given scale are weakly interpenetrating [11].

Finally, for numerical work we restricted our attention to the regime of linear response, in which the perturbation to the fluid motion from the presence of polymer can be calculated without accounting for the change in density profile of the chains arising in turn from the fluid motion itself. In principle, a nonlinear calculation is possible using our methods; however, such calculations would require prohibitive computation. To study the nonlinear regime, a scaling approach (along the lines given below for polymer brushes) is more appropriate [14, 15].

2.1. RESULTS

In view of the algebraic complexity of the theory, we give no further details here, but summarize some of the main results [11].

For the case of a single adsorbed homopolymer layer, one expects the hydrodynamic thickness R_H to scale as the gyration radius of the adsorbed chains. Within the "loop decomposition" picture, this becomes the gyration radius of the largest adsorbed loops. The expected scaling is reproduced satisfactorily, although there is a residual weak (logarithmic) dependence on the total density ρ of adsorbed polymer. This effect may be an artefact of the method.

More interesting is the flow profile near the wall. Here, one finds that the flow exponent y, defined as $u \sim z^y$ (where u is the flow velocity parallel to the wall and z the perpendicular distance from it), has a value $y = 2.0 \pm 0.1$ for a layer consisting entirely of loops. At the same time, the computed flow velocity obeys $u(z) \sim 1/\rho$ at fixed z. An interesting question is whether the exponent y is universal or whether it depends on details of the layer architecture. To test this, we studied a polydisperse "pseudobrush" layer consisting entirely of tails, as opposed to loops; we found the same ρ scaling but instead a flow exponent $y \simeq 2.5$. This suggests that the flow exponent depends on at least some of the details, though apparently these do not include the overall density ρ.

It is useful to compare the flow profiles arising from this theory with

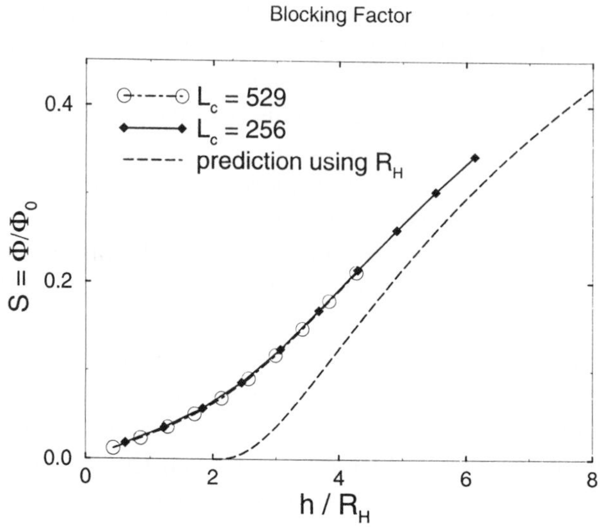

Figure 1. The blocking factor S for flow between parallel plates, as a function of h/R_H (plate separation over hydrodynamic thickness of a layer). The dotted line represents the prediction for impermeable layers of thickness R_H. Data for two different discretizations (L_c values) are shown, showing the convergence of the results.

those of the simplistic "porous medium" approaches, of which there are two [7, 8]. These differ in their estimation of the local porosity of the layer in terms of the local segment density. A careful comparison is made in [11], however, the main conclusion is that these theories predict extremely small flow velocities near the wall, compared to our non-local hydrodynamic theory, if parameters are chosen to make them equivalent in other ways (e.g. by matching R_H in the various models).

The multiple scattering theory can be adapted to study the flow of solvent through polymer layers in a confined geometry (two parallel plates at separation h coated with polymer). This arrangement is reminiscent of the "squeeze flow" geometry arising during a close collision of two colloidal particles, although here the plate separation is held fixed and the flow velocity assumed small. Figure 1 shows the total flux through such a parallel plate arrangement, normalized by that arising in the absence of polymer. This ratio is the "blocking factor" S and would be zero for $h < 2R_H$ if the layers were modelled by impenetrable slabs of thickness R_H (which correctly describes their effects on distant flow fields). Obviously in practice the flux remains finite and for $h/R_H \ll 1$ is expected to obey $S \sim (h/R_H)^{y-1}$. (This follows by simple scaling from the definition of y.) This behaviour is satisfactorily reproduced.

Finally, the model can be used to study frequency dependent effects.

These are expressed in terms of the complex mechanical impedance of the layers or equivalently the complex hydrodynamic thickness, $R_H^*(\omega)$. The real and imaginary parts of this quantity respectively determine the amplitude and phase of the velocity response, far from the layer, in response to a periodic shearing force. The real part, R_H' is found to decrease slowly with ω once ω exceeds the Zimm relaxation rate of the largest loops; it then falls faster once that of the *smallest* loops is reached. This is broadly in line with a scaling prediction of Sens et al. [14] that $R_H \sim (i\omega)^{-1/3}$ for the intermediate frequency range. On the other hand, the imaginary part R_H'' shows an increasing trend in the same range, which is not predicted. However, with the computer resources available we were unable to go far enough toward the scaling limit (very long chains) to expect to see this asymptote clearly.

3. Flow through Adsorbed Brushes

This topic has recently been reviewed elsewhere [16] and so will be summarized only briefly. Fuller accounts appear in [17-19].

Polymers attached by one end to a surface at high density form a polymer brush whose equilibrium properties are now well understood [20]. More recently, experimental and theoretical work has focussed on the dynamic response of a brush to strong shear flows and in other non-equilibrium conditions [22, 21]. Strong flow conditions can give rise to novel and unexpected brush behavior. For example, in the case of neutral polymer brushes subjected to shear flows, there is experimental evidence that brushes can swell in the normal direction [21]. Our recent theoretical work is based on semiquantitative scaling ideas; this allows relatively rapid progress to be made. It is related to previous treatments [23-25] but differs in that an element of self-consistency is introduced, by calculating the flow field and the brush deformation simultaneously. It shares with these treatments the simplifying assumption that all chains behave alike. Though false [20], this seems to be necessary at the present state of our understanding.

3.1. METHOD

The scaling analysis is based on estimating the "effective tension" $\mathbf{t}(s)$ of the chains in a brush, as a function of an arclength coordinate $0 < s < L$, under the supposition that all chains behave alike. This tension can be found by estimating the free energy of a short section of chain as the sum of an elastic and an osmotic contribution, and finding a derivative at fixed number of monomers with respect to variation in the end-to-end vector of the section. The result is

$$\frac{t_x(s)}{k_B T} = \frac{\sin\theta(s)}{\xi(s)} + \frac{\xi(s)}{\xi_0^2}\tan\theta(s) \tag{1}$$

$$\frac{t_z(s)}{k_B T} = \frac{\cos\theta(s)}{\xi(s)} - \frac{\xi(s)}{\xi_0^2}\left(\frac{2-\cos^2\theta(s)}{\cos^2\theta(s)}\right) \tag{2}$$

in a coordinate system where x is parallel to the wall, z is normal, and θ is the angle made by the local chain tangent vector to the vertical direction. Here various combinations of order unity prefactors are set to unity, and $\xi(s)$ is the "blob size" which parameterizes the local segment density in the brush. This effective tension is constructed so that, when it is everywhere zero (no external forces), the brush adopts a step function density profile of the Alexander de Gennes type [20], with $\xi(s) = \xi_0$ everywhere.

Under flow, the external tractions arising from the fluid must balance a non-zero value of the effective tension. More precisely, at each point along a chain, the difference in the effective tension $\Delta\mathbf{t}$ across a blob of size $\xi(s)$ is balanced by the total hydrodynamic force upon it. This drag force is estimated by simple arguments [18] as $\mathbf{f} = \eta\xi(s)\mathbf{v}(s)$ where η is solvent viscosity and $\mathbf{v}$ the local flow velocity. In shear flow, this requirement can be combined with the conservation of total chain length, and a formula for the polymeric contribution to the shear stress (which, added to the solvent stress, is constant through the layer), so as to fix *both* the velocity profile $v_x(z)$ *and* the blob size $\xi(z)$, for a brush subject to a shear flow. In the case of permeation flow things are easier, because $v = v_z$ is a constant specified in advance. (This case describes a brush subjected to a normal flow velocity of fluid, entering or leaving a porous support to which the polymer is attached.) In general, the calculations of the flow and segment density profiles follow from the above considerations by straightforward numerical integration [17, 18].

3.2. RESULTS

Results for a normal (permeation) flow are discussed in [17]. Here we summarize briefly the results for a shear flow which, in the absence of polymer, would have $\mathbf{v} = \dot{\gamma}z\hat{x}$. Results involve the dimensionless shear rate $\dot{\gamma}\tau$ where $\tau = \eta\xi_0^3/k_B T$ is of order the characteristic relaxation time of a Zimm blob of radius ξ_0. We find [18] that, roughly speaking, stretched chains arising at $\dot{\gamma}\tau \gg 1$ can be divided into two regions: (i) an interior region of essentially uniform stretching and tilt in which the fluid velocity is very small, and (ii) a boundary region of thickness $\Delta z \simeq \xi_0$ of weakly stretched and tilted chains in which $\xi(z)$ and $\theta(z)$ vary rapidly. The thickness of this boundary layer corresponds roughly to the hydrodynamic penetration depth calculated from the velocity profiles [18]. We note that there is both a qualitative similarity and a quantitative difference between our predictions and those made earlier by Barrat [24]; see [18] for a much fuller discussion.

3.3. CHAIN DESORPTION AND FRACTIONATION EFFECTS

The above ideas can be used to study the desorption of individual grafted chains in a brush subjected to shear flow. This issue has been considered in some detail by Aubouy [19]. In general, one expects the chain desorption rate to be increased by putting grafted chains under external tension. However, it is found that the enhancement factor is quite different for two main types of anchoring: that of a chain with a compact end-sticker, and that for a diblock copolymers grafted via an insoluble block. In the first case, the rate limiting step is a local desorption of the sticker and its motion over a distance of order one blob near the wall [26]. The flow induced tension reduces the timescale for this by a factor algebraic in $\dot{\gamma}$. In contrast, for block copolymers, desorption involves unwinding a long section of the anchor block into the solvent (creating a configurational energy barrier). This barrier is lowered by the hydrodynamic tension and the desorption rate is exponentially enhanced at high $\dot{\gamma}$.

So far, we have considered brush deformation and chain desorption processes assuming that all chains in a brush behave alike (the Alexander–deGennes ansatz). For equilibrium brushes, however, self-consistent field calculations [20] have shown that the free ends are in fact distributed throughout the layer, rather than concentrated at its extremity. One might expect this to also be the case for brushes subjected to solvent flows. The issue can be addressed, at least qualitatively, within a simple "dual chain" picture devised by Aubouy [19]. In this model, only two types of chains are present: a fraction f of the chains which are extended and tilted by the shear flow (the "dragged" chains), and the remaining fraction $1 - f$ which lie deeper in the adsorbed layer, where the flow is screened (the "quiescent" chains). The result is that f falls rapidly with increasing $\dot{\gamma}\tau$, approaching a plateau for $\dot{\gamma}\tau \gg 1$. In this final state, there are just enough chains participating in the dragged layer to ensure hydrodynamic screening of the quiescent one. Although its details are quite complex, the model suggests that strong shear flow may pick out a few chains and pull them very hard. This has some interesting consequences for the desorption behaviour; see Ref. [19].

4. Depletion Forces

We now turn to the case of depletion [1]. Here one is interested in the interaction between two flat surfaces (related, via the Derjaguin approximation, to that between two spherical colloids of radius R) mediated by *nonadsorbing* polymers (or another species such as small spheres or rods) of much smaller size. The "first order" physics has been known for some time [27]: given a dilute suspension of depletant, there is, for small plate separations h, an attractive force between them equal to the osmotic pressure of the solution. For small spheres of diameter σ, this sets in abruptly at $h = \sigma$; for depletion by rods, the onset is gradual. The force arises because the depletant particles are prevented (spheres) or hindered (rods) from entering the space between the plates; the osmotic pressure exerted on the external faces of the plates is not balanced by that arising within the narrow gap.

There is no doubt that this physics is right. However, it is based on a picture of noninteracting depletant, and is rigorous only in the limit of small concentrations. Recently we have extended the theories of Asakura and Oosawa [27] to higher depletant concentrations, for both spherical and rodlike depletants. The important new effect entering is that the depletant particles have to avoid not only the plates *but also each other.* Because of this interaction, the depletion force extends to a longer range than the particle size. An interesting issue, which is relevant to various aspects of colloid dynamics, is whether the depletion force can be repulsive at these longer distances. This could be important since the presence of a repulsive barrier of sufficient magnitude could lead not to destabilization (creaming of emulsions, flocculation of colloids) as normally expected from the depletion force, but to kinetic *stabilization.* This would require a free energy barrier much larger than k_BT; a sufficiently large barrier would prevent the attractive forces at close separations from every being reached.

Our conclusion is that the depletion force at separations beyond the particle size does indeed show a repulsive barrier, but that this can become large compared to k_BT only for extreme size ratios between the colloidal particles and the depletant species.

4.1. DEPLETION BY SMALL SPHERES

This has been studied using a perturbation expansion in powers of depletant concentration. For spherical depletants, this was done to second order by Walz and Sharma [28] and to third order by Mao et al. [29]. Figure 2 shows the interaction potential $W(h)$ between large colloidal spheres in a sea of small spheres at volume fraction $\phi = \pi/15$. The size ratio is 10 between large and small spheres, allowing comparison with recent simulation results [30]. For these parameters, an entropic barrier of order k_BT is present. The barrier height is linear in the size ratio, and, to second order in ϕ, its height is in fact $\Delta F = (12R\phi^2/5\sigma)k_BT$. To achieve a kinetic stabilization for reasonable ϕ, this requires both small depletants (say 5nm micelles) and

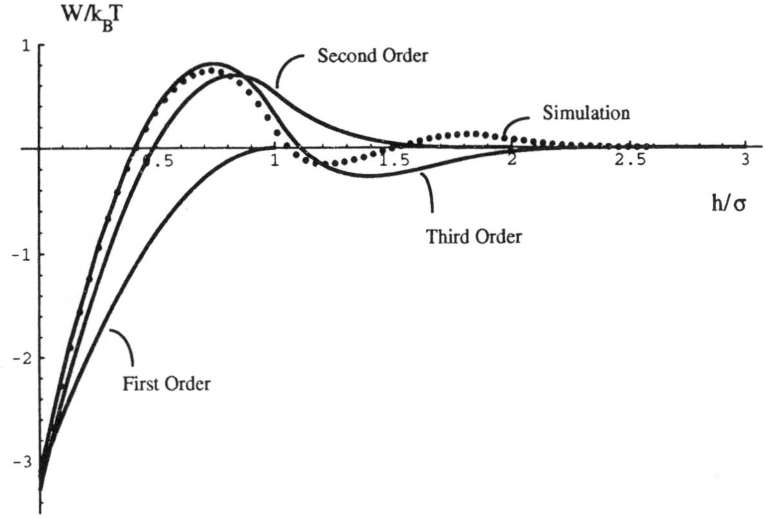

Figure 2. Interaction potential curves for $2R/\sigma = 10$ and $\phi = /15$. Simulation data from Ref. [30].

large colloids (say 10μm radius; this combination gives $\Delta F = 24k_BT$).

Although for more normal parameters the barrier is too low to allow kinetic stabilization, the physics behind it is quite interesting. The main point is that, for separations h just larger than σ, the osmotic pressure exerted within the plates is the result of very few particles, very far apart, having extremely numerous collisions with the walls. In this situation, the resulting pressure is ideal: these particles have negligible excluded volume interaction with each other. This must be compared with the pressure exerted by the particles outside the gap (which do have excluded volume interactions) *at the same chemical potential.* It is easily shown that the density (per unit volume available) of depletant between the plates is slightly higher than outside (exploiting the absence of interactions), and that this yields an extra osmotic pressure which more than balances the interaction terms to the pressure outside. These arguments rest on the fact that the region available to particle centres between the plates, for h slightly above σ, is an extremely thin slab. Accordingly the particles are very few per unit area (and hence noninteracting), although their voluminal number density is comparable to that outside the plates [29].

4.2. DEPLETION BY RODS

This case has been studied by two approaches: a perturbation theory to second order in rod concentration [31] and a self-consistent integral equation method, which becomes exact in the limit of rods whose length/diameter ratio L/D is extremely large [32].

At the second order level, one again finds a barrier whose height now scales as $\Delta F = \alpha k_B T c_b^2 R/D$. Here α is a numerical constant, and c_b is the reduced concentration of rods ($c_b = nDL^2$ where n is the number density), which is of order 4 at the isotropic-nematic transition. For a semidilute solution of rods, values of c_b up to of order unity are reasonable, and on this basis in Ref. [31] we proposed that barrier heights large compared to $k_B T$ could be expected for $R/D \gg 1$. Since D is the rod *diameter*, not length, very large values of this ratio are in principle achievable. For example, with $R/L = 10, L/D = 20$ and $c_b = 2$ we found $\Delta F = 25 k_B T$. However, the use of the second order formula for this value of c_b certainly involved an element of extrapolation.

Subsequently, the asymptotically exact treatment [32] showed that this extrapolation was in fact not justified for $c_b \simeq 2$. Instead, the maximum barrier height is $\Delta F = a(c_b) k_B T R/D$ where the function $a(c_b)$ has a maximum of order 10^{-3} for $c_b \simeq 1$. (The perturbative approach on which the previous estimate was based is found to break down at around $c_b = 0.5$.) Accordingly, a barrier large enough to cause kinetic stabilization never occurs for realistic size ratios. This negative result is based on the extraordinary smallness of $a(c_b)$ which, on dimensional grounds, should be of order unity for $c_b \simeq 1$. The fact that it is so small is something that could not have been surmised without actually doing this rather demanding calculation. Along the way, the approach yields a lot of other interesting information concerning the surface tension of rod solutions, density and order parameter profiles, etc., all of which are asymptotically exact for large L/D and therefore form an important baseline for future studies of interacting particles in confined spaces. For example, in Figure 3 we show the density of rod midpoints as a function of the normal coordinate between plates for $c_b = 2$ and $h = 1.8L$ (taken from Ref.[33]). As well as the exact asymptotic theory, simulations are shown for aspect ratio 10 and 20 showing the extent of the corrections present for realistic ratios. The depletion potential is, of course, also subject to finite L/D corrections, but even for $L/D = 10$ there is no departure from the previous conclusion that kinetic "depletion stabilization" does not arise in practice, in solutions of mutually avoiding rods.

Acknowledgements

The work reviewed above was brought about (and mostly financed) by the DTI Colloid Technology Programme. Thanks are due to the industrial partners, especially Alex Lips of Unilever, firstly for shouldering much of the

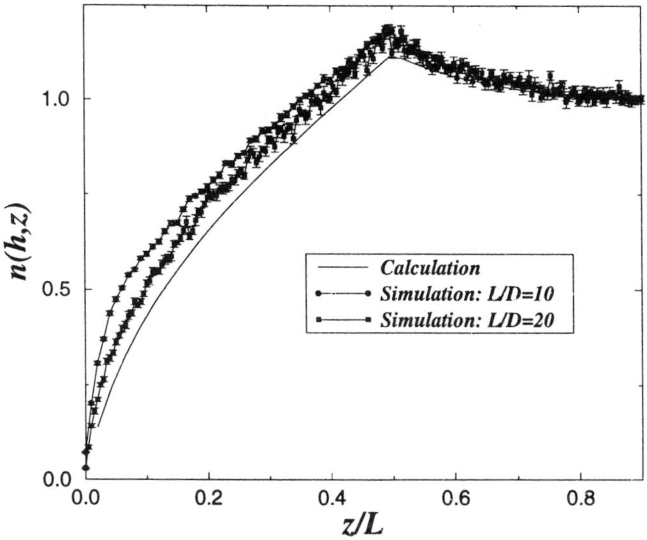

Figure 3. Normalized density of rod midpoints $n(h,z)$ as function of z (coordinate normal to plates) for $h = 1.8L$.

burden of running the programme, and secondly for many helpful scientific interactions along the way. I also thank collaborators David Wu, Jim Harden, Miguel Aubouy, Yong Mao, Henk Lekkerkerker and Peter Bladon whose work is described above, and colleagues Robin Ball and John Melrose for their valuable input.

References

1. Fleer, G. J. , Cohen Stuart, M. A., Scheutjens, J. M. H. M., Cosgove, T. and Vincent, B. (1993) "Polymers at interfaces", Chapman & Hall, London.
2. Lips, A., Campbell, I. J. and Pelan, E. G. (1991) in E. Dickinson (ed.), *Food Polymers, Gels and Colloids*, RSC London, special publication 82.
3. See the article of J. R. Melrose, J. H. van Vliet, L. E. Silbert, R. C. Ball and R. Farr, in this volume.
4. Poon, W. C. K. and Pusey, P. N. (1994) in Baus, M., Rull, L. F. and Ryckaert, J. P. (eds.) *Observation, Prediction and Simulation of Phase Transitions in Complex Fluids*, ; Kluwer, Dordrecht.
5. Russel, W. B., Saville, D. A. and Schowalter, W. R. (1989) *Colloidal Dispersions*, Cambridge University Press.
6. Doi, M. and Edwards, S. F. (1986) *The Theory of Polymer Dynamics*, Clarendon, Oxford.
7. de Gennes, P.-G. (1981) *Macromolecules* **14**, 1637; (1980) *Macromolecules* **13**, 1069; (1983) *C.R. Acad. Sci. Paris* **297**, 883.
8. Cohen Stuart, M. A., Waajen, F. H. W. H. and Dukhin S. S. (1984) *Colloid Polym. Sci.* **262**, 423; Cohen Stuart, M. A, Waajen, F. H. W. H., Cosgrove, T., Vincent, B. and Crowley, T. L. (1984) *Macromolecules* **17**, 1825.
9. Wu, D. T. and Cates, M. E. (1993) *Colloids and Surfaces* **71**, 4142.
10. Wu, D. T. and Cates, M. E. (1993) *Phys. Rev. Lett.* **71**, 4142.
11. Wu, D. T. and Cates, M. E. (1996) *Macromolecules* **29**, 4417.
12. Edwards, S. F. and Freed, K. F. (1974) *J. Chem. Phys.* **61**, 1189; Freed, K. F. and Edwards, S. F. (1974) *J. Chem. Phys.* **61**, 3626.
13. Guiselin, O. (1992) *Europhys. Lett.* **17**, 225.
14. Sens, P, Marques, M. and Joanny, J.-F. (1994) *Macromolecules* **27**, 3812.
15. Aubouy, M. unpublished
16. Harden, J. L and Cates, M. E. (1997) to appear in M. Adams (ed.) *Proceedings of Indo-UK forum on Dynamics of Complex fluids*, Imperial College Press.
17. Harden, J. L. and Cates, M. E. (1995) *J. Phys. II France* **5**, 1093; erratum ibid., 1757.
18. Harden, J. L. and Cates, M. E. (1996) *Physical Rev. E* **53**, 3782.
19. Aubouy, M., Harden, J. L. and Cates, M. E. (1996) *J. Phys. II France* **6**, 969.
20. Milner, S. T. (1991) *Science* **251**, 905.
21. Klein, J. (1996) *Annu. Rev. Mater. Sci.* **26**, 581.
22. Cohen Stuart, M. A. and Fleer, G. J. (1996) *Annu. Rev. Mater. Sci.***26**, 463.
23. Rabin, Y. and Alexander, S. (1990) *Europhys. Lett.* **13**, 49.
24. Barrat, J.-L. (1992) *Macromolecules* **25**, 832.
25. Kumaran, V. (1993) *Macromolecules* **26**, 2464.
26. Wittmer, J., Johner, A., Joanny, J.-F. and Binder, K. (1994) *J. Chem. Phys.* **101**, 4379.
27. Asakura, S. and Oosawa, F. (1954) *J. Chem. Phys.* **22**, 1255; (1958) *J. Polym. Sci.* **33**, 183.
28. Walz, J. Y. and Sharma, A. (1994) *J. Colloid Interface Sci.* **168**, 485.
29. Mao, Y., Cates, M. E. and Lekkerkerker, H. N. W. (1995) *Physica A* **222**, 10.
30. Biben, T. and Bladon, P. to be published.
31. Mao, Y., Cates, M. E. and Lekkerkerker, H. N. W. (1995) *Phys. Rev. Lett.* **75**, 4548 (1995).
32. Mao, Y., Cates, M. E. and Lekkerkerker, H. N. W. (1997) *J. Chem. Phys.* In press.
33. Mao, Y., Bladon, P., Cates, M. E. and Lekkerkerker, H. N. W. (1997) Molec. Phys. In press.

THEORY OF VIBRATION-INDUCED DENSITY RELAXATION OF GRANULAR MATERIAL

S.F.EDWARDS and D.V.GRINEV
Polymers and Colloids Group, Cavendish Laboratory,
University of Cambridge, Madingley Road, Cambridge CB3 OHE,
UK

We propose a simple theory which describes the settling of loosely packed, cohesionless granular material under mechanical vibrations. Using thermal analogies and basing the theory on an entropic concept, formula are derived for how the density of a powder depends on its history. Comprehensive data from the Chicago group show how an initially deposited powder changes its density under carefully controlled packing to reach a terminal density at a terminal tapping rate. A reduction in the tapping rate moves the system to a higher density, but whereas the first measurements follow an irreversible curve, the second measurements have established a reversible curve. These two regimes of behaviour are analyzed theoretically and a qualitative understanding of them is offered.

1. Introduction

There is an increasing interest in applying the methods of statistical mechanics and of transport theory to granular systems where processes are dominated by friction, and have possesion of vital memories [1]. In this paper we propose a simple theory which gives a qualitative explanation of experimental data obtained by Chicago group [2]. They have shown that external vibrations lead to a slow, essentially logarithmic approach of the packing density to a final steady-state value. Depending on the initial conditions and the magnitude of the vibration acceleration, the system can either reversibly move between steady-state densities or can become irreversibly trapped into metastable states.

A granular material is a system with a large number of individual grains. If we hypothesise that it may be characterised by a small number of parameters (analagous to temperature etc.) and that this system has properties which are reproducible given the same set of extensive operations (i.e. operations acting upon the system as a whole rather than upon individual grains) then we may apply the ideas of statistical mechanics to these granular systems [3] (however, the interactions between the grains are dominated by frictional forces). We briefly review these ideas here. In conventional statistical mechanics the microcanonical ensemble gives us the probability that the system is found in the state with energy E:

$$P = e^{-S}\delta(E-H), \qquad \int P = 1, \quad e^S = \int \delta(E-H)\, d(all) \tag{1}$$

R. H. Ottewill and A.R. Rennie (eds.), Modern Aspects of Colloidal Dispersions, 185–192.

where H is the Hamiltonian and S is the entropy. Further one defines temperature $T = \frac{\partial E}{\partial S}$ and can transfer to the canonical ensemble with:

$$P = e^{\frac{F-H}{kT}}, \qquad e^{-\frac{F}{kT}} = \int e^{-\frac{H}{kT}}\, d(all) \tag{2}$$

where $E = F - \frac{\partial F}{\partial T}$ and $S = -\frac{\partial F}{\partial T}$.
We now introduce a volume function W (assuming the simplest case that all configurations of a given volume are equally probable; in many cases of cause the mechanism of deposition will leave a history in the configuration but this will not be considered here, (see e.g. [4]) of the coordinates of the grains which specifies the volume of the system in terms of the positions and orientations of its constituent grains. Averaging over all the possible configurations of the grains in real space gives us a configurational statistical mechanics describing the random packing of grains. Thus we have a microcanonical probability distribution:

$$P = e^{-S}\delta(V - W), \qquad e^S = \int \delta(V - W)\, d(all) \tag{3}$$

Further we define the analogue of temperature $X = \frac{\partial V}{\partial S}$. This parameter is the *compactivity* and measures the packing of the granular material. It may be interpreted as being characteristic of the number of ways it is possible to arrange the grains in the system into volume ΔV such that disorder is ΔS. Consequently the two limits of X are 0 and ∞, corresponding to the most and least compact arrangements. Then we have the canonical probability distribution:

$$P = e^{\frac{Y-W}{\lambda X}} \tag{4}$$

where λ is a constant which gives compactivity the dimension of volume, Y we call the effective volume; it has the role of free energy:

$$S = -\frac{\partial Y}{\partial X}, \qquad V = Y - X\frac{\partial Y}{\partial X} \tag{6}$$

To illustrate this theory consider the simplest example of a W:each grain can occupy a space which as a minimum of ν_0 and a maximum of ν_1, so that there is a degree of freedom, μ say, such that:

$$W = \nu_0 + \mu(\nu_1 - \nu_0) \tag{7}$$

One can think of μ as $(1 - \cos\theta)/2$, where θ is an "angle of orientation" of a grain. So when $\mu = 0(\theta = 0), W = \nu_0$ then the grain is "well oriented" and when $\mu = 1(\theta = \pi), W = \nu_1$ then the grain is "not well oriented". The simple calculation of Y and V gives us:

$$Y = N\nu_0 - N\lambda X \ln\left\{\frac{\lambda X}{\nu_1 - \nu_0}(1 - e^{-\frac{\nu_1-\nu_0}{\lambda X}})\right\} \tag{8}$$

$$V = N(\nu_0 + \lambda X) - \frac{N(\nu_1 - \nu_0)}{e^{\frac{\nu_1-\nu_0}{\lambda X}} - 1} \tag{9}$$

Thus we have two limits: $V = N\nu_0$, when $X \to 0$ and $V = N(\nu_0 + \nu_1)/2$ when $X \to \infty$ (N is a number of grains). Note that the maximum V is *not* $N\nu_1$ just as in the thermal system with two energy levels E_0 and E_1 one has $E = E_0$ when $T \to 0$ and $E = (E_0 + E_1)/2$ when $T \to \infty$.

We study systems dominated by friction which will be 'shaken' or tapped but do not acquire any non ephemeral kinetic energy. Any such powder will have a remembered history and in particular can have non-trivial stress patterns, but we will confine the analysis of this paper to systems with homogeneous stress which will permit us to ignore it.

The fundamental assumption is that under shaking a powder can return to a well defined state, independent of its starting condition. Thus in the simplest system, a homogeneous powder, the density characterises the state.

It is possible to imagine a state where all the grains are improbably placed, i.e. where each grain has its maximum volume ν_1. In a thermal analogy this would be like a magnet below the Curie temperature where the magnetic field is suddenly reversed. Such a system is highly unstable and equilibrium statistical mechanics doesn't cover the case at all. It will thermalize consuming the very high energy with establishing the appropriate temperature. Powders however are dominated by friction, so if one could put together a powder where the grains were placed in high volume configuration, it will just sit there until shaken; when shaken it will find its way to the distribution (4). It is possible to identify physical sates of the powder with characteristic values of volume in our model. The value $V = N\nu_1$ corresponds to the "deposited" powder, i.e. the powder is put into the most unstable condition possible, but friction holds it. When $V = N\nu_0$ the powder is shaken into closest packing. The intermediate value of $V = (\nu_0 + \nu_1)/2$ corresponds to the minimum density of the reversible curve. We discuss this further and show that these values of powder's volume are the limiting values of density packing function in experimental data [2].

2. Dynamical consideration

The terms "shaken" or "tapped" have been used above and we must try to make this more precise. It is sensible to seek the *simplest* algebraic model for our calculation and to this end we think of a grain described by two variables μ_1 and μ_2, roughly speaking describing the orientation such that our volume function is:

$$W = \nu_0 + (\nu_1 - \nu_0)(\mu_1^2 + \mu_2^2) \tag{10}$$

so that we use a square of our previous parameter μ : $\mu^2 = \mu_1^2 + \mu_2^2$ and the variables of integration are: $d\mu_1 d\mu_2 = \mu d\mu$. The point of this assignment is to aim for the simplicity of a Fokker-Planck equation:

$$\left(\frac{\partial}{\partial t} - \frac{\partial}{\partial \mu} D\left(\frac{\partial}{\partial \mu} + \frac{1}{\lambda X}\frac{\partial W}{\partial \mu}\right)\right) P = 0 \tag{11}$$

This equation can be derived from the Langevin equation:

$$\frac{d\mu}{dt} + \frac{1}{\nu}\frac{\partial W}{\partial \mu} = \sqrt{D} f \tag{12}$$

where $\langle f(t)f(t')\rangle \simeq 2\delta(t-t')$ and the derivation gives the analogue of the Einstein relation that $\nu = (\lambda X)/D$. If we identify f with the amplitude of the force a used in the tapping, the natural way to make this dimensionless is to write the "diffusion" coefficient as :

$$D = \left(\frac{a}{g}\right)^2 \frac{\nu\sigma^2}{v} \tag{13}$$

and

$$\lambda X = \left(\frac{a}{g}\right)^2 \frac{\nu^2\sigma^2}{v} \tag{14}$$

where v is a typical volume of a grain, σ a typical frequency of a tap and g the gravitational acceleration. The Fokker-Planck equation (11) has right- and left-hand eigenfunctions P_n and Q_n and eigenvalues ω_n such that:

$$\omega_n P_n = \frac{\partial}{\partial \mu} D\left(\frac{\partial}{\partial \mu} + \frac{1}{\lambda X}\frac{\partial W}{\partial \mu}\right) P_n \tag{15}$$

$$\omega_n Q_n = \left(-\frac{\partial}{\partial \mu} + \frac{1}{\lambda X}\frac{\partial W}{\partial \mu}\right) D \frac{\partial}{\partial \mu} Q_n \tag{16}$$

or equivalently a Green function:

$$G = \sum_n P_n(\mu) Q_n(\mu) e^{-\omega_n t} \tag{17}$$

It follows that if we start with a non-equilibrium distribution:

$$P^{(0)}(t=0) = \sum_{n=1}^{\infty} A_n P_n, \qquad A_n = \int Q_n P^{(0)} \mu \, d\mu \tag{18}$$

and it will develop in time like:

$$P^{(0)}(t) = A_0 P_0 + \sum_{n\neq 0}^{\infty} A_n P_n e^{-\omega_n t} \tag{19}$$

where $\int P^{(0)} \mu d\mu = A_0$. This coefficient is determined by a number of grains present in the powder, hence must be a constant. Suppose now that deposition produces a highly improbable configuration, indeed the most improbable configuration which is $\mu^2 = 1$ and the mean volume function is $\bar{W} = \nu_1$, where:

$$\bar{W}(X,t) = \int P^{(0)}(X,t) W \mu \, d\mu \tag{20}$$

The general solution of the Fokker-Planck equation (19) goes to its steady-state value when $t \to \infty$ so we can expect $\bar{W}(X,t)$ to diminish (as the amplitude of tapping increases) until one reaches the steady-state value $\bar{W}(X)$. The formula (9) can be obtained using (20) when $t \to \infty$ and represents a reversible curve in experimental data of [2]: altering a moves one along the curve $\rho = \frac{v}{\bar{W}(X)} = \rho(a)$. We can identify time with the number of taps, so wherever we start with any initial $\rho^{(o)}$ and a, successive tapping takes one to reversible curve $\rho(a)$. Or, if one decides on a certain number of taps, $t \neq \infty$, one will traverse a curve $\rho_t(a)$, where $\rho_\infty(a) = \rho(a)$. Notice that the simple result lies with the crudety of our Fokker-Planck equation approach. The general problem will not allow us to think of X as $X(a)$ independent of the development of the system. The thermal analogy is this: if the Brownian motion in an ensemble of particles is controlled by a random force f which is defined in terms of its amplitude and time profile, this random force defines the temperature in the system. Our problem is like a magnetic system where magnetic dipoles are affected by a constant magnetic field, being random at high temperature , and increasingly oriented by the external field as the temperature falls. The algebraic expressions are rather complex and are relegated to Appendix.

3. The experimental situation

The physical picture presented in section 2 is consistent with everyday knowledge of granular materials: when poured they take up a low density but when shaken settle down, unless violently shaken when they return to low density. These effects are much more pronounced in systems with irregularly shaped grains than with fairly smooth uniform spheres, indeed the more irregular a grain is, the more the discussion above describes big differences between $\rho^{(0)}$ and ρ. The experimental data of [2] show the packing density dependence on parameter $\Gamma = a/g$ (fig. 1). A loosely packed bead assembly first undergoes irreversible compaction corresponding to the lower branch of $\rho(\Gamma = \frac{a}{g})$. The settling behavior becomes reversible only once a characterestic acceleration has been exceeded. Our crude formula that three points $\rho(X = 0)$, $\rho(X = \infty)$ and $\rho(t = 0, X = 0)$ are in the ratio: $1/\nu_0$, $2/(\nu_0 + \nu_1)$, $1/\nu_1$ is clearly not worked out and this is to be expected given the crudness of the model. It is a difficult problem to decide whether embarking on a vast amount of algebraic work that a superior mode would entail is worthwhile. Rather a study of stress in the Chicago group's experiments should be more fruitful, both experimentally and theoretically. A final point is that we find the lower (irreversible) curve build up to the upper (reversible) curve exponentially in time. One can always expect tails of the Vogel-Fulcher type i.e. whereas we have (see Appendix):

$$V(t) = V_{\text{initial}}\, e^{-\omega t} + V_{\text{final}}(1 - e^{-\omega t}) \tag{21}$$

one can expect slower tails to be present like:

$$V(t) = V_f + (V_i - V_f)e^{-\omega t} + Tt^{-\epsilon} \tag{22}$$

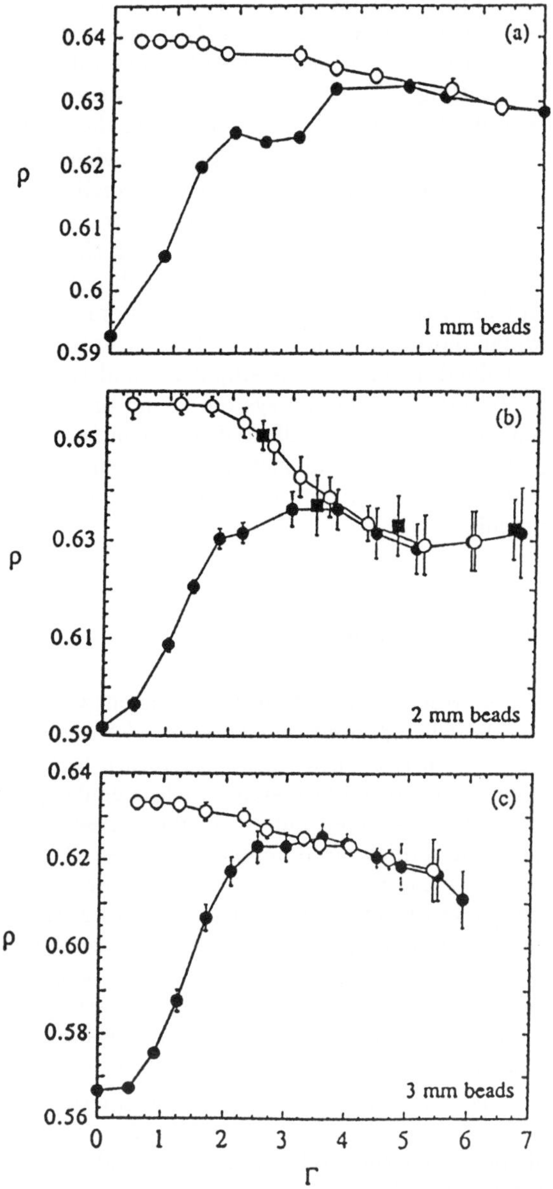

Figure 1: The dependence of the volume packing fraction ρ on the vibration history for various bead sizes. The beads are first prepared in the low density state. The acceleration amplitude was then slowly, and successively, first increased (solid symbols) and then decreased (open symbols). For each curve, the upper branch that has the higher density is reversible, and upon subsequently raising the value of $\Gamma = a/g$ again, ρ retraces the value on the downward trajectory.

where ϵ is large. In fact, identifying t with the number of taps n, the law seems to be even slower at $(\ln t)^{-1}$. Our analysis is clearly inadequate to obtain such a result which is quite outside the straightforward "eigenfunction" expansion method. However there is an argument by P.G. de Gennes [5] which argues that a Poisson distribution can provide this logarithmic behaviour.

4. Appendix

The model adopted is that a grain has two degrees of freedom μ_1, μ_2 and the volume function is:

$$W = \nu_0 + (\nu_1 - \nu_0)(\mu_1^2 + \mu_2^2) \tag{23}$$

The eigenfunctions are taken to be functions of the polar variable: $\mu^2 = \mu_1^2 + \mu_2^2$ and the Fokker-Planck operator becomes:

$$\hat{L}_{FP} = \frac{\partial}{\partial \mu}\frac{\lambda X}{\nu}\left(\frac{\partial}{\partial \mu} + \frac{2\mu(\nu_1 - \nu_0)}{\lambda X}\right) \tag{24}$$

and its ground-state eigenfunction (which is the steady-state solution of the eq.(11)) is:

$$P^{(0)}(t \to \infty) = \frac{e^{-\frac{(\nu_1-\nu_0)\mu^2}{\lambda X}}}{\int_0^1 e^{-\frac{(\nu_1-\nu_0)\mu^2}{\lambda X}}\,\mu\, d\mu} \tag{25}$$

One can easily check that using (11) he will get the result (9). The Fokker-Planck operator (24) has a complete orthogonal set of eigenfunctions:

$$P_n = H_n\left(\sqrt{\frac{\nu_1 - \nu_0}{\lambda X}}\,\mu\right) e^{-\frac{(\nu_1-\nu_0)\mu^2}{\lambda X}} \tag{26}$$

where H_n are Hermite polynomials and $\mu \in (0, \infty)$. In our case $\mu \in (0, 1)$. One can avoid this mathematical difficulty, taking into account the crudety of our model and constructing the "first excited state",say, $P_2 = (a\mu^2 + b)\, e^{-\frac{(\nu_1-\nu_0)\mu^2}{\lambda X}}$ orthogonal to the ground-state eigenfunction P_0. This eigenfunction describes the i nitial state of our system i.e. loosely packed deposited powder. Therefore it is easy to see the initial non-equilibrium distribution (18) depends on how the the powder is deposited. Constants a and b can be defined from the orthonormality relations. By using:

$$P_n = Q_n\, P^{(0)}(t \to \infty), \quad Q_0 = 1 \tag{27}$$

and:

$$\int_0^1 Q_2\, \hat{L}_{FP}\, P_2 \mu\, d\mu = \omega_2 \tag{28}$$

one can easily verify that the eigenvalue ω_2 (which corresponds to P_2 and gives us the decay rate of our nonequilibrium distribution) is a constant dimensionless number.

5. Acknowledgements

This research was carried out as part of the DTI Colloid Technology Link Project supported by Unilever, ICI, Zeneca and Schlumberger. D.V.G. acknowledges a Research Studentship from Shell (Amsterdam).

6. References

1. Mehta, A. (ed.) (1993) *Granular Matter: An Interdisciplinary Approach*, Springer-Verlag, New York.
2. Nowak, E. R., Knight, J. B., Povininelli, M., Jaeger, H. M. and Nagel, S. R. (1997) to be published.
3. Edwards, S. F., and Oakeshott, R. B. S. (1989) *Physica A*, **157**, 1080
4. Shapiro, A. A., Stenby, E. H. (1996) *Physica A*, **230**, 285.
5. de Gennes, P. G. (1997) to be published.

THE EFFECT OF PARTICLE SIZE AND COMPOSITION ON THE REACTIVITY OF ENERGETIC POWDER MIXTURES.

RICHARD BUSCALL, RAMMILE ETTELAIE, WILLIAM J. FRITH[++] and DAVID SUTTON.
Centre for Particle Science & Engineering, ICI Technology
P0 Box 90 Wilton, Middlesbrough, Cleveland, TS90 8JE, UK.

[++]*Current Address: Unilever Research, Colworth House Shambrook, Nr Bedford, UK.*

Pyrotechnic compositions are examples of reactive particulate mixtures. Because of statistical fluctuations in composition, solid mixtures comprising two or more particulate reactants will be non-stoichiometric when examined on a microscopic scale, even when the mixture is stoichiometric overall. The reaction zone in pyrotechnic mixtures is commonly of order a few particle diameters in size. As a consequence, such mixtures do not react with 100% efficiency. The effect of compositional fluctuations due to granularity is analysed for the case where oxidizer and fuel have equal particle sizes. In this case analytical results for the extent of reaction can be obtained.

1. Introduction

The determination of the relation between the bulk properties of heterogeneous materials and their microscopic structure is a long standing problem of great importance in many fields of science and technology[1-3]. One such problem, which has not received too much attention in the past, is the influence of particle size and packing on the extent of solid state reactions in energetic particulate mixtures (pyrotechnics).

The pyrotechnic compositions of interest here comprise a mixtures of two particulate components [4] which react without evolving any gas and without melting. The particle sizes are usually somewhere between 1-100μm. Particulate mixtures can be expected to be inhomogeneous on length scales comparable with the particle size and one is caused to ask whether this is a potential source of variation in the bulk properties of pyrotechnics, e.g., of the burning rate. Intrinsic fluctuations in density and composition arising from the granularity can be expected to depend on a number of factors such as particle size, quality of mixing and the scale on which they are examined. When a sufficiently large volume of the particle mixture is considered the concentrations of the two components in

R. H. Ottewill and A.R. Rennie (eds.), Modern Aspects of Colloidal Dispersions, 193–203.

the mixture will be the same as the overall composition. However, if the volume is small then there is an increasing likelihood that the concentration will deviate from the overall value. An obvious question to ask is how the length scale for compositional fluctuations compares with that associated with the chemical reaction. It is the aim of the present work to address this question. The results discussed here should be of relevance to the pyrotechnic synthesis of ceramics, cermets and intermetallics, also perhaps to problems like that of incorporating cross-linking agents into powder-coating formulations.

Examples of energetic compositions of the type we have in mind are (W and $K_2Cr_2O_7$) or (Sb and $KMnO_4$). In these two cases it is believed that the reaction is diffusion-controlled, i.e. the activation energy is close to zero. The lack of an activation energy raises the question of why these systems are stable at room temperature. Laye and Nelson [5] have proposed an explanation of this. At room temperature, they suggest the diffusion of reactants is negligible so that a reaction cannot be sustained. As the temperature is increased however some change, possibly in the phase of the material, occurs which allows diffusion to take place. This temperature (T_{ig}) is the point at which reaction begins. The reaction is thought to proceed as follows: If we consider some point *A* in the reactant mixture, then as the reaction front approaches *A* the temperature at this point will begin to rise until T_{ig} is reached. At this stage the reactants begin to diffuse and mix and subsequent reactions at point *A* produce heat which sustains the propagation of the reaction front.

It is evident that in the absence of any evolved gas, the main mechanism for heat transfer will be conduction. The heat generated at a point *A* has to diffuse in order to reach neighboring points. The diffusion coefficient for this is given as usual by

$$D = \frac{\kappa}{\rho c} \tag{1}$$

Here κ is the thermal conductivity, ρ is the density and c is the specific heat capacity of the mixture. Let us now assume that the combustion front has just arrived at point *A* in the system and that T_{ig} has been reached, initiating a reaction at *A*. After a further time t the combustion front would have traveled a distance vt ahead of *A*, where v is the propagation rate. During the same time the heat from the reaction at *A* would have diffused a distance of the order $\sqrt{Dt}$. For short times $\sqrt{Dt} > vt$. Thus, the heat generated at point *A* initially moves ahead of the combustion front, affecting the subsequent propagation. Over a longer period however, the front catches up and moves far ahead. At this stage, whatever happens at *A* has no further influence on the propagation of the combustion front. As a result there is only a limited time, once T_{ig} is reached, during which reactions at point *A* can affect the propagation. This influence time (τ) is given by the cross over point $\sqrt{D\tau} \approx v\tau$ which leads to

$$\tau \approx \frac{D}{v^2} = \frac{\kappa}{c\rho v^2} \tag{2}$$.

Equation (2) expresses the influence time in terms of the propagation rate v. In its derivation we have assumed that v is approximately constant over the period τ. When this is not the case, one can still define τ through the relation

$$\sqrt{D\tau} \approx \int_0^\tau v(t)\,dt$$

where t is taken to be equal to zero at the ignition time. This new value of τ will vary from region to region in the pyrotechnic material.

The existence of the influence time, i.e. the fact that there is only a limited time during which reactions at a given point in the system can influence the combustion front, has major implications for the overall rate of propagation within this type of system. A small value for τ means that the reactants have only a short time to diffuse and mix if they are to affect the combustion front. Indeed, there is a characteristic distance L_r over which they will be able to do so. This distance is of the order

$$L_r \sim \sqrt{D_{com}\tau} \tag{3}$$

where D_{com} is the diffusion coefficient of the reacting components. If we assume [5] that the reaction is diffusion limited and that the volume defined by L^3_r has a uniform composition equal to the stoichiometric value, then the reaction will go to completion within a time of order τ. That is to say that all the matter within the volume L^3_r would react during the period τ. If however, the composition is inhomogeneous on length scales comparable to L_r then there will be local deviations from the stoichiometric value, within a volume L^3_r. Any imbalance means that there is not enough time for all the matter within the reaction zone to react fully. Thus, there would remain some residual unreacted matter. In so far as the propagation of the flame front is concerned then, the extent of reaction is less than complete. This reduction in the extent of reaction in turn will affect the propagation rate of the combustion front, leading to a decrease in v. In addition, it may also be that such inhomogeneities will cause the break-up of the combustion front and lead to irreproducible burning rates.

In this work the model descried above in concept has been used to assess the effect of powder composition and associated variables on the efficiency of reaction in pyrotechnic powder mixtures. A full report of the work will be given elsewhere [7]. In this short paper we shall outline the general formalism for calculating the extent of reaction and describe one limiting case where analytical results can be obtained. However, analytical results are only to be had in certain special limiting cases and, therefore, resort to computer simulations has to be made for more general situations [7].

2. General theory

It is assumed then that the composition ratio of the reacting components (hereafter referred to as A and B) within a volume of the size of L^3_r, , is the crucial quantity determining the extent of reaction in time τ. When A and B are both present in the correct stoichiometric ratio then, all of A will react with all of B and the reaction would be complete. It can be shown that the mean value of A:B (in a region of any size) will always be the same as the overall composition of the system. It might be supposed then that for maximum reaction one should choose the composition ratio to be the same as the stoichiometric ratio. Indeed, if the fluctuations on the length scale of L_r were unimportant this would be true. However, as will be shown later, the situation is altered when the fluctuations are no longer negligible. The effect of fluctuations is to cause the local composition ratio to deviate from its mean value. This produces an excess of A in some regions and a shortage in others. In either case the extent of reaction is reduced as a certain amount of matter is left unreacted. Thus, far from averaging each other out, fluctuations reinforce to give an overall macroscopic effect on the burning rate. The residual matter will be of type B where there is a shortage of A, and of type A if we have a lack of B.

From the above discussion, it is clear that the extent of reaction in any region is determined by whichever reactive component is deficient. Let us take the volume fraction of A and B components, within a volume of size L_r^3 to be ϕ_A and ϕ_B, respectively. At a place where we have an excess of A, all of B will react. Furthermore, it will do so with an amount $\lambda\phi_B$ of the component A, where the stoichiometric ratio of A:B is λ. The total volume fraction of the reacted matter is therefore equal to $\phi_B(1+\lambda)$. Similarly, if A was deficient, we would find the total volume fraction of the reacted matter to be $\phi_A(1+\lambda^{-1})$. To summarise

$$E_r^v(\phi_A,\phi_B) = \left\{ \begin{array}{ll} \phi_B(1+\lambda) & if \ \frac{\phi_A}{\phi_B} \geq \lambda \\ \\ \phi_A(1+\lambda^{-1}) & if \ \frac{\phi_A}{\phi_B} < \lambda \end{array} \right\} \qquad (4) \ ,$$

where E^v_r denotes the total volume fraction of the matter which has reacted. We should point out that, in the following analysis, we have taken the densities of A and B to be identical. Generalisation of (4) to cases where this might not be so is rather trivial but it does not help the clarity of the discussion and will not be considered further.

The value of E^v_r for a particular region is not of much interest in itself. Instead, one requires the average value of this quantity taken throughout a macroscopic sample. Representing such averages by < >, the extent of reaction for the system is defined as

$$E_\tau = \frac{< E_\tau(\phi_A, \phi_B) >}{< \phi_A + \phi_B >} = \frac{< E_\tau(\phi_A, \phi_B) >}{< \phi >} \tag{5}$$

i.e, as the ratio of the average volume fraction of the reacted matter to the average volume fraction of the total matter, within the reaction zone. Note that if A and B are mixed in the stoichiometric ratio, and fluctuations are insignificant, equation (5) gives E_τ=1 as expected.

The geometry of the particle packing enters equation (5) through the averaging process. This can be expressed in a more explicit form by defining $P(\phi_A,\phi_B)d\phi_A d\phi_B$ as the probability of A and B having volume fractions in ranges ϕ_A to $\phi_A + d\phi_A$ and ϕ_B to $\phi_B + d\phi_B$ within the reaction zone. Using $P(\phi_A,\phi_B)$, equation (5) becomes

$$E_\tau = \frac{\int E^v_\tau(\phi_A, \phi_B) P(\phi_A, \phi_B)\, d\phi_A\, d\phi_B}{\int (\phi_A, \phi_B)\, P(\phi_A, \phi_B)\, d\phi_A\, d\phi_B} \tag{6}$$

In defining $P(\phi_A,\phi_B)$ we have made the important assumption that the system is statistically homogeneous. This at once excludes such considerations as segregation due to gravity or imperfect mixing processes since these can induce inhomogeneities on a macro scale. Such effects are taken into account formally by rewriting $P(\phi_A,\phi_B)$ as $P(\phi_A,\phi_B,\mathbf{r})$ to indicate the dependence of the probability function on the position of the reaction zone. Major complications then arise. Here we are interested only in intrinsic microscopic fluctuations due to finite particle size effects.

When the full structural information is obtainable, for example through computer simulations, $P(\phi_A,\phi_B)$ can be deduced and the problem of finding E_τ reduces to one of performing the integrals in (3). This is the approach that can be applied generally and which is taken in reference [7]. For a restricted class of problems however equation (6) simplifies, allowing one to directly calculate E_τ. For these type of problems there exists a very strong correlation between the fluctuations of ϕ_A and those of ϕ_B. This strong dependence allows one to express the volume fraction of one of the components in terms of that of the other. Writing this relation as $\phi_B = f(\phi_A)$ we have

$$P(\phi_A, \phi_B) = P_A(\phi_A)\delta(\phi_B - f(\phi_A)) \tag{7}$$

where δ is the Dirac delta function. Substituting (7) in (6) and performing the integral in ϕ_B one gets

$$E_\tau = \frac{\int E^v_\tau(\phi_A, f(\phi_A)) P_A(\phi_A)\, d\phi_A}{\int (\phi_A + f(\phi_A))\, P_A(\phi_A)\, d\phi_A} \tag{8}$$

This represents a considerable simplification in the calculations. Expression (8) for E_τ now only requires the knowledge of the structure of one of the components. Thus, a binary mixture problem has now been reduced to a single component problem. Furthermore, for sufficiently large values of the reaction length L_r (as compared to the radius of the largest particles), $P_A(\phi_A)$ is expected to approach a Gaussian distribution. Once the mean and the variance of this distribution are specified the value of E_τ is easily calculated, using (8).

In the following section we shall present one of two important limiting cases which fall in the above class. For the sake of simplicity, we shall represent powder particles as impenetrable spheres interacting with each other via hard core potentials. For such hard-sphere mixtures, it has been shown elsewhere[7] that when A and B particles are either equal or very different in size, equation (7) becomes a good approximation. We will refer to these two limiting situations as 1:1 and ∞:1 cases, respectively. These two systems excepted, in the majority of other cases the degree of correlation between ϕ_A and ϕ_B is not strong enough to allow one to write a unique relationship between these two quantities. This is easy to see for the general case of impenetrable hard spheres. Varying the value of ϕ_A alters the amount of space available to the B particles within the reaction zone. There has to be a corresponding change in the probability of finding a certain value of ϕ_B. This however, does not mean that a particular value of ϕ_A will force ϕ_A to take up a unique value. ϕ_B can still fluctuate within the restrictions imposed by a given value of ϕ_A. Simplifications such as (7) are therefore not applicable in these situations. Strictly speaking, all practical cases of interest fall in this group. However, as mentioned before 1:1 and ∞:1 cases satisfy (7) sufficiently well as to allow for the use of (8). For intermediate size ratios no such justifications can be given.

3. Equal-sized particles

When both species of particles are the same size it is obvious that one can interchange an A particle with B, without affecting the overall structure. The structure of the mixture can therefore be thought of as one constructed out of a set of identical particles. Labels A and B are assigned to each particle later according to the probability given by the composition ratio. For a set of identical particles one expects the adopted structure to be the same as or at least close to the random close-packing. For a randomly closed packed system of spherical particles we have the average occupied volume fraction within the reaction zone $<\phi>=<\phi_A>+<\phi_B>\approx 0.64$. As one moves from one region to another ϕ varies around its mean value of 0.64. On top of this, for each value of ϕ, both ϕ_A and ϕ_B can vary as well, provided of course $\phi_A + \phi_B = \phi$. An analysis of the computer simulation data of reference [7] reveals that the second set of fluctuations are much more significant than the first. For L_r greater than three times the particle radius, the size of the relative fluctuations in either ϕ_A or ϕ_B were at least an order of magnitude larger than those in ϕ. For example at L_r=3.2 (in units of particle radius), and for a composition ratio A:B=1:4, the relative fluctuation in ϕ_A was 0.238 as compared to 0.018 for ϕ. At L_r=8.0 the respective values were 0.068 and 0.004. From these results it seems reasonable that, at least to a first approximation, the fluctuations in the value of ϕ can be ignored. This amounts to taking ϕ equal to $<\phi> \approx 0.64$ everywhere, leading to the following simple relation between ϕ_A and ϕ_B

$$\phi_B \approx <\phi> - \phi_A \quad (9) .$$

At this stage it is more convenient to work with the actual number of particles within the reaction zone instead of the volume fractions. This change of variables is easily accomplished with the help of the following equations

$$N_A = \phi_A \left(\frac{V_r}{V_p} \right) \quad , \quad N_B = \phi_B \left(\frac{V_r}{V_p} \right) \quad (10) ,$$

where N_A and N_B are the number of particles of A and B type in the reaction zone and V_p and V_r are the volumes of a single particle and that of the reaction zone, respectively. Equation (9) now becomes

$$N_B = <N> - N_A \quad (11) .$$

Making a similar change of variables in equation (8) gives

$$E_\tau = \frac{\int_0^{<N>} E_\tau^\nu(N_A, <N> - N_A) P_A(N_A)\, dN_A}{\int_0^{<N>} <N>\, P_A(N_A)\, dN_A}$$

$$= \frac{1}{<N>} \int_0^{<N>} E_\tau^\nu(N_A, <N> - N_A) P_A(N_A)\, dN_A \quad (12) .$$

Equation (11), together with the composition ratio of the mixture $C'_A : C'_B$ (where C'_A and C'_B are normalised to give $C'_A + C'_B = 1$), give a binomial distribution for N_A with a mean of $<N_A> = C'_A <N>$ and a standard deviation of $\sigma_A = \sqrt{<N>\, C'_A C'_B}$. Note that in assuming this, one is neglecting the fact that some particles might only partially be included in the reaction zone. Such particles would have to be within one particle diameter of the surface of the reaction zone. That is to say that their relative number is expected to drop as L^{-1} compared to the total number of particles. Therefore, for not too small a value of L_r the error introduced in using particle numbers rather than volume fractions should be tolerable. Furthermore, for N>10 the binomial distribution can be approximated by a Gaussian one

$$P_A(N_A)\,dN_A = \frac{1}{\sqrt{2\pi <N> C'_A C'_B}} \exp\left[\frac{-(N_A - C'_A <N>)^2}{2<N> C'_A C'_B}\right] dN_A \qquad (13)\ ,$$

which is more convenient to use in the calculations. The value of E_τ can now be calculated by substituting (13) into equation (12) and specifying the stoichiometry of the system. Taking this to be C_A:C_B (not necessarily the same as the composition ratio C'_A:C'_B) in terms of C_A and C_B the quantity λ in equation (4) reads $\lambda = C_A/C_B$, leading to

$$E^{\nu}_{\tau}(N_A, <N>-N_A) = \frac{N_A}{C_A} \quad \text{when} \quad \frac{N_A}{<N>-N_A} < \frac{C_A}{C_B} \qquad (14)$$

$$E^{\nu}_{\tau}(N_A, <N>-N_A) = \frac{<N>-N_A}{CB} \quad \text{when} \quad \frac{N_A}{<N>-N_A} > \frac{C_A}{C_B} \qquad (15)\ .$$

Once again C_A and C_B are normalised such that $C_A + C_B = 1$. Using (13),(14) and (15), the integral in equation (12) can be evaluated to give

$$E_\tau = \frac{-\sqrt{C'_A C'_B}}{\sqrt{2\pi <N>}\, C_A C_B} \exp\left(\frac{-(C_A - C'_A)^2 <N>}{2C'_A C'_B}\right)$$

$$+\frac{C'_A}{2C_A}\left[2 - erfc\left(\frac{(C_A - C'_A)\sqrt{<N>}}{\sqrt{2C'_A C'_B}}\right)\right] + \frac{C'_B}{2C_B} erfc\left(\frac{(C_A - C'_A)\sqrt{<N>}}{\sqrt{2C'_A C'_B}}\right) \qquad (16)\ .$$

where the erfc(x) function is defined as usual:

$$erfc(x) = \frac{2}{\sqrt{\pi}} \int_x^{\infty} \exp(-y^2)\, dy$$

Simplifications in (16) occur when C'_A:C'_B = C_A:C_B . For a stoichiometric overall mixture, putting $C'_A = C_A$ and using the fact that erfc(0)=1, expression (16) reduces to

$$E_\tau = 1 - \frac{1}{\sqrt{2\pi <N> C_A C_B}} \qquad (17)\ ,$$

As expected, equation (17) shows that for a stoichiometric mix the extent of reaction approaches 1 as <N> increases. For large values of L_r, and hence <N>, the influence of fluctuations become insignificant resulting in $E_\tau=1$.

With the aid of equation (16) it is also possible to predict the mixture ratio for which E_τ attains its maximum value. To do so we need to differentiate (16) with respect to C'_A. This is not difficult to do, although the resulting expression is rather long and will not be listed in here. It suffices to say that for the stoichiometric mixture ($C'_A = C_A$), one finds that

$$\frac{dE_\tau}{dC'_A} = \frac{(C_B - C_A)}{2C_A C_B}\left[1 - \frac{1}{\sqrt{2\pi <N> C_A C_B}}\right] \qquad (18) .$$

This is a particularly useful equation. It indicates at once that if $C_A \neq C_B$ (i.e stoichiometric ratio not 1:1) then the gradient of E_τ at the point $C'_A = C_A$ is non-zero. In other words, the maximum value for the extent of reaction does not occur at the stoichiometrically mixed composition. In fact, since

$$\frac{1}{\sqrt{2\pi <N> C_A C_B}} \quad < \quad 1$$

(see equation (17)), the sign of the gradient depends on the sign of ($C_B - C_A$). When A is the minor component the gradient will be positive. This indicates that increasing C'_A will increase E_τ. Consequently, E_τ is expected to reach its maximum value for a composition ratio richer in A then the stoichiometric mix. With B as the minor component the situation is reversed. It is interesting that this observation has been experimentally known for a long time. For example Spice and Stavely [6] have listed twenty different formulations for which this behavior has been demonstrated.

A comparison of the results of simulations [7] and those calculated using (16) are presented in figure 1. This shows the extent of reaction plotted against L_r for a number of different compositions. In each case the discrete points represent the simulation data, and lines the calculated values. The stoichiometric ratio A:B of the system was 1:4. Provided $L_r > 4$, the degree of agreement between the two sets of data is quite good. For smaller values of L_r however, clear discrepancies between the simulation and the calculated results are apparent. This is not surprising. For a reaction zone with a linear dimension smaller than say three times the particle radius many of the assumptions made in the calculations simply do not hold. In particular, the replacement of the binomial distribution with the Gaussian distribution (13) becomes invalid. An interesting feature of the figure 1 is the point of cross over for the curves with $C'_A = 0.2083$ (almost a stoichiometric mix) and $C'_A = 0.25$ (a mixture rich in the minor component). It is seen that the calculations are in good accord with the simulation results on predicting the correct cross over point.

More generally [7], examples involving Systems with different stoichiometries and various composition ratios have shown that the difference between the simulation results and the predictions of equation (16) are less than 1% at values of $L_r>5$.

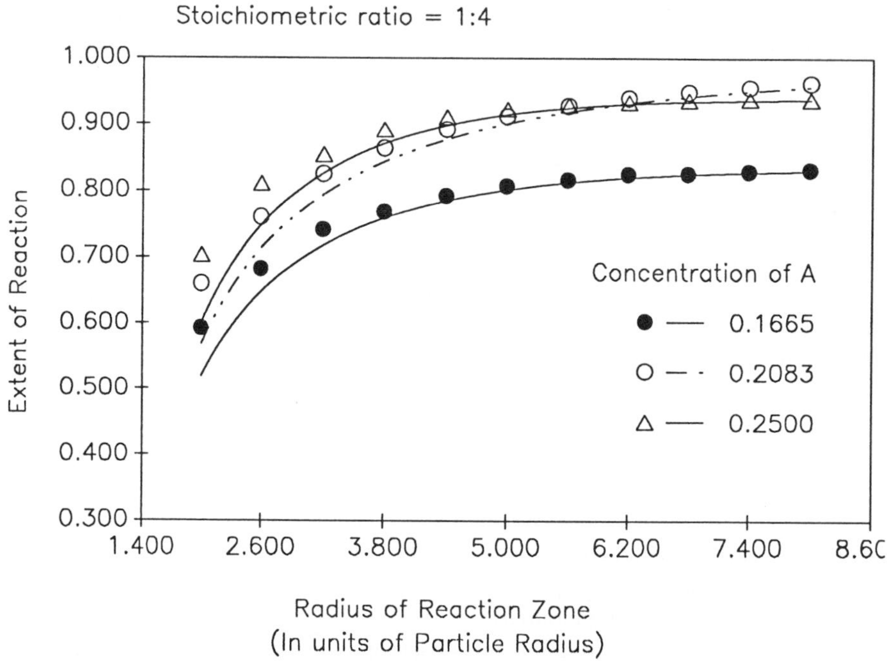

Figure 1: The extent of reaction as a function of reaction zone radius for three different composition ratios. The stoichiometry of the system is A:B=1:4 (by volume). The lines are from equation 16, the points are the results of computer simulations and are taken from reference [7].

4. Conclusions

The reaction zone size in typical pyrotechnic mixtures is estimated to be a few particle diameters and the chemical reaction sees the intrinsic fluctuations in powder composition. These cause the reaction to be less than 100% efficient overall. In the simplest case where oxidizer and fuel particles have the same particle size, analytical results for the overall extent of reaction can be obtained. These show inter alia that the maximum extent of reaction is not had with a mixture which is stoichiometric overall, rather, in one which is enriched somewhat in whichever component is the minor in terms of volume-fraction.

References

1. Beran, M. J. (1968) *Statistical Continuum Theories*, Wiley, NewYork.
2. Jeffrey, D. *J.* and Acrivos, A. (1976) *AIchE J.* **22**, 3.
3. Kirkpatrick, S. (1973) *Rev. Mod. Phys.* **45**, 574.
4. Beck, M. W. (1984) ph. D. Thesis, Rhodes University.
5. Laye, P. G. and Nelson, D. C. (1992) *Thermochimica Acta.* **197**, 123.
6. Spice, J. E. and Stavcly. L. A. K. (1949) *J. Soc. Chem. Ind.* **68**, 348.
7. Buscall, R. , Ettelaie, R. , Frith, W. J. and Sutton, D. (1997) to be submited to *J.Phys.Chem.*

FLUIDISATION, SEGREGATION AND STRESS PROPAGATION IN GRANULAR MATERIALS

S. WARR[1], J. M. HUNTLEY[2] AND R. C. BALL[1]
1. University of Cambridge,
Cavendish Laboratory, Madingley Road, Cambridge CB3 0HE
2. Loughborough University,
Department of Mechanical Engineering, Loughborough LE11 3TU

The paper reviews recent experimental investigations into three areas of the dynamic and static behaviour of model granular materials. In the first, digital high speed photography and computer image processing are used to study the fluidisation of a two-dimensional model powder undergoing vertical vibration. The granular temperature is deduced from measured velocity distribution functions, and the scaling dependence on vibration frequency, amplitude and number of grains is compared with simulations and a simple analytical model. The system has also been used to investigate the size-ratio and acceleration dependence of particle size segregation behaviour for a single intruder in a two-dimensional bed of monodisperse particles at low accelerations. Particle trajectory maps show that, at all base accelerations, the intruder and surrounding particles move upwards at the same speed. Unlike recent suggestions in the literature, there appears to be no fundamental difference in segregation mechanism between the so-called intermittent and continuous regimes. The third area concerns the propagation of stress within a three-dimensional conical pile. A simple elasto-optical method has been used to measure the normal force distribution at the pile base. Experiments on sandpiles confirmed the existence of counter-intuitive pressure dips at the centre. Variations in surface roughness were found to have little effect on the pressure profiles.

1. Introduction

In recent years a large body of literature has emerged on the physics of granular materials [1], and in particular, the vertical vibration of granular materials has been extensively studied. A wide range of unusual phenomena is experimentally observed [2], including heaping and convection rolls [3] and size segregation [4-6]. At larger vibration amplitudes, period doubling instabilities lead to both standing waves, [7], and travelling waves, [8], on the free surface, and eventually the system can become fully fluidised. Under quasi-static conditions, stress can propagate within granular materials in counter-intuitive ways, leading for example to a pressure dip under the peak of a conical pile of

R. H. Ottewill and A.R. Rennie (eds.), Modern Aspects of Colloidal Dispersions, 205–214.

sand [9]. This paper reviews some of the research carried out at the Cavendish Laboratory over the past four years into some of these areas.

2. Instrumentation

The fluidisation and segregation experiments were performed using an electromagnetically driven shaker (Ling Dynamics Systems Model V650) driven by sine waves from a low-distortion signal generator (Farnell DSG2) and 1 kW power amplifier (LDS PA1000). The moving part of the shaker is a platform 156 mm in diameter which can attain a maximum peak to peak displacement of 25.4 mm and a maximum velocity and acceleration of 1.06 m s^{-1} and 70 g (g = 9.81 m s^{-2}) respectively. A cell made up of two glass plates 165 mm wide by 285 mm high was mounted on the moving platform. The width between the plates was controlled by spacers of varying thicknesses, to a resolution of 0.05 mm. By adjusting the plate spacing to exceed the particle size by 0.05 mm a close approximation to an idealised two-dimensional model powder was obtained. Vertical accelerations were monitored using piezoelectric accelerometers, one attached to each side of the support, and displacements of the vibrating cell were measured using a calibrated laser displacement meter (Nippon Automation LAS-5010V). This allowed the horizontal acceleration to be checked at various points. The horizontal acceleration was found to be less than 2% of the vertical acceleration for the working range used in this paper, indicating an essentially one-dimensional acceleration field.

A Kodak Ektapro 1000 high speed video camera system is used to record the two-dimensional motion of granular particles, with the resulting images stored digitally on processor boards within the camera. When images are stored to the full size of 239 x 192 pixels, a maximum framing rate of 1000 frames per second can be achieved giving 1.6 s of recording time. The framing rate can be increased by decreasing the image size with an upper limit of 12,000 frames per second. Figure 1 shows a typical image from a size segregation experiment.

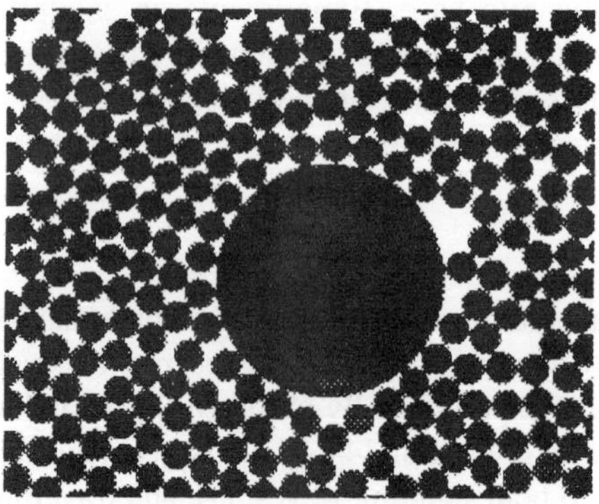

Figure 1. Typical grey-scale image from the Ektapro high speed video camera.

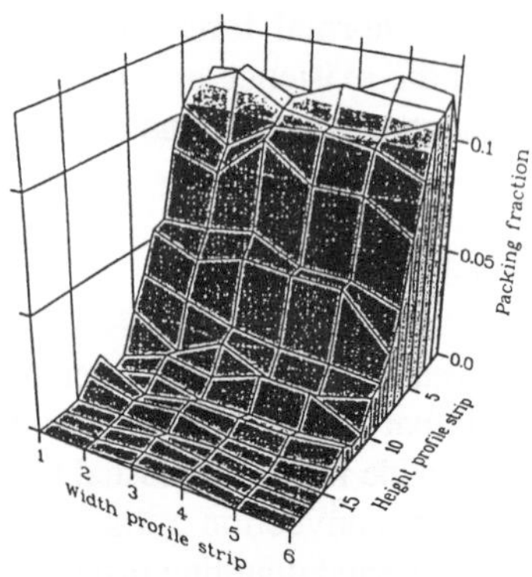

Figure 2. Packing fraction distribution in fluidised powder (N = 90, A_0 = 2.12 mm, f = 50 Hz).

Images are transferred by GPIB interface to a Sun IPX SPARCstation where the in-plane position of particles are located using digital image processing software. There are three main steps to the image analysis: (a) the edges of the particles are detected by means of a Sobel filter; (b) the positions of the centres are then detected by Hough transformation of the resulting image; and (c) the position of all particles within the field of view are then tracked from frame to frame, from which speed and velocity distribution functions can be calculated. These operations are described in detail in [10].

3. Fluidisation

The fluidisation of granular materials has recently been investigated numerically by molecular dynamics and event-driven simulation techniques. Luding et al. [11, 12] obtained the conditions required to observe the condensed and fluidised regimes and showed scaling relations for the height of the centre-of-mass for the system of beads in the fluidised regime in one- and two-dimensional powders. Previous experimental studies in two-dimensional systems have only considered surface fluidisation whereby a condensed phase and fluidised phase coexist [13].

Fluidisation behaviour can be described using the kinetic theory concept of granular temperature that is widely used in theories of rapid granular flow. The reviews by Campbell [14] and Savage [15], on computer simulation and theoretical studies respectively, discuss the concept of granular temperature and its relevance to granular materials.

3.1 TWO-DIMENSIONAL RESULTS

The instrumentation described in section 2 was used to investigate the fluidisation of model powders consisting of 5 mm diameter steel spheres [16]. The number of spheres in the cell, N, took the values 27, 40,60 and 90, and for each of these, experiments were carried out with amplitudes A_o of 0.5, 1.123, 1.84 and 2.12 mm. The base frequency was 50 Hz throughout. Camera runs were carried out at three different heights for each of these combinations of N and A_o, resulting in a total of nearly 80,000 frames of data.

Figure 2 shows the time-averaged three-dimensional packing fraction surface for the case $A_o = 2.12$ mm and N = 90. No significant side-wall effects are evident. Averaging the data across the width of the cell improves the signal-to-noise ratio; results for all four N-values at $A_o = 2.12$ mm are shown in Figure 3(a). The N = 90 data are replotted on log-linear axes in Figure 3(b), together with a linear fit to the exponential tail of the distribution. The gradient of the best fit line provides one measure of the granular temperature, E_o. E_o can also be estimated from the width of the measured velocity distribution functions. Figure 4 shows the distribution for the horizontal component of velocity at a height in the cell of 65 mm, where N=90 and A_o =2.12 mm. Crosses correspond to data points and the line is the best fit Gaussian curve.

The simulations carried out by Luding et al. [11, 12] indicated that E_o scales with A_o and N as

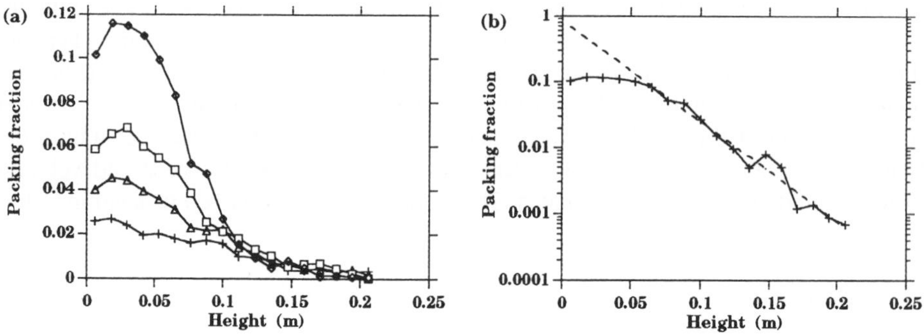

Figure 3. Effect of system size on the average packing fraction profiles. Crosses, triangles, squares and diamonds correspond to N=27, 40,60 and 90 respectively.

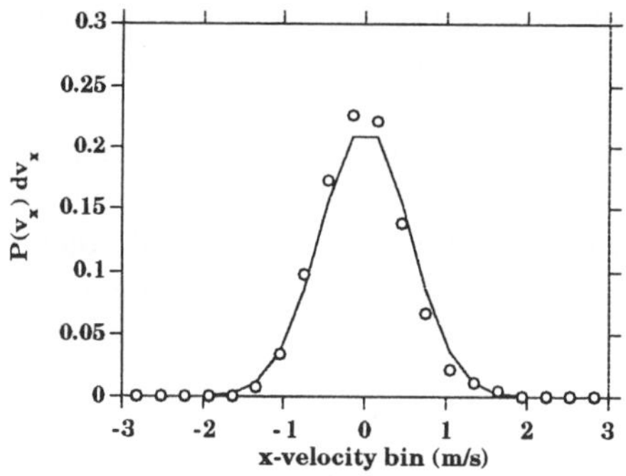

Figure 4. Horizontal velocity distribution: experimental (discrete points) with best fit Gaussian.

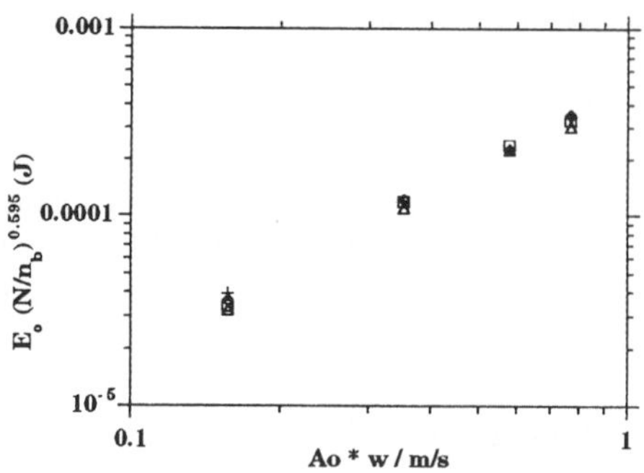

Figure 5. Granular temperature as a function of peak base velocity.

$$E_o \propto (A_o\omega)^{\alpha}[N(1-\varepsilon)]^{-\beta} \tag{1}$$

where ω is the angular frequency of the base and ε is the restitution coefficient. The exponents α and β were found to take the values 2.0 and 1.0, respectively, for a one-dimensional powder, and the values 1.5 and 1.0 in two dimensions. Figure 5 shows the variation of E_0 with $A_0\omega$ as calculated from the velocity distribution functions. The effects of system size have been scaled out using the average exponent $\beta = 0.60$. The datapoints fall on a line with a gradient $\alpha = 1.41 \pm 0.03$ which is close to the value $\alpha = 1.5$ calculated from simulations.

3.2 ONE-DIMENSIONAL RESULTS

The simplest possible one-dimensional powder consists of a single particle bouncing on a vibrating base. In [17] we derived an exact expression for the rate of energy input into the system for the case of sawtooth excitation and in the limit of high excitation (vibration frequency >> collision rate). By assuming a Gaussian velocity distribution function of the particle the expression was evaluated numerically. This resulted in the same scaling law as Equation (1) with $\alpha = 2.0$ and $\beta = 1.0$. In [18] the analysis was extended using a Langevin-type equation to calculate the form of the rebound velocity, v_o, particle velocity, v, and height, h, probability density functions (PDF's) for a single particle on a sinusoidally oscillating base. Very good agreement between the model and results of simulations was obtained. Scaling laws for the particle granular temperature with peak base velocity and particle-base restitution coefficient, determined from previous work, can also be predicted from the PDF. A fine scale "spiky" structure in the rebound velocity PDF was found, using numerical simulations, to be a consequence of resonance phenomena between the particle and vibrating base.

4. Size segregation

The segregation of powders under vibration according to particle size is a well-known phenomenon, and can cause significant problems during powder handling. A number of possible microscopic mechanisms have been presented in the literature. In the low amplitude and high frequency regime convection rolls, driven by particle-wall friction, control segregation [4]. The intruder is seen to be carried upwards at the same velocity, leading to a continuous ascent. At very low accelerations, recent experiments on two-dimensional systems have identified a transition from a continuous to intermittent, step like, motion as the particle size decreases below a critical size ratio [5]. The experiments have indicated that segregation in this regime is no longer driven by convection, with the larger particles rising relative to the background particles. An arching effect model has been proposed [6] to explain this transition.

The ability to measure and track particles with the Ektapro camera has been exploited to investigate this transition between the continuous and intermittent regimes [19]. The experiments were carried out using a monolayer of approximately 5000 oxidised duralumin spheres (diameter = 2 mm) in the test cell. A set of intruders were made from 1 mm thick duralumin discs of various diameters, through which three 2 mm diameter chrome steel spheres were pressed to form an equilateral triangle. Γ is used to denote the peak acceleration normalised by g, the acceleration due to gravity. The vibration frequency throughout was 10 Hz.

Figure 6 shows three frames of an intruder with size ratio, Φ=7, taken from the low acceleration, Γ=1.17, regime. Regions of disorder often appear around the intruder (a). Small gaps may open up below the intruder which particles can be pushed into by collective block motion, however large gaps and avalanche events are not observed. In (b) a slip plane is captured below the intruder resulting in the upper block of particles moving upwards. Horizontal slip planes are also often observed (c).

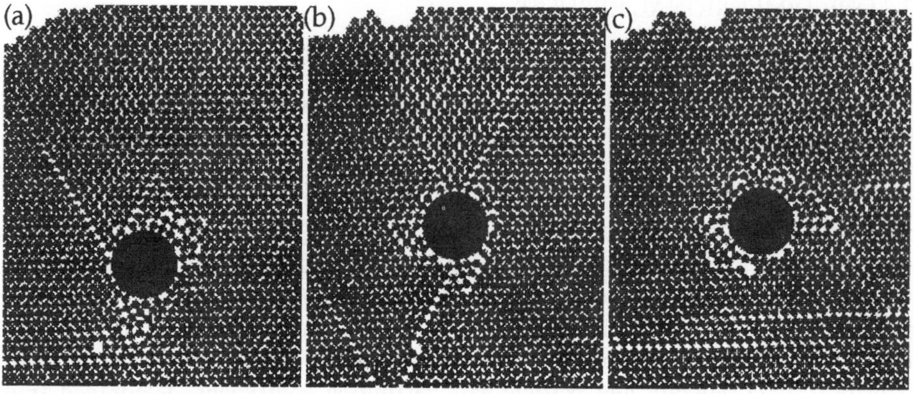

Figure 6. Images from size segregation experiments (see text for details).

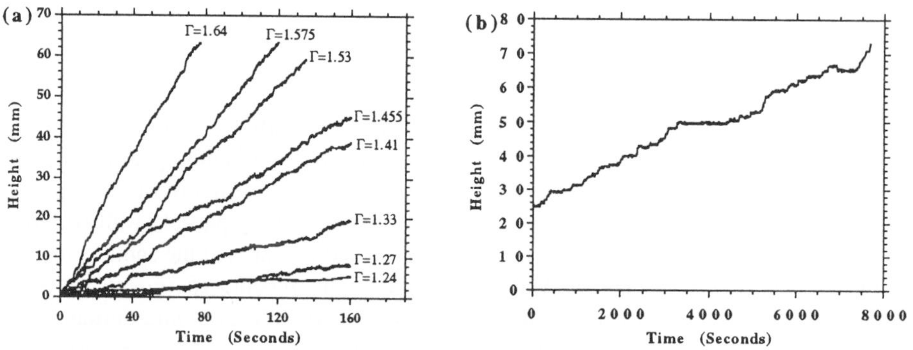

Figure 7. Intruder ascent diagrams for the (a) continuous and (b) intermittent regimes.

In Figure 7 the measured position of the intruder is plotted against time for a range of reduced accelerations. These plots are at a constant size ratio Φ=7. The results in (a) correspond to the continuous regime, whereas in (b) the acceleration is lower, (Γ=1.17) and the intermittent motion is clearly visible. The results in (a) were obtained by filming at one frame per cycle; in (b) a frame was recorded every 48 cycles.

The use of Hough transforms and the particle tracking routines allowed trajectory maps of all the particles in the field to be calculated. Figure 8(a) shows results from one experiment with a size ratio Φ=7, and an acceleration of Γ=1.65. This corresponds to a regime of quite strong convection; the intruder moved over the field of view in 266 frames as indicated by its trajectory. The other trajectories all show the position of the background particles **relative to the intruder** disc. A similar plot is shown in (b), but in this case the acceleration was reduced to Γ=1.32 so that we are in the regime with intermittent rise characteristics. The trajectories still resemble those of Figure 8(a)

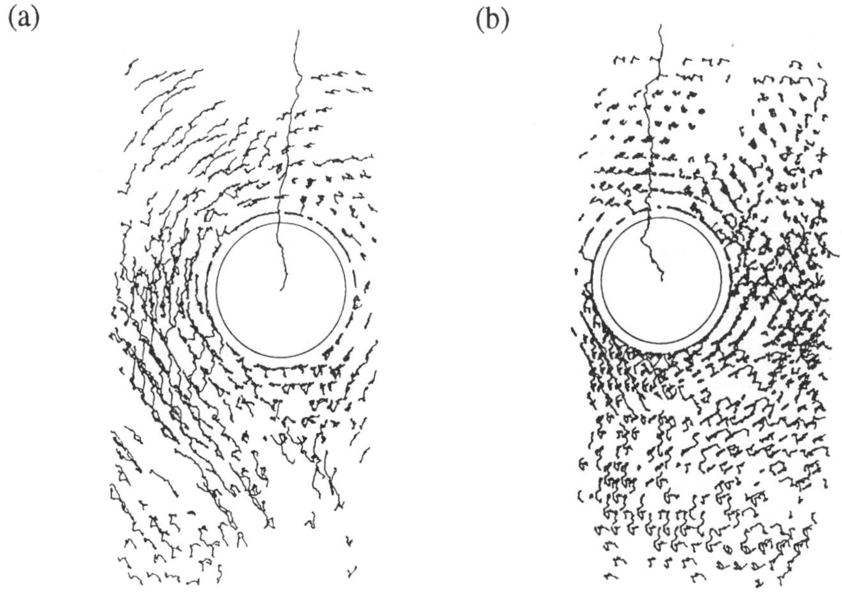

Figure 8. Trajectory maps for intruder and background particles: (a) $\Gamma = 1.65$; (b) $\Gamma = 1.32$.

showing that the intruder and background particles rise at the same rate in a collective motion; the intruder moved over the field of view in 1355 frames.

The main conclusion from these experiments is that the intruder disc rises at the same speed as the background particles over the entire range of accelerations studied. If we call such a phenomenon convective then the mechanism of segregation is driven by convection rolls over all accelerations. Previous studies which indicate that convection is absent at low accelerations (i.e. upwards motion of the intruder is not accompanied by upwards motion of the surrounding particles) and that geometrical effects strongly influence the segregation mechanism are not supported by our observations.

5. Contact force distribution beneath a three-dimensional granular pile

The propagation of stress within a granular material under quasi-static conditions is currently not well-understood. One interesting example is the pressure distribution beneath a conical sandpile, in which intuitively one expects the pressure to be maximal where the total height of material is also maximal, i.e. at the centre. However, experimental measurements have shown a significant local dip at this point [9]. This section describes an elasto-optical technique and its application to the measurement of vertical force profiles beneath loosely packed conical piles. It provides a much higher spatial resolution than the previous experimental work [9] and by using a powder of particle size directly comparable to this resolution, measurements of the fluctuations in the individual grain contact forces were also obtained.

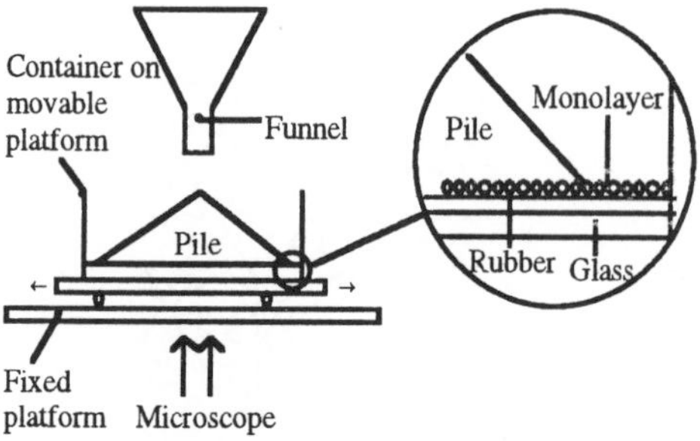

Figure 9. Experimental apparatus to measure the force profile beneath the test piles.

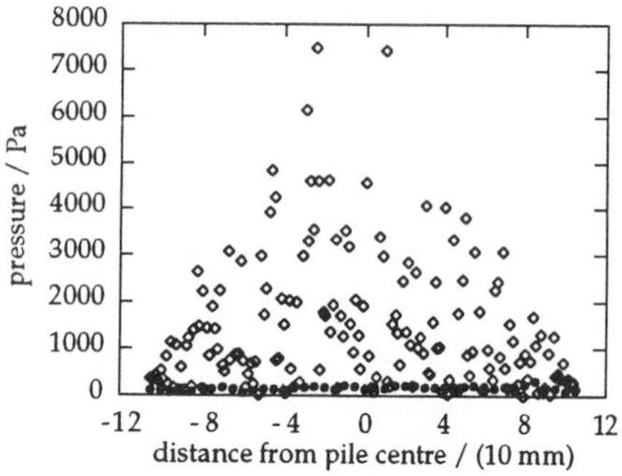

Figure 10. Pressure profiles from a lead shot pile. Circles: ball bearing monolayer; diamonds: lead shot.

The force distribution was determined by measuring the elastic deformation of a transparent silicone rubber surface, on which the pile was constructed (see Figure 9). The rubber was cast as a uniform layer, 2.25 mm thick, on top of a glass flat. A hexagonal-close-packed monolayer of 2.5 mm diameter steel ball-bearings, placed on top of this rubber, acted as a blanket of pressure sensors: the circular area of contact between a given ball and the rubber was measured using an optical microscope and was related directly to the normal force acting on the ball.

The powders used were chosen for their variation in physical and geometrical properties. Lead shot had the highest density, and the mean bead diameter was chosen to match that of the ball bearings to allow force measurements to be made on the same scale as the individual particles. Sand was used for comparison with the results of previous investigators [9]. Glass beads (180 and 560 μm diameter) were used to investigate the effect of particle size distribution; the effect of friction coefficient was studied by etching the larger glass beads with a chemical etchant. The etching resulted in increases in the angle of repose of up to 4°.

The results of some of the pressure measurements made under 22 different piles of diameter ~ 200 mm is given in graphical form in Figures 10 and 11. Figure 10 shows a typical lead-shot pile profile as measured from a single experiment. Each point represents a single contact diameter reading. The force values show large variations from point to point; since the spatial resolution is equal to the bead diameter, this therefore represents the true fluctuations in contact forces in such a pile. The contact forces were found to follow a probability density function that was approximately negative exponential in form.

The experiment was repeated several times with each material (typically 3 times; 5 times with the lead shot) to reduce the effect of the natural contact force fluctuations. All the results for a given material were then averaged; further averaging was also done within discrete radial ranges (bins). The results are shown in Figure 11, in which the error bars represent the standard deviation in the mean of the pressure values averaged.

Significant pressure dips were found to occur with sand (Figure 11(b)), and to a lesser extent with the small glass beads (Figure 11(d)). An increase in glass bead diameter by a factor of three results in almost complete suppression of the dip, regardless of whether

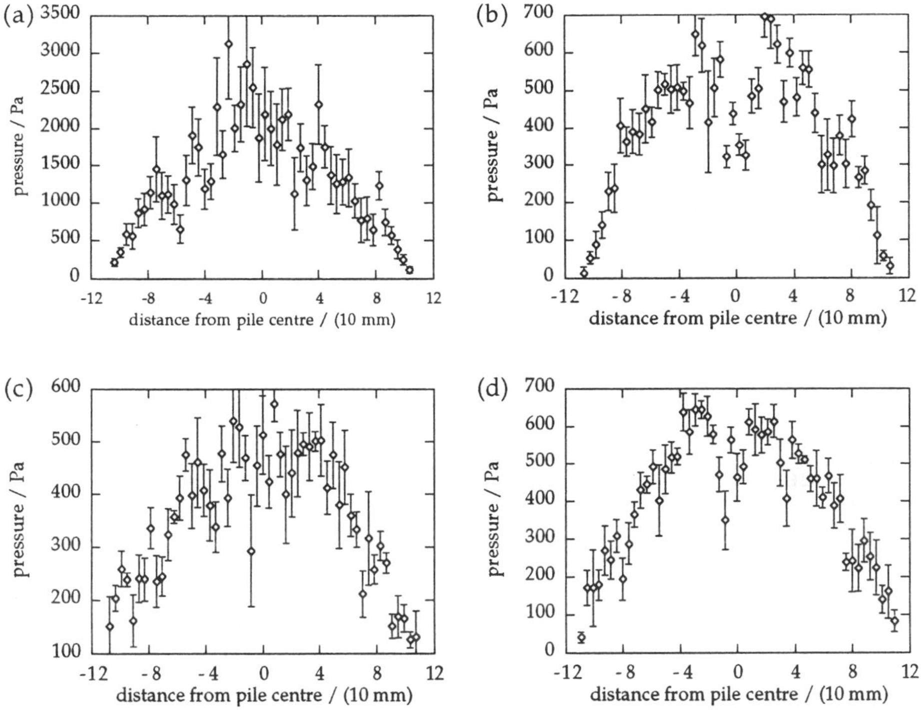

Figure 11. Pressure profiles averaged over several experiments (number of piles averaged in parentheses). (a) lead shot (5); (b) sand (3); (c) large glass beads (3); (d) small glass beads (3).

the beads were smooth (Figure 11(c)) or etched. Likewise, the much larger lead shot (Figure 11(a)) showed no central minimum.

6. Acknowledgements

The research summarised in this paper benefited from experimental and programming input from R. Brockbank, W. Cooke, H. T. Goldrein and G.T.H. Jacques; technical support from D. Johnson, R. Flaxmann, and P. Bone; and useful discussions with C. C. Mounfield, S. F. Edwards, J. E. Field and E. M. Terentjev. Funding from the Department of Trade and Industry, Unilever Plc, ICI Plc, Zeneca Plc and Schlumberger Cambridge Research is also gratefully acknowledged.

7. References

1. Jaeger, H. M. and Nagel, S. R.(1992) *Science* **255**, 1523.
2. Evesque, P.(1992) *Contemporary Physics* **33** 245.
3. Evesque, P. and Rajchenbach, J.(1989) *Phys. Rev. Lett.* **62**, 44.
4. Knight, J. B., Jaeger, H. M. and Nagel, S. R.(1993) *Phys. Rev. Lett.* **70**, 3728.
5. Duran, J., Mazozi ,T., Clement, E. and Rajchenbach J.(1994) *Phys. Rev.* E **50**, 5138.
6. Duran ,J., Rajchenbach, J. and Clement, E.(1993) *Phys. Rev. Lett.* **70**, 2431.
7. Douady, S., Fauve, S. and Larouche, C.(1989) *Europhys. Lett.* **8**, 621.
8. Pak, H. K. and Behringer, R. P.(1993) *Phys. Rev. Lett.* **71**, 1832.
9. Smid, J. and Novosad, J.(1981) IChE Symp. **63,** D3/V/1.
10. Warr, S., Jacques, G. T. H. and Huntley, J. M.(1994) *Powder Technol.* **81,** 41.
11. Luding, S., Clement, E., Blumen, A., Rajchenbach, J. and Duran,(1994) *J. Phys. Rev. E* **49**, 1634.
12. Luding, S., Herrmann, H. J. and Blumen, A. (1994) *Phys. Rev. E* **50**, 3100.
13. Clement, E. and Rajchenbach, J. (1991) *EuroPhys. Lett*, **16**, 133.
14. Campbell, C. S. (1990) *Ann. Rev. Fluid Mech.* **22**, 57.
15. Savage, S. B. (1984) *Adv. Appl. Mech.* **24**, 289.
16. Warr, S., Huntley, J. M. and Jacques, G. T. H.(1995) *Phys. Rev. E.* **52,** 5583.
17. Warr, S. and Huntley, J. M.(1995) *Phys. Rev. E.* **52,** 5596.
18. Warr, S., Cooke, W., Ball, R. C. and Huntley, J. M.(1996) *Physica A.* **231,** 551.
19. Cooke, W., Warr, S., Huntley, J. M. and Ball, R. C.(1996) *Phys. Rev. E.* **53**, 2812.

MEASUREMENT OF FLOC STRUCTURE BY SMALL ANGLE LASER LIGHT SCATTERING

A. LIPS and D. UNDERWOOD
Unilever Research
Colworth Laboratory
Sharnbrook, Beds MK44 1LQ

SALLS, based on a customised use of the Malvern Mastersizer, enables structure to be probed over three decades of scattering vector K; typically for aggregates of 0.5μ spheres simultaneous access is possible to the Guinier, fractal and onset of the Porod region. When applied to the aggregation of initially monodisperse latex spheres, intra-particle scattering form factors can be decoupled from the measured intensities to yield an absolute measurement of the inter-particle structure factor S(K). This together with the wide K range provides a wealth of information on floc formation from a single technique. The method however is limited to the study of dilute dispersions and to a size range of primary particles, optimally 0.5 to 2μ, for which aggregate formation can be strongly sensitive to convective factors. We can observe *shear consolidation* of flocs formed from particles interacting with fully screened electrostatics. The evolution of floc kinetics in high shear (>1000 s^{-1}) for particles with varying strengths of electrostatic repulsions shows an unusual electrolyte dependence implicating a dependence of floc kinetics on electrostatic *force* rather than *potential. Thermal consolidation* of flocs formed from particles by diffusion-limited aggregation can be a relatively fast process when the colloid interaction is mediated by a weakly adsorbing polymer bridging agent.

1. Introduction

SALLS is a well established technique for the measurement of particle size distributions, with a range typically 0.3 to 600 μ. Commercial instruments involve an inversion analysis of the measured angular distribution of scattered light in terms of Mie scattering theory for a polydisperse set of spheres. Malvern Instruments, under a confidentiality arrangement, have given us sufficient detail on the collection optics of their instrument to enable us to circumvent the Mie inversion and to use the apparatus as an absolute scattering photometer. As such it has proven to be a remarkably powerful and reliable instrument. A particularly attractive feature is the wide and well spaced range of scattering wave vector K ($=4\pi\sin(\theta/2)/\lambda$) with θ the scattering angle and λ the wavelength in the dispersion medium) offering at least three decades of reciprocal space. When applied to well characterised model latex spheres an absolute measurement of the inter-particle structure factor S(K) is possible. As will be shown the approach is well suited to studies of floc structure in convective fields.

R. H. Ottewill and A.R. Rennie (eds.), Modern Aspects of Colloidal Dispersions, 215–223.

2. Experimental

2.1 MATERIALS

We used highly monodisperse polystyrene latex standards, supplied by Polysciences Warrington, PA, with mean particle radii a=0.46 and 0.48μ. To avoid multiple scattering, the disperse phase volumes ϕ needed to be extremely low, typically 0.000015.

Diffusion-limited aggregation in a range of shear regimes was effected by the addition of 0.1M lanthanum chloride which ensured efficient screening of electrostatics. For the study of orthokinetic coagulation of latex, in the highest available shear field, a range of concentrations of sodium chloride, from 10^{-6} to 10^{-1} M, was used as the coagulant to provide a decreasing range of electrostatic barrier.

Thermal consolidation studies involved a combination of polymeric flocculant, Dextran T500, at a nominal adsorption density of 0.5 mg.m^{-2}, and 0.01 M sodium chloride. These conditions provided for rapid diffusion-limited aggregation but into a *bridging* rather than van der Waals minimum, the latter remaining counteracted by residual unscreened electrostatics [1].

2.2 METHODS

To obtain an absolute measurement of the inter-particle structure factor S(K) we first measured the angular distribution of scattered light for the latex in its unaggregated state (S(K)=1) as a divisor for corresponding distributions measured for aggregated latex. (Taking account of the scattering geometry of the Malvern Mastersizer, we confirmed consistency of the scattering from the unaggregated latex with Mie predictions of the

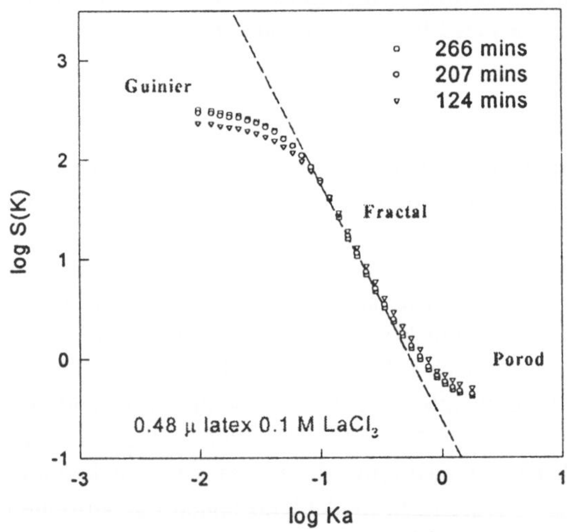

Figure 1 *'Near Brownian'* Diffusion-limited Aggregation

intra-particle scattering factor P(K) for the EM sizes and refractive index of the latex particles. Details will be presented elsewhere.) The orthokinetic studies were done using the stirring facitities of the Malvern Mastersizer. Unfortunately these are not well quantified. We describe three situations of convective flows: *'near-Brownian'* where the growing flocs were occasionally minimally stirred in a tank arrangement to counteract settling, *'moderate'* shear involving the minimum available continuous shear from a pumping device option of the instrument and *'high'* shear corresponding to the maximum setting for the pumping device. (From other studies we estimate an effective shear rate G of ca 4000 s^{-1} for the *'high'* shear regime but with the complication of some extensional flow).

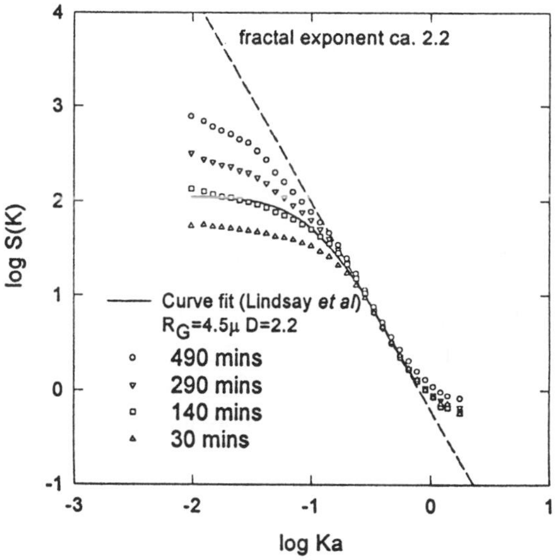

Figure 2 'Zero Barrier' Aggregation in *'Moderate'* Shear

3. Results and Discussion

Figures 1 to 3 show the dependence of S(K) on the reduced inverse interparticle separation Ka (with a the monomer radius) at different times of floc growth. In all cases 0.1 M lanthanum chloride was used to ensure effective screening of electrostatics. The Guinier, fractal and transition to the Porod region can be clearly demarcated. Interestingly, for the weaker convective regimes, there is convergence of the curves at different times for Ka>0.1 indicating an approximately time invariant 'fine structure' on length scales smaller than a few particle diameters; for the *'high'* shear case, by contrast, there appears substantial shear consolidation extending down to those length scales. For all cases, the power law regions are narrow in part reflecting finite cluster size effects, the weight average number of particles in flocs $DP_W(=S(0))$ being only of order 10^3. The fractal exponents d inferred either directly from the slopes of the power law region or from an approach suggested by Lindsay et al. [3] (see below) decrease from 2.5 ± 0.1 to 2.0 ± 0.1 with increasing shear intensity.

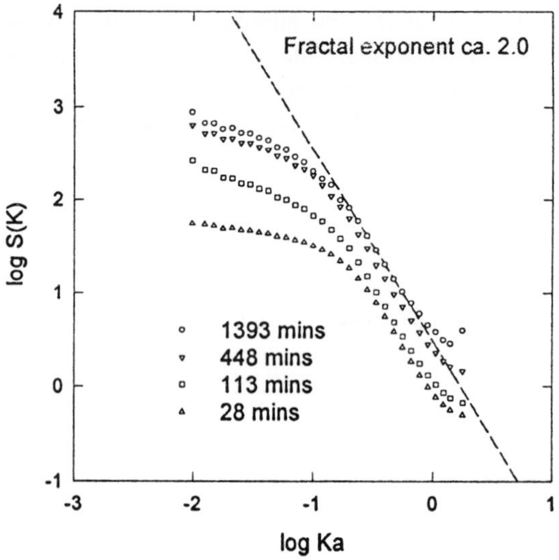

Figure 3 'Zero Barrier' Aggregation in '*High*' Shear

This could be consistent with DLA for the '*near Brownian*' case though convection-limited aggregation is also predicted to yield 2.5 in three dimensions [2]. Using the Guinier expression $S(K)= DP_W(1+<S^2>_Z K^2/3)^{-1}$ we can derive the z average floc radius of gyration $<S^2>_Z^{(1/2)}$ (also referred to as R_G).

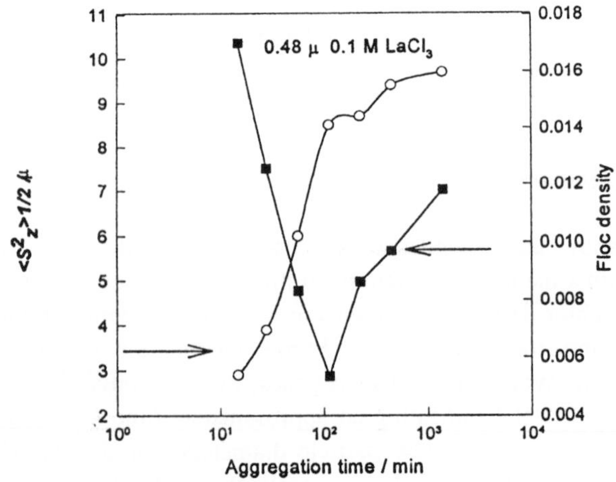

Figure 4 '*High*' Shear Floc Consolidation

Figures 4 and 5 chart the growth in floc molecular weight DP_W, floc density, and average floc radius for the *'moderate'* and *'high'* shear case. For the latter, a limiting size is established within which the density of particles increases with time, floc density at first decreases but then increases with shear consolidation.

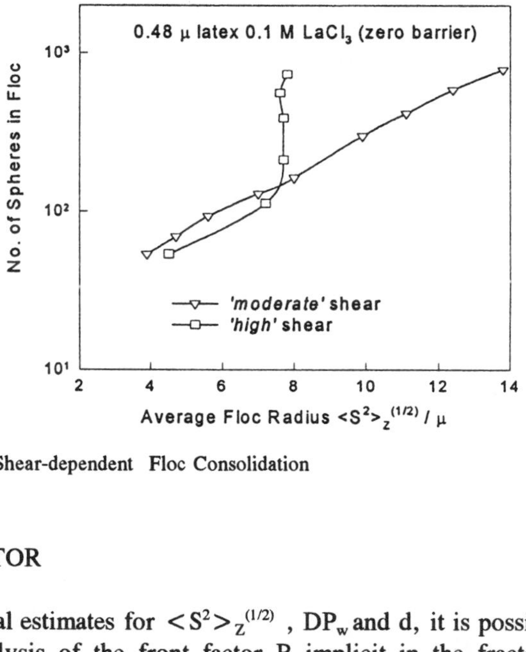

Figure 5 Shear-dependent Floc Consolidation

3.1 FRONT FACTOR

From the individual estimates for $<S^2>_Z^{(1/2)}$, DP_w and d, it is possible to undertake a more detailed analysis of the front factor B implicit in the fractal scaling relation $DP_W=B(<S^2>_Z^{(1/2)}/2a)^d$. The inferred values for B depend strongly on shear and are respectively 0.4,1.0 and 5.5 for the *'near Brownian'*, *'moderate'* and *'high'* shear case.

For centro-symmetric growth, B should indicate a nearest neighbour coordination number if the 'chemical' length scale equals the monomer size. Here the low value of B for the *'near Brownian'* case suggests a strong non central element in the growth of flocs. We could consider modelling the flocs as connected stiff linear chains using the well known blob model [4] ('sliding rod') according to which $B=[2(1+1/d)(1+2/d)]^{2/d}k^{(1-d)}$ (with d the fractal exponent and k the 'persistence length' expressed in terms of monomer diameters). On this basis we infer 'persistence lengths' k of 6 ± 2, $3\pm1$and 1.2 ± 0.4 respectively for the *'near Brownian'*, *'moderate'* and *'high'* shear case.

3.2 KRATKY PLOTS

Kratky plots, in Figures 6 and 7, weighting higher K contributions to S(K) and therefore short range structural features more strongly, also support the prediction of the front factor analysis with similar values of derived persistence lengths.

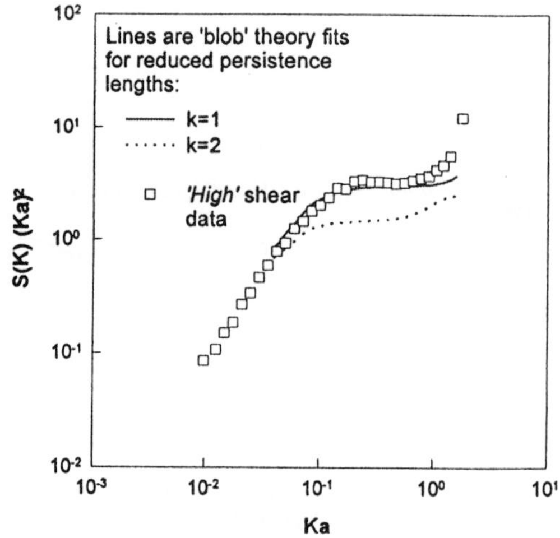

Figure 6 Kratky Plot for '*High*' Shear Aggregation

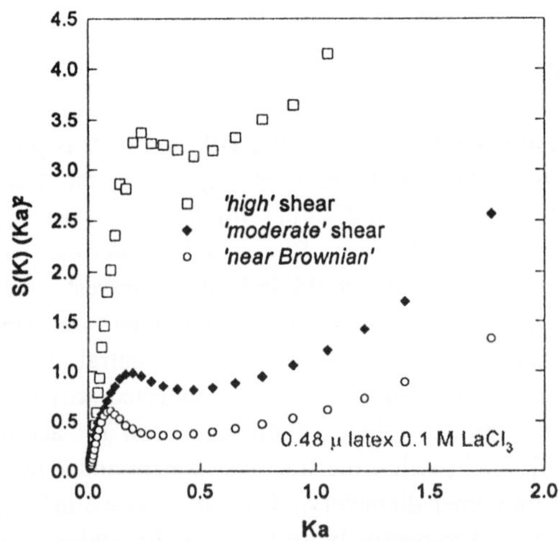

Figure 7 Kratky Plots for different Shear Regimes

3.3 REACTION LIMITED ORTHOKINETIC COAGULATION

Figure 8 illustrates a perhaps surprising insight that electrolyte addition can lead to increased stability of charge-stabilised particles in high shear (G ca 4000 s^{-1}). As can be seen there is apparently greater kinetic stability at intermediate concentrations of electrolyte; at very low salt the particles are relatively unstable to orthokinetic coagulation. Our suggestion is that orthokinetic behaviour then couples more directly with the interparticle colloidal force rather than the potential. DLVO theory predicts an increase in the maximum electrostatic *force* with increasing electrolyte but with a concomitant decrease in range. At intermediate salt levels the latter effect is insufficient to counteract the increase in maximum force.

A second aspect is the pronounced acceleration in the kinetics of aggregate formation for the case of no added electrolyte. We believe this reflects a transition from initial reaction control to collision control as higher order aggregates can more easily overcome the electrostatic barrier by virtue of their greater shear-induced kinetic energy.

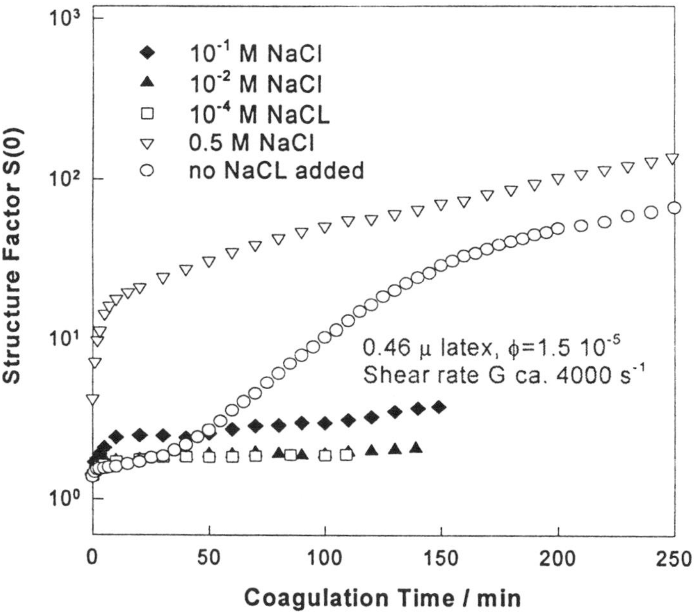

Figure 8 Orthokinetic Coagulation in *'High'* Shear

3.4 THERMAL CONSOLIDATION

Figures 9 and 10 show an example of colloid aggregation with fractal dimension d= 3. The special feature is the use of the nonionic polymeric flocculant dextran T500. We believe this acts as a bridging flocculant. Judicious choice of electrolyte concentration ensures an adequate barrier against van der Waals forces at short range so *diffusion-limited aggregation proceeds into a bridging minimum at long range.* It seems that dextran is rather special either in providing a relatively shallow but wide bridging minimum or through weak adsorption facilitating maximum bridging interaction by 'condensation' in the pores of a close packed, regular structure of flocculated latex particles. We have demonstrated elsewhere [1] that, for *dilute* polysyrene latex flocculated with dextran, this process of thermal consolidation can be fast compared to times for diffusion-limited encounters.

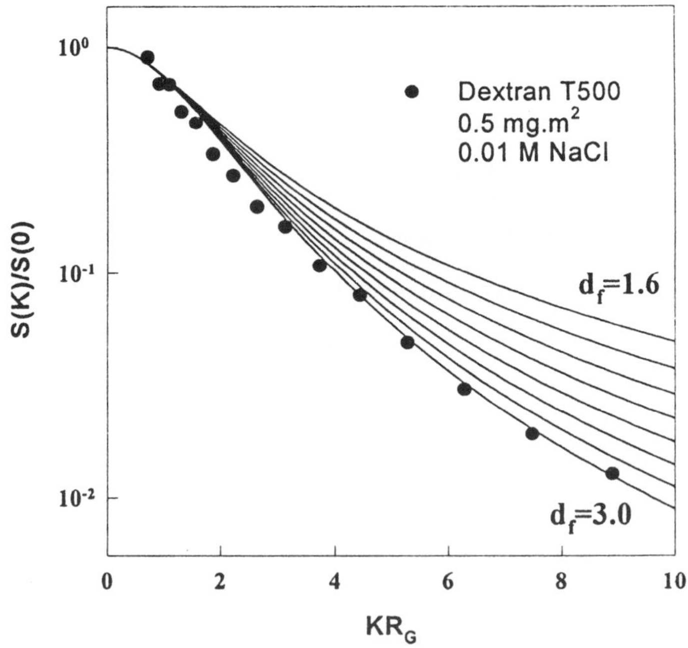

Figure 9 Bridging-mediated Diffusion-limited Aggregation with Thermal Consolidation

Figures 9 and 10 represent P(K) as the ratio S(K)/S(0) plotted against KR_G , R_G (or $<S^2>_Z^{(1/2)}$) having been determined from a Guinier analysis of the low angle scattering data (as above). The lines are theoretical curves for different values of fractal dimension based on an equation proposed by Lindsay et al [3], viz: $P(K)=S(K)/S(0)=(1+2K^2$

$<S^2>_Z/3d)^{d/2}$. Conformity with d=3 for the dextran system is well illustrated, Figure 10 also shows this type of representation for the data in Figures 1 and 3.

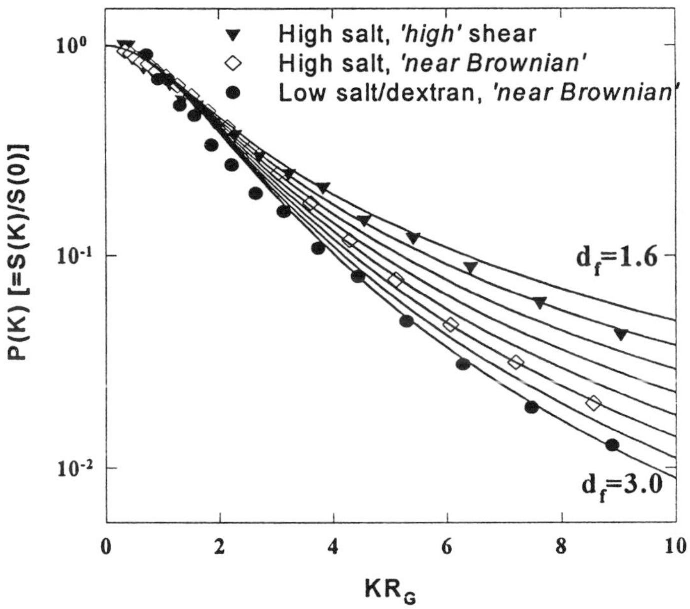

Figure 10 Examples of Shear and Thermal Consolidation in Diffusion-limited Aggregation

4. References

1. Giles D. and Lips A., *J. Chem. Soc., Faraday Trans. I,* (1974), **74**,733.
2. Warren P. B., Ball R. C. and Boelle A., (1995) *Europhys. Lett.*, **2**, 339.
3. Lindsay H.M., Lin M.Y., Weitz D.A., Cheng P., Klein R. and Meakin P., (1987), *Faraday Discuss. Chem. Soc.*, **83**, 153.
4. de Gennes P.-G., (1976), *Macromolecules*, **9**, 594.

THE LOW SHEAR RHEOLOGY OF CONCENTRATED O/W EMULSIONS

P.TAYLOR, R.PONS*
Jealott's Hill Research Station, Bracknell Berkshire, RG42 6ET
* *Centro de Investicion y Desarrollo. C.S.I.C. c/Jordi Girona18-26.*
08034 Barcelona Spain

The rheology of concentrated O/W emulsions is reported for a range of emulsifier systems. It was found that the rheology is approximated by the Princen treatment, and is marginally better described by the scaling laws suggested by Mason et al.. The importance of the method of determing the interfacial tension within the emulsion is discussed, and it is found that by using the aqueous phase of the emulsion for such measurements leads to a universal scaling of the low shear modulus above a critical volume fraction. It is demonstrated that the rheology may be controlled by varying the emulsifier or by using mixtures of emulsifiers.

1. Introduction

Emulsions may be regarded as dispersions of deformable particles, however, unlike more complex microgel particles, their rheology is relatively simple. This paper will consider the universality of the rheology of concentrated emulsions when considered in relation to the physical characteristics of the emulsion, the important parameters being the interfacial tension between the two phases, the droplet size and the dispersed phase volume fraction. Concentrated O/W emulsions are extensively used in both the pharmaceutical and cosmetic industries as skin cream formulations. The feel of such emulsions is dependent upon their rheological properties. It is only within the last decade or so that the rheological properties of concentrated emulsions have been studied with any degree of detail [1-12]. One of the earliest investigations was by Princen et al. who found that the static shear modulus of these systems could be related to the droplet size and the interfacial tension within the systems once a critical oil volume fraction (ϕ_0) had been reached [2]. At this volume fraction the emulsion droplets begin to distort as they interact with each other and the low shear rheology is governed by the dilation of the interface during shear. Similarly, the steady state properties of the emulsion could be scaled using the capillary number. Solans, Pons et al. have made a systematic study of the properties of 'gel-emulsions', emulsions having a very high volume fraction of dispersed aqueous phase [9]. The low shear rheology of these emulsions were also found

R. H. Ottewill and A.R. Rennie (eds.), Modern Aspects of Colloidal Dispersions, 225–234.

to scale with the interfacial tension, though there was some degree of scatter in the coefficients obtained. Pons et al. have recently reported the effect of temperature on the low shear rheology of decane in water emulsions stabilized by Synperonic PE L62 or P94 [4].

2. Experimental

Decane was supplied by Aldrich, Isopar M (a paraffinic oil, Exxon UK) was used as received. The Isopar M was also cleansed with activated charcoal and sorbsil silica (giving an O/W tension close to 50 mN m^{-1}). The polymers Synperonic PE F108, P105, P104, P103 (PEO/PPO/PEO block co-polymers each with an average PPO chain length of 3250, and, on average, 80 50, 40 and 30% of ethylene oxide on the total molecular mass respectively) and Atlox 4913 (a comb polymer with a PMMA backbone with grafted PEO chains of molecular weight 750). were supplied by ICI Surfactants while Mowiol 4-88 (88% hydrolyzed PVA with a molecular weight of around 28000) was supplied by Harco. Sodium dodecyl sulphate (SDS) was was supplied by BDH and was of special Biochemical grade and Synperonic A7 (an ethoxylated synprol alcohol, with an average of 7 ethylene oxide units) was supplied by ICI Surfactants. All were used as received. Water was twice distilled from an all Pyrex still,

Most of the emulsions were prepared, at a decane or Isopar M volume fraction of 0.6 and an emulsifier concentration of 2%(w/w) in the aqueous phase, by high shear mixing of the oil into the aqueous phase using an Ystral high shear mixer. The oil was added slowly to the aqueous phase under relatively low shear. Once all of the oil had been introduced the speed of the mixer was increased to full power for a period of 10 minutes. During this period the temperature of the system reached around 50 °C, emulsions based on P103 and P104 were cooled during shear.

The stock emulsions were centrifuged at 10000 rpm for 20 minutes to concentrate the emulsion, the aqueous phase serum was carefully seperated from the concentrated cream. The oil volume fraction in this cream was determined by dilution with water and then measuring the density of the diluted emulsion with a pyknometer. For the rheological measurements the concentrated emulsions were diluted with the aqueous phase which was seperated from the stock emulsion to give a range of oil volume fractions, which were calculated from the volume fractions of the concentrated emulsions.

In some cases emulsions were prepared using the emulsification technique described above but at an oil volume fraction of 0.8 and an emulsifier concentration of 4% in the aqueous phase. In these experiments the oil used was the unpurified Isopar M and the dilutions were made with water.

The low shear oscillatory rheological measurements were made on a Bohlin VOR rheometer (Bohlin Rheologi, Cirencester, UK) using the cone and plate platens (cp 5/30). A strain sweep was initially made at a frequency of 1Hz to establish the linear viscoelastic region, this was stopped immediately the critical strain was reached. A frequency sweep was then performed in the range 0.01-20 Hz at a strain within the linear viscoelastic region (the upper frequency limit was dependent on the particular torsion element used to avoid problems of resonance). A strain sweep was then performed in which the strain was taken past the onset of nonlinear behaviour. The moduli reported here are the storage moduli (G') taken from within the frequency independent part of the frequency sweeps.

The interfacial tensions within the emulsions were determined using the static Wilhelmy plate described earlier [8]. The aqueous phases separated from the emulsions were measured against fresh samples of oil.

The droplet sizes of the emulsions were determined using a Malvern Mastersizer. Emulsions were initially diluted into a 2% solution of the appropriate copolymer and then further diluted in the feed tank of the instrument. The average of 5 repeat runs on each dilution was taken using a very polydisperse model and an oil refractive index of 1.42. No significant coalescence was seen in the emulsions during the period of the measurements.

Results

The compositions of the emulsions used are listed in table 1, along with the droplet sizes and interfacial tensions. The centrifugation techniques allowed oil volume fractions of up to ca. 0.96 to be achieved, though this value depended on the rheological properties of the systems.

Table 1: Composition and characteristics of the emulsions used in this study. Emulsions A-F were prepared at a volume fraction of 0.6 whilst G was prepared at a volume fraction of 0.7. P denotes purified oil while U denotes unpurified oil. γ denotes the oil/ water interfacial tension whilst R_{32} is the volume/ surface mean radius of the emulsion droplets.

Emulsion	Stabilizer	[Stabilizer]/%	Oil	γ/ mNm^{-1}	R_{32}
A	F108	2	decane	14.4	2.05
B	P105	2	decane	3.35	1.78
C	P104	2	decane	2.42	1.80
D	P103	2	decane	1.40	1.90
E	SDS	2	Isopar M (P)	7.67	0.98
F	PVA	2	Isopar M (P)	27.5	0.70
G	Atlox 4913	4	Isopar M (U)	11.1	0.64
H	F108	2	decane	14.4	1.06

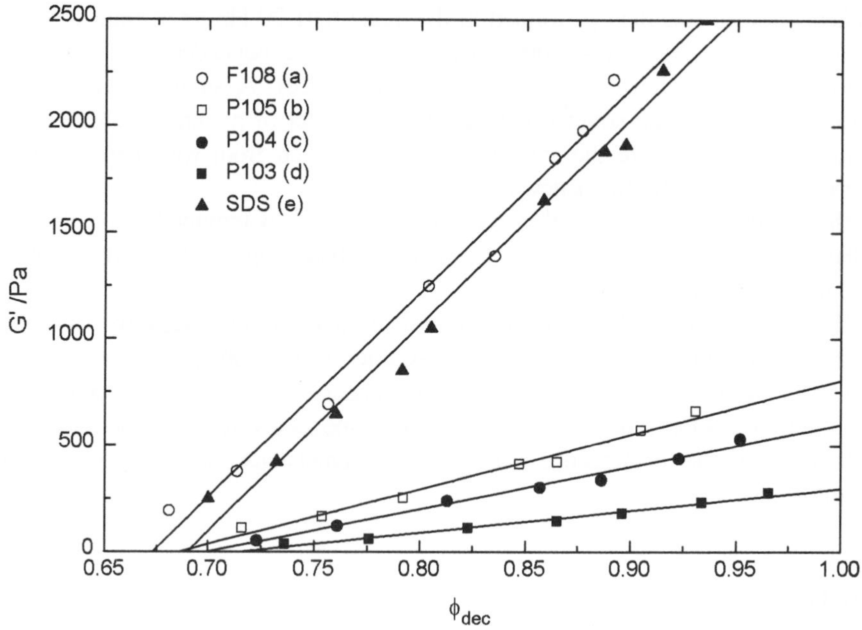

Figure 1: Storage moduli versus oil volume fraction for emulsions stabilized by the Synperonic surfactants and by SDS

The rheology of the emulsions varied considerably between the different stabilizers used. A typical set of results are shown in figure 1 in which storage modulus is plotted as a function of oil volume fraction for emulsions stabilized by the Synperonic PE surfactants. Most of the samples behave as viscoelastic solids, showing relatively little variation in modulus with applied frequency, as shown in an earlier publication [8]. All samples were highly elastic.The systems stabilized by F108 showed the largest moduli, reaching values of around 3000 Pa. The moduli fell with decreasing ethylene oxide with P103 stabilized systems showing the lowest moduli.. The plots appear linear at first sight, as has been found in earlier work, however, they are slightly curved upwards. Mason et al. [11] have recently reported a non-linear response in concentrated monodisperse emulsions. Monodispersed systems are expected to give non-linear behaviour [10,11], though polydispersity masks out this curvature to give a linear response. Similar plots were obtained for the systems stabilized by PVA, SDS and Atlox 4913.

An interesting variation between stabilizers was seen when the storage moduli of these emulsions were plotted as a function of applied strain for each of the highest oil volume fractions used. At low strains the moduli were independent of strain, the so-called linear viscoelastic region. At a certain strain, the distorted droplets are able to hop from one configuration to another of lower energy, allowing the strain to be relaxed

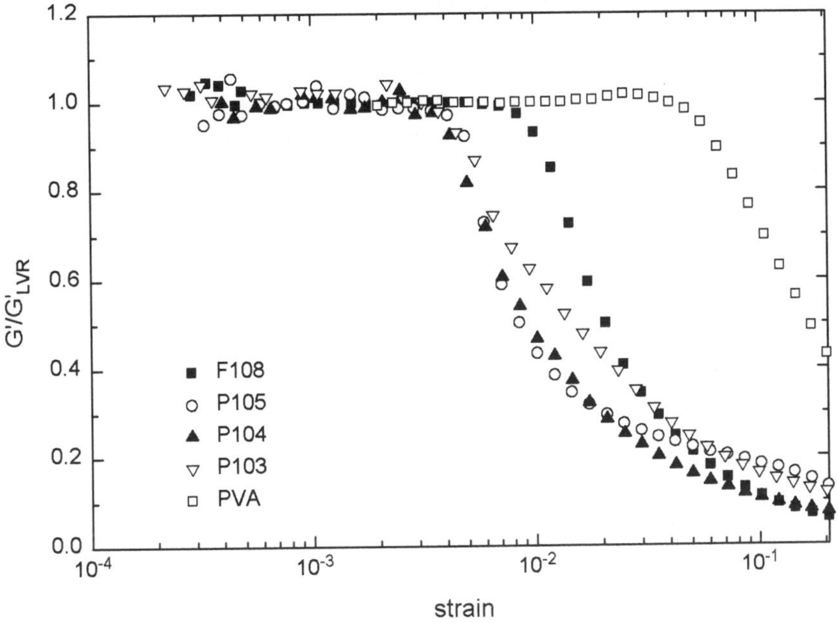

Figure 2: Moduli (normalized to the linear visco-elastic region value) for emulsions stabilized by the Synperonic surfactants and by PVA (each at the highest volume fraction used)

resulting in a reduced modulus. Above this critical strain, the emulsions became more liquid like with increasing strain. An increase in molecular mass of the emulsifier generally resulted in longer linear viscoelastic regions as shown by the Synperonic polymers in figure 2. The differences between these polymers were relatively small, but further support was given by the extended linear region shown by emulsions stabilized by PVA. This was also noted by Pons et al. [13] where the SDS and Atlox 4913 stabilizers were compared. The longer chain polymers may form a more coherent layer at the interface that is more resistant to deformation in that it can be strained further before the deformed droplets jump to a lower energy configuration. Other factors may also affect the critical strain, such as the droplet- droplet interaction potential or polydispersity in the droplet size [13,14]. The latter appears to be the case; Mason [11] found critical strains for SDS concentrated O/W emulsions of 0.08-0.1 at an effective volume fraction of 0.8 compared to 0.008 found in the current work.

The critical strain was also found to increase with increasing oil volume fraction for all of the stabilizers. This effect has been noted before and may arise from a change in the interaction potential with droplet separation. Alternatively, it may reflect the increasingly cooperative nature of the process by which the deformed droplets jump from one configuration to another.

Much of the work on concentrated emulsion rheology analyses the data in terms of Princen's model [2,4,8,13]. At low volume fractions the droplets are well separated and are undeformed. However, at a certain volume fraction (ϕ_0) the droplets become close-packed. At volume fractions above this point the droplets have to deform in order to pack. When a stress (or strain) is applied to these deformed emulsions the interfaces are dilated within the shear field and the energy (and hence modulus) associated with this dilation is dependent upon the interfacial tension and interfacial area. The area increases with decreasing droplet size and so the modulus varies inversely with size. Princen considered the deformation of a hexagonally close packed array of cylinders. Adapting this model to a three dimensional system, he obtained a semi-empirical equation relating the static shear modulus, G_0, to the droplet size (surface mean radius), interfacial tension, γ, and droplet volume fraction, ϕ. He obtained the following equation [2],

$$G_0 = \frac{A\ \gamma\ \phi^{1/3}}{R_{32}}(\phi - \phi_0) \qquad (1)$$

where A had a value of 1.76 and ϕ_0 (the maximum packing fraction of the undeformed droplets) a value of 0.71. The static shear modulus may be replaced by the elastic modulus provided it shows little frequency dependence. This treatment works well for polydisperse emulsions, however Mason et al. [11] have reported that the modulus of monodisperse emulsions scaled as $(G'R_{32}/\gamma)=A'\phi(\phi-\phi_0)$ in contrast to the $\phi^{1/3}(\phi-\phi_0)$ dependence found by Princen. Plots of $G'R_{32}/\phi^{1/3}\gamma$ (Princen) and $G'R_{32}/\phi\gamma$ (Mason) against ϕ are shown in figures 3 and 4. The use of the Mason scaling slightly reduced the curvature in the scaled plots and so was a slightly more accurate representation of the data. The emulsions were, by no means, monodisperse and it was surprising that the Mason scaling law fitted the data so well. The values obtained for A and A' and ϕ_0 for both scalings are listed in table 2.,

Table 2: Parameters obtained from the Princen or Mason scaling relationships for the various emulsions.

Emulsion	Stabilizer	Princen		Mason	
		A	ϕ_0	A'	ϕ_0
A	F108	1.49	0.661	1.41	0.669
B	P105	1.38	0.665	1.37	0.677
C	P104	1.49	0.680	1.48	0.694
D	P103	1.39	0.695	1.40	0.708
E	SDS	1.22	0.673	1.33	0.704
F	PVA	1.38	0.669	1.56	0.665
G	Atlox	1.66	0.644	1.48	0.649
H	F108	1.46	0.636	1.40	0.649

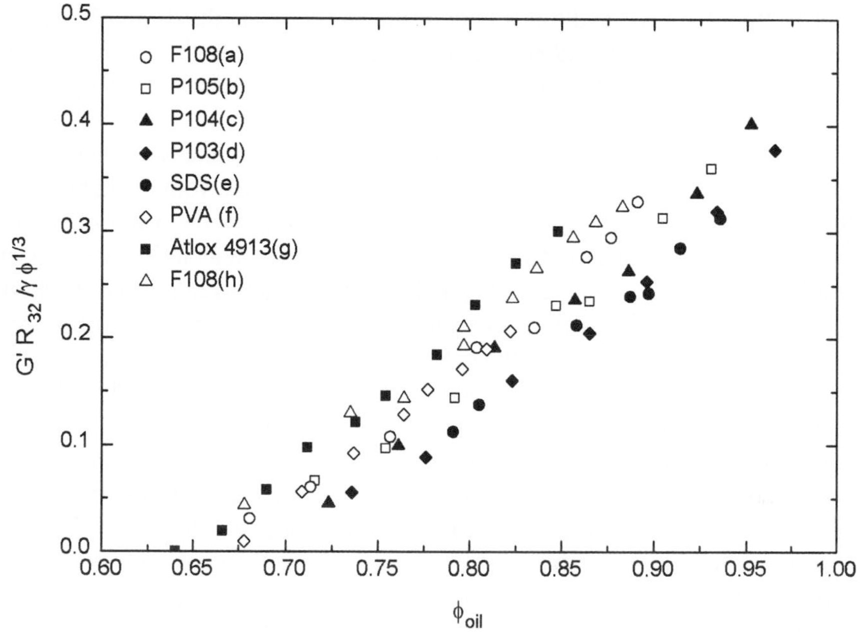

Figure 3: Storage moduli of the emulsions as a function of volume fraction scaled according to Princen's equation.

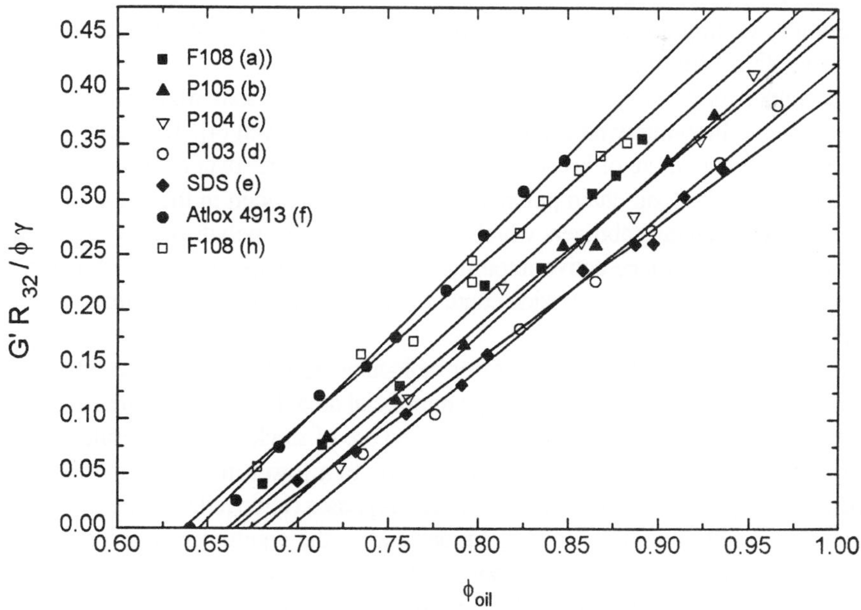

Figure 4: Storage moduli of the emulsions as a function of volume fraction scaled according to Mason's equation.

It is clear from figures 3 and 4 that both treatments correlate the data well, with only minor shifting of the curves along the volume fraction axis. Both A and A' lie in the region of 1.2-1.6, which in view of the errors in the experiment are not significantly different.

These results suggest that the rheology of concentrated emulsions is universally governed by the interfacial tension, droplet size and effective volume fraction. There appears to be little molecular mass effect in the parameter A, the major effect of the stabilizer appears to be in the extent of the linear viscoelastic region. The term effective volume fraction refers to the fact that both the adsorbed stabilizer layer on the droplets and the thin aqueous liquid film between the compressed droplets contribute to the volume fraction of the emulsion. The extent of this effect will depend upon the ratio of the thickness of the stabilizer layer and the thin film to the droplet radius. The effect is relatively small (but measurable) in large emulsions such as those used here (as shown by the shifting of the curves to higher or lower volume fractions), and the values listed in table 2 show that the longer chain molecules, such as PVA or F108 generally result in smaller values of ϕ_0. Increasing the molecular weight of the Synperonic polymers resulted in progressively lower critical volume fractions. This decrease was too large too be accounted for by the thickness of the polymer layer alone showng that the thickness of the thin liquid film was, indeed, a contributing factor. Changing the droplet radius appears to have little effect on the slope of the plots. The two results for Synperonic F108 both show the same slope, the only difference being that the curve for the smaller droplet emulsion, h, (radius 1.05μm compared to that of 2.05μm for emulsion a) is shifted to lower volume fractions in agreement with the concept of the polymer + thin film thickness to droplet radius ratio.

The values for the slope A from the Princen equation are somewhat lower than those reported in the literature where values as high as 13 are given [4]. The values obtained here apparently are more or less independent of the stabilizer over a wide range of type (ionic surfactant to long chain polymer). The differences seen in the value of A lie in the use of commercial stabilizers which show considerable polydispersity in molecular mass and possibly surface active impurites [15]. In the Synperonic series for instance, molecules with lower ethylene oxide content will be more surface active than those with greater contents.

In an emulsion the effect of the lower molecular weight molecules and the surface active impurities is diluted over the large interfacial area. However, the area in a typical interfacial tension measurement is small, and so use of stock solutions results in higher interfacial concentration of the impurites leading to a lower interfacial tension. In the experiments reported here, the emulsion aqueous phases (removed with the droplets still intact) were used rather than stock solutions and so the equilibrium concentrations of the impurities were present giving a more representative value for the interfacial tension than may be otherwise measured. For instance, the emulsion aqueous phase for F108 gave a tension of $14.4 mNm^{-1}$ whilst a fresh solution gave a value of the order of 4-$5 mNm^{-1}$.

It is clear that the rheology of emulsions may be controlled by the judicious use of surfactants, polymers or their mixtures. In a series of experiments the effect of presence of a non-ionic surfactant (Synperonic A7) in emulsions stabilized by Atlox 4913 was studied [16]. The presence of the surfactant always resulted in a lower modulus compared to that found for the polymer alone, figure 5. The surfactant adsorbs more readily than the polymer at the interface resulting in a lower interfacial tension (off-setting any reduction in size of the droplets due to the use of a mixed emulsifier system).

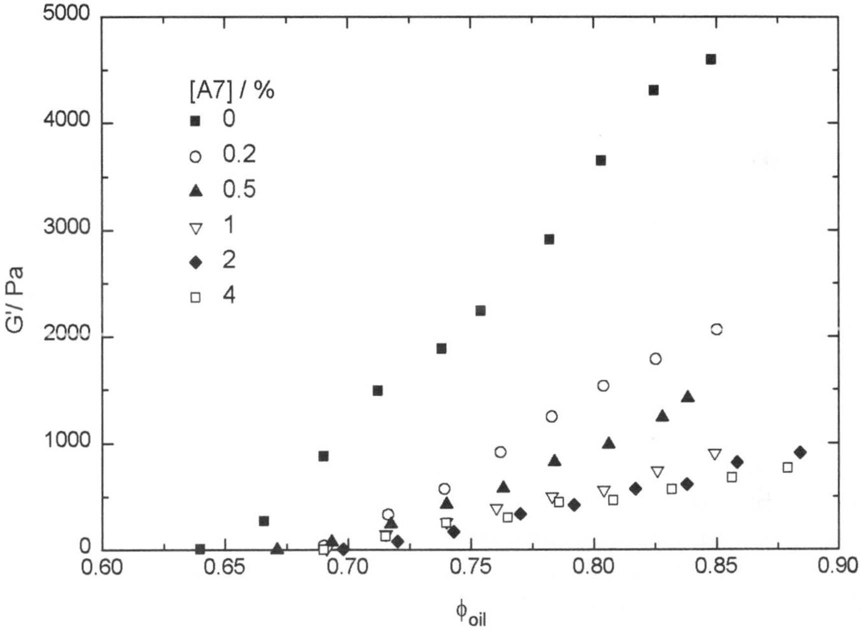

Figure 5: Plots of the storage moduli as a function of oil volume fraction for emulsions stabilized by Atlox 4913/ Synperonic A7 mixtures.

Due to the partitioning of the A7 between the two phases (an accurate value of the interfacial tension could not be obtained) it was impossible to carry out a full analysis of the data in terms of the Princen or Mason scaling. Synperonic A20 (which contains on average 20 ethylene oxide units per molecule rather than the 7 in A7) gave similar results to Synperonic A7 but the reduction in moduli was less. The latter is less surface-active than the A7 due to its greater water solubility and reduces the interfacial tension to a lesser extent. These results (and others reported in the literature for mixtures of Atlox 4913 and SDS and for PVA and sodium dodecyl benzene sulphonate) demonstrate that the rheology of concentrated emulsions may be controlled by the use of surfactant mixtures [13,8].

3. Conclusions

The rheology of concentrated O/W emulsions may be approximated by either Princen's or Mason's approaches. Both analyses suggest that the slope of the reduced modulus versus volume fraction plots is independent of the emulsifier. The variation in stabilizer leads to differences in the extent of the linear viscoelastic region and the oil volume fraction at which the droplets come into contact and begin to deform. The importance of the use of the emulsion aqueous phase to determine the interfacial tension is highlighted by the results. The rheology of concentrated emulsions is readily controlled by the use of mixed emulsifiers or by the judicious use of single emulsifiers.

Acknowledgements:

R.P. acknowledges the Human Capital Mobility Programme of the EC (contract no. ERB CHB ITC 920177). Th.F.Tadros is thanked for useful discussions. We remember the late Helene Moysan and appreciate her carrying out some of the experiments.

References

1. Princen, H. M. (1979) *J. Colloid Int. Sci.*, **71**, 201
2. Princen, H. M. and Kiss, A. D. (1986) *J. Colloid Int. Sci.*, **112**, 427
3. Schwartz, L. W. and Princen, H. M. (1987) *J. Colloid Int. Sci.*, **118**, 201
4. Pons, R. P., Solans, C. and Tadros, Th. F. (1995) *Langmuir*, **11**, 1966
5. Otsubo, Y. and Prud'homme, R. K. (1994) *Rheol. Acta*, **33**, 29
6. Otsubo, Y. and Prud'homme, R. K. (1994) *Rheol. Acta*, **33**, 303
7. Ravey, J. C., Stebe, M. J. and Sauvage, S. (1994) *Colloids Surf.*, **91**, 237
8. Taylor, P. (1996) *Colloid Polym. Sci.*, **274**, 1061
9. Pons, R. P., Erra, P., Solans, C., Ravey, J. C., and Stebe, M. J. (1993) *J. Phys. Chem.*, **97**, 12320
10. Buzza, D. M. and Cates, M. (1993) *Langmuir*, **9**, 2264
11. Mason, T. G., Bibette, J., and Weitz, D. A. (1995) *Phys. Rev. Lett.*, **75**, 2051
12. Lacasse, M. D., Grest, G. S., Levine, D., Mason, T. G. and Weitz, D. A. *Phys. Rev. Lett*, in press
13. Pons, R. P., Taylor, P. and Tadros, Th. F., accepted for publication in *Colloid Polym. Sci.*
14. Pons, R, P., Rossi, P. and Tadros, Th. F. (1995) *J. Phys. Chem.*, **99**, 12624
15. Hancock, R. I. (1984) in Tadros, Th. F. (ed.), *Surfactants*, Academic Press London, pp. 287-322
16. Pons, R. P., Taylor, P., Moysan, H. M. and Tadros, Th. F., in preparation

INTERACTION BETWEEN AN OIL DROP AND A PLANAR SURFACE IN AN AQUEOUS ENVIRONMENT

M. S. NADIYE-TABBIRUKA, R. H. OTTEWILL,
School of Chemistry, University of Bristol,
Bristol BS8 1TS, U. K.

A. LIPS,
Unilever Research, Colworth House, Bedford,
Beds. MK44 1LQ, U. K.

Image analysis techniques have been used to investigate the interaction of a macroscopic oil drop (dodecane) with a molecularly smooth mica surface in an aqueous environment containing a surfactant. Results are described for experiments using hexadecyltrimethylammonium bromide as the surfactant at concentrations below and above the critical micelle concentration.

1. Introduction

A basic question of fundamental interest is whether interaction of an oil-in-water emulsion drop with a solid surface leads to flocculation, with consequential breaking of the emulsion and wetting of the surface with oil, or whether the oil drop retains its stability with an aqueous film between the drop and the surface. In either case it is important to understand the nature of the forces which control the interactions and explain the phenomena observed. In view of the ubiquitous occurrence of emulsions in this situation and whether stability or breakage is required, this problem is relevant to many areas of research, such as, oil recovery from rock strata [1], separation of minerals using oil droplets [2], road surfacing with bitumen emulsions [3], the preparation and storage of food emulsions [4] and many others.

The purpose of the present work was to investigate the interaction between a macroscopic droplet of dodecane in an aqueous phase containing hexadecyltrimethylammonium bromide, HTAB, and a molecularly smooth mica surface. Measurements using a computerised image analysis technique [5] have allowed the thickness and stability of the thin aqueous film to be assessed as a function of the HTAB concentration in the aqueous phase. This method provided an advance on the previous procedure which used a video camera to examine the dynamics of wetting [6].

Although extensive research has been carried out on the stability of

R. H. Ottewill and A.R. Rennie (eds.), Modern Aspects of Colloidal Dispersions, 235–246.

surfactant foam films [7] and on the interaction of air bubbles with surfaces [8], the interaction between oil droplets and solid surfaces appears to have been much less thoroughly investigated.

A point of major interest was the behaviour above the c.m.c. of HTAB when interaction occurs between the dodecane drop and the mica plate in the presence of charged micelles. In this region HTAB is known to form non-spherical micelles [9].

2. Experimental

2.1 MATERIALS

All water used was double distilled. Dodecane was Fluka Puriss grade material (>99.5%) which was filtered through a Nucleopore filter before use.

HTAB was obtained from the Sigma Chemical Company and was recrystallised from acetone and ethanol and then dried in a vacuum desiccator before use.

The mica used was a high quality muscovite mica supplied by the British Mica Company. A small plate ca. 1 x 3 cm was cut from a large sheet. Then from the small plate a fresh sheet was cleaved from the middle of the plate. This process was carried out under water and gave a clean molecularly-smooth surface as shown in previous work [6,10,11].

2.2 MEASUREMENT OF AQUEOUS FILM THICKNESS

The equipment used for simultaneously studying the drainage dynamics and obtaining the equilibrium film thickness was similar in design to that used previously [5]. However, the video recording system was replaced by a CCD camera which was focussed on the fringes formed by the film between the curved surface of the oil drop and the planar surface of the mica, using a low power objective and an eyepiece. The resulting analogue voltage from the light sensors of the CCD camera was scanned and converted into a digital signal by a frame grabber in a personal computer and then displayed on a calibrated monitor with a display area of 512 x 512 pixels. Since the digital voltage was proportional to the grey levels registered by the camera the light intensity on the fringe pattern was also proportional to the grey levels [4]. A schematic diagram of the interphase region and a sample of the fringes observed is shown in Figure 1.

2.3 INTERFACIAL TENSION MEASUREMENTS

Measurements of surface tension at the air-water interface were made using a Du Nouy tensiometer and those at the oil-water interface using a spinning drop-tensiometer [12].

2.4 ELECTROKINETIC MEASUREMENTS

The electrophoretic mobilities of the dodecane droplets were measured using a Pen Kem 3000 electrophoretic analyser. Zeta-potentials, ζ in mV, were calculated using the Smoluchowski equation in the form,

$$\zeta = 1.177 \times 10^9 u \quad (1)$$

with u the mobility expressed in $m^2V^{-1}s^{-1}$ at 30°C.

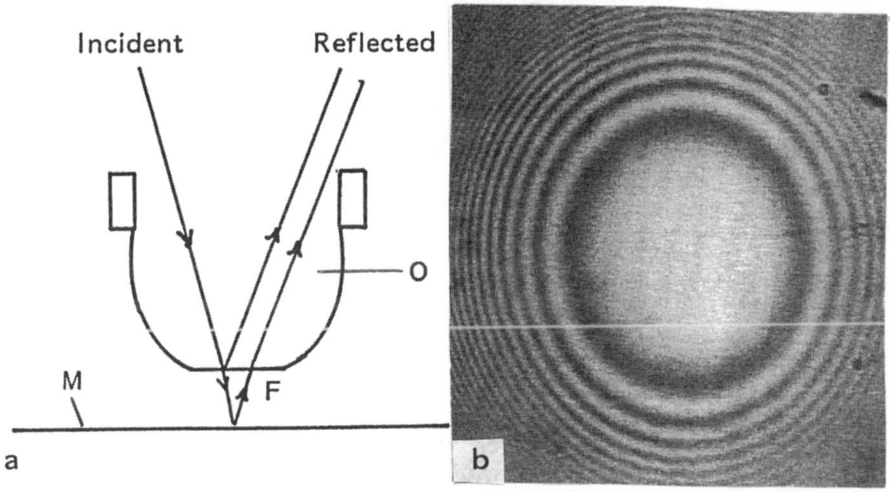

Figure 1: a) Sketch of interphase region showing oil droplet, O, aqueous film, F, and mica plate, M;
b) Fringes from the interphase region.

3. Results

3.1 ADSORPTION OF HTAB AT THE DODECANE-WATER AND AIR-WATER INTERFACES

Figure 2 shows the results for the interfacial tension of aqueous HTAB solutions against air at 25.0°C [13] and against dodecane at 30.0°C; the Krafft temperature of HTAB was taken as ca 28°C [14]. The critical micelle concentrations, c.m.c., against air was found to be 9.20×10^{-4} mol dm^{-3} and against dodecane 6.31×10^{-4} mol dm^{-3} ; these values are in good agreement with those in the literature [15]. The small difference in values could well be an indication of a small solubility of HTAB in dodecane.

A major effect is that over the concentration range from 10^{-6} mol dm^{-3} to the c.m.c, the interfacial tension at the oil-water interface is substantially lower than that at the air-water interface. Hence, the capillary pressure (section 4.3) exerted by an oil drop on the film is always lower than that of an air bubble. This indicates the possibility of thicker aqueous films being formed with oil drops than with air bubbles.

The surface excesses of HTA^+ ions close to the cmc were calculated from the Gibbs' equation in the form [16,17,18],

$$-d\gamma = 2\Gamma_{HTA+}RT(\ 1\ -\ 0.57\sqrt{C_{HTAB}}\ )\ d\ ln\ C_{HTAB} \qquad (2)$$

with C_{HTAB} the concentration of the HTAB solution in mol dm^{-3}. The values of Γ_{HTA+} close to the c.m.c. were found to be 2.54×10^{-6} mol m^{-2} and 2.08×10^{-6} mol m^{-2} at the air-water and oil-water interfaces respectively. This gave 0.66 nm^2 and 0.80 nm^{-2} for the areas per charged head group.

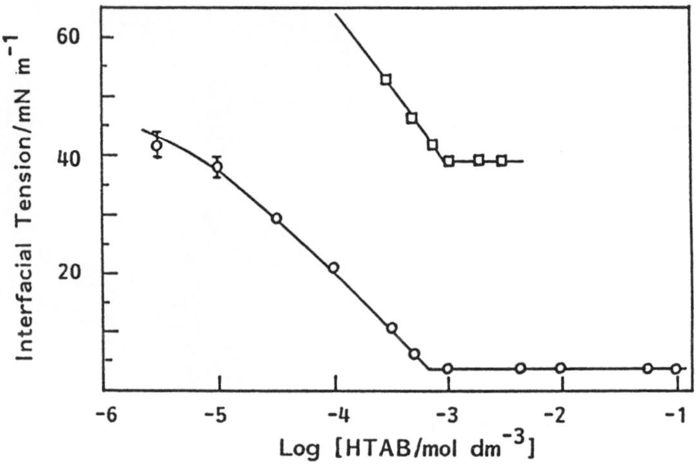

Figure 2: Interfacial tensions at the air-water and oil-water interfaces as a function of HTAB concentration. □, air-water; O, oil-water.

3.2 A COMPARISON OF THE ELECTROKINETIC PROPERTIES OF MICA-WATER AND DODECANE-WATER INTERFACES

The results obtained for the zeta-potentials at the dodecane-water interface and the mica-water interface as a function of HTAB concentration are shown in Figure 3. Those for the mica-water system are from earlier work [19] and indicate that charge reversal of the surface, as a consequence of HTA^+ adsorption, occurs at 5.62×10^{-6} mol dm^{-3} HTAB. In the case of the dodecane-water interface this occurs at 1.78×10^{-7} mol dm^{-3}. Thus beyond 5.62×10^{-6} mol dm^{-3} both surfaces become positively charged.

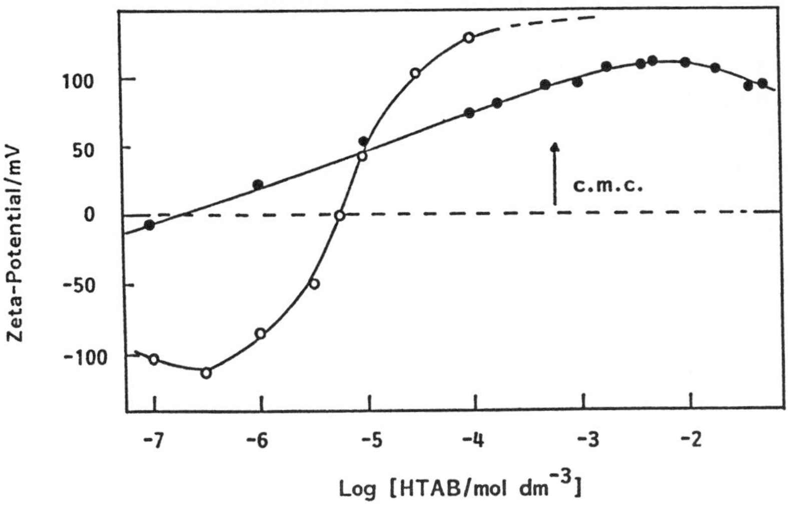

Figure 3: Zeta-potentials at the mica-water interface and the dodecane-water interface as a function of HTAB concentration. O, mica-water; ●, dodecane-water.

3.3 AQUEOUS FILM THICKNESS MEASUREMENTS BELOW THE C.M.C.

The results showing the thickness of the aqueous film formed over a range of HTAB concentration from 10^{-6} mol dm^{-3} to 10^{-3} mol dm^{-3} are shown in Figure 4.

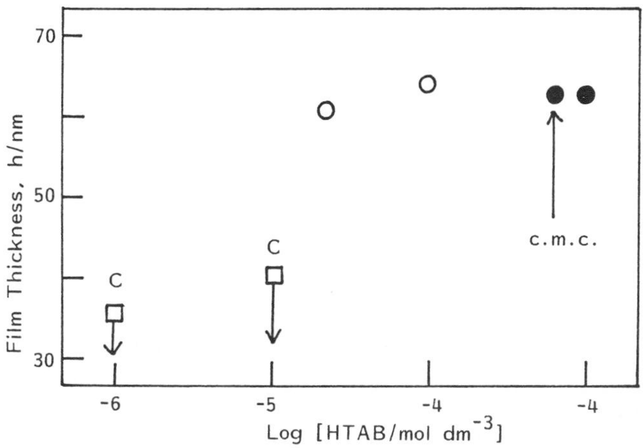

Figure 4: Aqueous film thickness, h/nm, against log[HTAB/mol dm^{-3}]; □, unstable film; O, metastable film; ●, stable film; ↓ C, collapsing film.

At a concentration of 10^{-6} mol dm^{-3} HTAB initially a film of ca. 35 nm thickness was formed, which gave a bright image in reflection. However, after about 10 s the bright image became dark and the film collapsed; this was considered to be an unstable film.

A similar behaviour was observed at 10^{-5} mol dm^{-3} HTAB where the film also had a short lifetime of stability, of the order of 10 s, during which time it was possible with the image analysis system to "capture" an image of the fringes and to obtain a thickness measurement of ca. 40 nm, before collapse occurred. This was also considered to be an unstable film.

At HTAB concentrations of 2 x 10^{-5} and 10^{-4} mol dm^{-3} films were formed with thicknesses of 60$\pm$2 nm and 64$\pm$2 nm respectively. These were quite stable with a low capillary pressure from the drop on the film where the capillary pressure, P_γ, was defined by [20]:

$$P_\gamma = 2\gamma_{ow}R/(R^2 - r^2) \tag{3}$$

with γ_{ow} the interfacial tension at the dodecane-water aqueous HTAB solution interface, R the internal radius of the drop holder and r the radius of the planar region of the film (see Figure 1a). Provided that R>>r this reduces to,

$$P_\gamma = 2\gamma_{ow} /R \tag{4}$$

At 10^{-4} mol dm^{-3} the film formed was stable but if the applied pressue was increased manually the film collapsed. This suggested that the internal stabilising pressure in the film had increased with increase in HTAB concentration. The films in this region were considered to be metastable in a similar manner to those described by Kolarov et al [21] for interaction of an air bubble with a solid surface under similar conditions.

In the region of the c.m.c. (6.31 x 10^{-4} mol dm^{-3}) the aqueous film thickness remained at 62$\pm$2 nm. The films in this region were very stable and did not collapse at the maximum capillary pressure, 4.9 N m^{-2}, which could be applied.

3.4 AQUEOUS FILM THICKNESS ABOVE THE CMC

The thickness of the aqueous films formed when the concentration of HTAB in the aqueous phase was above the c.m.c. are plotted in Figure 5 as a function of the concentration of HTAB present in the micellar form, namely as [C_T - C_0] with C_T the total concentration of surfactant and C_0 the c.m.c. All the films in this region were very stable and the film thickness increased over the region from 60$\pm$2 nm at 10^{-3} mol dm^{-3} to 130$\pm$2 nm at 0.05 mol dm^{-3} HTAB. These results indicate that the internal pressure in the film increases as the concentration of the micellar species increases. As can be seen from Figure 2 the interfacial tension at the dodecane-water interface above the c.m.c. is 3.5 mN m^{-1} and this remains constant with increase in micellar concentration; hence, the capillary pressure exerted in this region is essentially constant at 4.7 N m^{-2}. However, since the concentration of co-ions increases with the concentration of HTAB this means that together with a small contribution from the counter-ions [22] the ionic strength is increasing over the region.

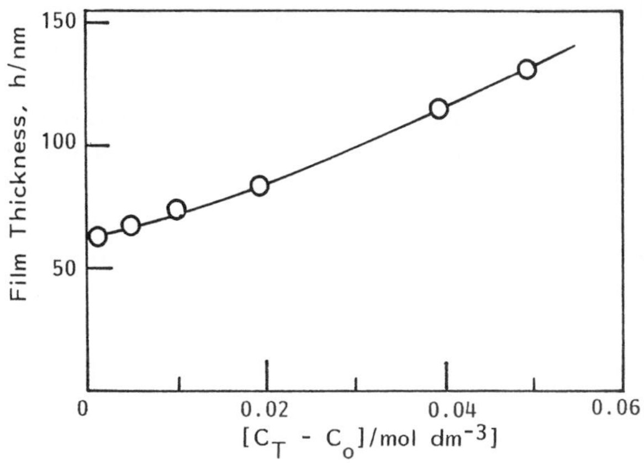

Figure 5: Film thickness, h/nm, against concentration of micellar species, C_T -C_0.

A further series of experiments was carried out at an HTAB concentration of 0.005 mol dm^{-3} in which the film thickness was measured at different applied capillary pressures. The results are illustrated in Figure 6.

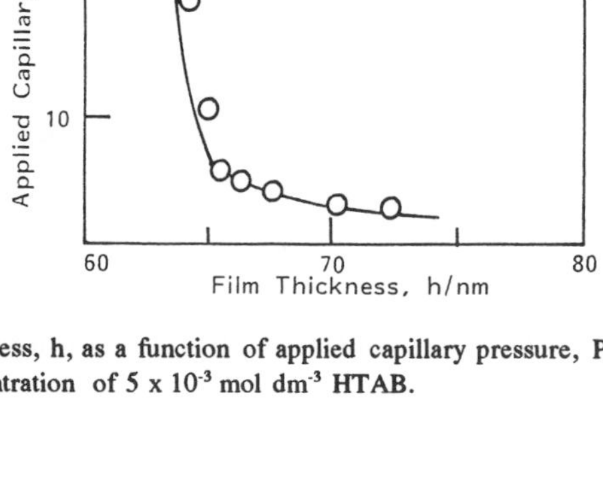

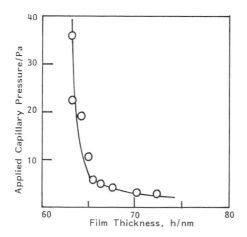

Figure 6: Film thickness, h, as a function of applied capillary pressure, P_γ, at a concentration of 5 x 10^{-3} mol dm^{-3} HTAB.

4. Some Theoretical Considerations

4.1 ELECTROSTATIC INTERACTIONS

In the present work it was established that when the system came to equilibrium the central region of the dodecane droplet was planar; this was confirmed from the fringe pattern. Hence, the system examined consisted of a planar dodecane-water-mica interphase region. In the presence of solutions of HTAB each interface has a charge determined by the surfactant concentration. As a consequence of the charge each surface has a surface potential and these can either be the same or different. The potential at the mica-water interface will be designated ψ_1 and that at the dodecane-water interface ψ_2. The distance between the interfaces will be designated h, so that schematically the potential profile will be as shown in Figure 7. Both potentials can be positive or both can be negative; also, one can be positive and the other negative.

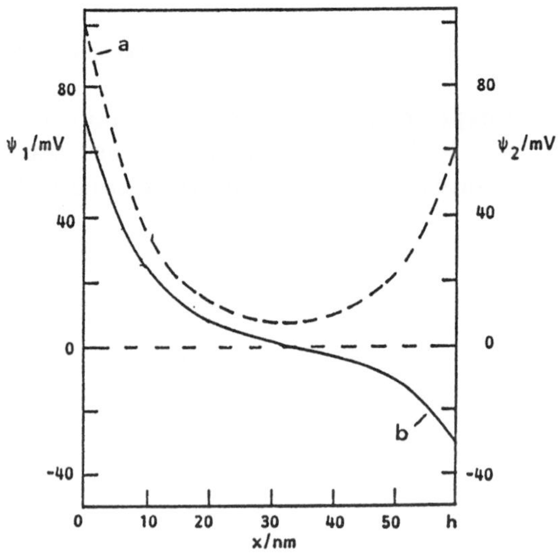

Figure 7: Potential profile between two interfaces:-
a) both potentials positive b) one positive and one negative.

For this situation the variation of potential with distance, x, across the film is given by [23]:-

$$\psi(x) = \psi_1 \cosh(\kappa x) + [(\psi_2 - \psi_1 \cosh(\kappa h))/\sinh(\kappa h)] \sinh(\kappa x) \quad (5)$$

For the purpose of the present article the electrostatic potential energy of interaction will be used in the form given by Hogg, Healy and Fuerstenau [24] in the form:-

$$V_{el}(h) = \epsilon_r\epsilon_0\kappa\ [(\psi_1^2 + \psi_2^2)\text{cosech}(\kappa h) - 2\psi_1\psi_2\text{cosech}(\kappa h)]/2 \qquad (6)$$

with ϵ_r the relative permittivity of water, ϵ_0 the permittivity of free space and κ the Debye-Hückel electrical double layer "thickness" parameter given by,

$$\kappa^2 = 2n_0v^2e^2\ /\epsilon_r\epsilon_0\ kT \qquad (7)$$

with n_0 = the number of ions of each type per unit volume, v = the basic ion charge, e = the fundamental charge on the electron, k = Boltzmann's constant and T = the absolute temperature.

It follows from equation (6) which gives the energy of interaction per unit area of plate, that the pressure due to electrostatic interaction is given by,

$$P_{el} = (\ dV(h)/dh)_{T,P}$$

$$= -\epsilon_r\epsilon_0\kappa^2[(\psi_1^2 + \psi_2^2)\text{cosech}(\kappa h) - 2\psi_1\psi_2\text{coth}(\kappa h)]/2\sinh(\kappa h) \qquad (8)$$

Equation (8) is based on using the linear approximation of the Poisson-Boltzmann equation so will not be completely accurate at predicting P_{el} at high surface potentials. However, a comparison [24] with the full computer solutions of Devereaux and de Bruyn [25] indicated that the deviations were acceptable for a semi-quantitative analysis.

4.2 LONG RANGE ATTRACTION

Following Verwey and Overbeek [23] the potential energy of the long-range attraction between the dodecane droplet and the mica plate will be taken as,

$$V_A(h) = -\ A_c/12\pi h^2 \qquad (9)$$

with A_c the composite Hamaker Constant given by,

$$A_c = (\sqrt{A_{11}} - \sqrt{A_{22}})(\sqrt{A_{33}} - \sqrt{A_{22}}) \qquad (10)$$

with A_{11}, A_{22} and A_{33} respectively the Hamaker Constants for mica, water and dodecane. These were taken as $1.0\text{x}10^{-20}$ J, $3.7\text{x}10^{-21}$ J and $5.0\text{x}10^{-20}$ J [26] which gave a value for A_c of $3.87\text{x}10^{-21}$ J; it follows therefore that A_C is positive. Since in the present work the attractive effect was small, retardation effects [23] were not considered.

From equation (9) the attractive pressure, $P_A(h)$, is given by,

$$P_A(h) = A_c/6\pi h^3 \qquad (11)$$

4.3 TOTAL PRESSURE EXERTED BY THE FILM

Considering only the electrostatic and attractive pressures gives for the total internal pressure within the film,

$$P_T = P_{el}(h) + P_A(h) \tag{12}$$

The pressure applied to the film by the droplet is essentially the capillary pressure, P_γ, as defined above. In the absence of other interactions within the film then,

$$P_\gamma = P_T \tag{13}$$

5. Discussion

5.1 AQUEOUS FILM BEHAVIOUR BELOW THE C.M.C

In the HTAB concentration range between 10^{-6} and 6.3×10^{-4} mol dm^{-3} several changes of film behaviour were observed (Figure 4).

Between 10^{-6} and 10^{-5} mol dm^{-3} films were formed but these only existed for a short time and then collapsed; this led to wetting of the mica substrate by the dodecane. From the results in Figure 3 it is clear that the dodecane-water interface in this region has a small positive charge; this is consistent with the interfacial tension data (Figure 3) which shows only a small lowering of the dodecane-water interfacial tension by HTAB. On the other hand the mica-water interface is negatively charged up to 5.62×10^{-6} mol dm^{-3}. This combination provides an electrostatic attraction between the two surfaces in this region.

At 10^{-5} mol dm^{-3} both surfaces have a net positive zeta-potential of ca. 40 mV which, in principle, should lead to film stability. However, instability was observed. It is possible that in this region, by analogy with other surfaces[27], that adsorption of the cationic surfactant could lead to mixed adsorption states, with some HTA^+ molecules head group up and some with the hydrocarbon tail up; unfortunately, adsorption isotherms of HTAB on molecularly smooth mica do not seem to be available. Thus patchy adsorption of HTAB could contribute to the instability of the film.

The changing positive zeta-potentials with concentration of HTAB between 10^{-5} and 10^{-4} mol dm^{-3} would seem to provide a reasonable explanation of the formation of metastable films in this region.

Between 10^{-4} mol dm^{-3} and the c.m.c., as judged by the zeta-potential data, the positive charge at both the mica-water and the dodecane-water increases more slowly; this was also observed by Churaev et al [28] with silica surfaces. At the c.m.c. saturation adsorption of both surfaces appears to have occurred. The interfacial tension data suggest a loose monolayer (area per chain = 0.80

nm^2) at the dodecane-water interface but whether there is a bilayer on the mica or a closely packed monolayer with some molecules head-up and some head-down is again not known.

The fact that both surfaces are highly positively charged in the c.m.c. region is consistent with strong electrostatic repulsion and film stability. In addition, the decreasing interfacial tension of the dodecane-water interface to 3.55 mN m^{-1} at the c.m.c. reduces the external capillary pressure on the film. It should also be noted that the long-range van der Waals' attraction is quite small at distances of the order of the film thickness.

Overall, it appears reasonable to conclude therefore that at HTAB concentrations up to the c.m.c. the film stability is dominated by electrostatic interaction between the surfaces.

5.2 AQUEOUS FILM BEHAVIOUR ABOVE THE C.M.C.

It is clear from Figure 3 that above the c.m.c. both the mica-water and the dodecane-water interfaces have high positive zeta-potentials, of the order of 140 mV and 100 mV respectively. Moreover, as a consequence of the formation of micelles, essentially only the co-ions contribute to the increase in ionic strength. Thus strong electrostatic repulsive forces between the two basic surfaces continue to operate throughout the region, but would be expected to decrease with ionic strength leading to a decrease in film thickness.

The near linear increase in film thickness with micellar concentration shown in Figure 5 is an interesting feature of the results. Also, the results shown in Figure 6 appear to indicate an exponential dependence of film thickness on capillary pressure, which at the larger distance is greater than expected from equation (12). This indicates that the electrostatic repulsive forces acting between the cationically charged micelles also act to thicken the film. The additional repulsive force is osmotic in nature and strongly dependent on the second virial coefficient, B, of the micellar species, as usually written in the form,

$$\Pi = RT(C_T - C_0)/M_{mic} + B(C_T - C_0)^2 \qquad (14)$$

where Π is the osmotic pressure and M_{mic} the micellar weight. This gives for the internal pressure above the c.m.c.,

$$P_T = \Pi + P_{el}(h) + P_A(h) \qquad (15)$$

This effect has been termed by Napper [29] depletion stabilisation since it provides an additional stabilising force in a situation where the gap between the surfaces is much greater than the size of the micelles and the molecular nature of both macroscopic surfaces is similar to that of the micelles. This aspect will be discussed in more detail in a later publication.

6. ACKNOWLEGEMENTS

We offer our sincere thanks to Dr Tony Makepeace for considerable help with the image analysis procedures. We also acknowledge with thanks support from the DTI Colloid Technology Project, including ICI, Schlumberger, Unilever and Zeneca.

7. REFERENCES

1. Anderson, W. G. (1986) *J. Petroleum Technology*, **38**, 1123-1144.
2. Sutherland, K.L. and Wark, I.W. (1955) *Principles of Flotation*, Australian Inst. Min. and Met.
3. Lane, A.R. and Ottewill, R.H. (1976) in Smith, A.L. (ed.) *Theory and Practice of Emulsion Technology*, Academic Press, London, 157-177.
4. Dickinson, E. and Stainsby, G. (1982) *Colloids in Food*, Applied Science Publishers, London.
5. Makepeace, A. and Nadiye-Tabbiruka, M.S. (1997) to be published.
6. Debacher, N.A. and Ottewill, R.H. (1991) *Colloids and Surfaces*, **52**, 9-161.
7. Buscall, R. and Ottewill, R. H. (1975) in Everett, D.H. (ed.) *Colloid Science*, The Chemical Society, London, **2**, 191-245.
8. Fisher, L.R., Mitchell, E.E., Hewitt, D. Ralston, J. and Wolfe, J. (1991) *Colloids and Surfaces*, **52**, 163-174.
9. Debye, P. and Anacker, E.W. (1951) *J. Phys. and Colloid Chem.*, **55**, 644-655.
10. Israelachvili, J.N. and Adams, G.E. (1978) *J. Chem. Soc. Faraday Trans.* I, **74**, 975-1001.
11. Lubetkin, S.D., Middleton, S.R. and Ottewill, R.H. (1984) *Phil. Trans. Roy. Soc. London*, **A311**, 353-368.
12. Newton, R., (1984) Ph. D. Thesis, University of Bristol.
13. Debacher, N.A. (1991) Ph. D. Thesis, University of Bristol
14. Adam, N.K. and Pankhurst, K.G.A. (1946) *Trans. Faraday Soc.*, **42**, 523-526.
15. Mukerjee, P. and Mysels, K.J. (1971) *Critical Micelle Concentrations of Aqueous Surfactant Systems*, Nat. Stand. Ref. Data Ser., Nat. Bur. Stand. (US) 1-222.
16. Cockbain, E.J. (1954) *Trans. Faraday Soc.*, **50**, 872-881.
17. Pethica, B.A. (1954) *Trans. Faraday Soc.*, **50**, 413-421.
18. Downes, N., Ottewill, G.A. and Ottewill, R.H. (1995) *Colloids and Surfaces*, **102**, 203-211.
19. Debacher, N.A. and Ottewill, R.H. (1992) *Colloids and Surfaces*, **65**, 51-59.
20. Scheludko, A., Platikanov, D. and Manev, E. (1965) *Discuss. Faraday Soc.* **40**, 253-265.
21. Kolarov, T., Yankov, R., Esipova, N.E., Exerowa, D. and Zorin, Z.M., (1993) *Colloid and Polym. Sci.*, **271**, 519-520.
22. Beresford-Smith, B., Chan, D.Y.C. and Mitchell, D.J. (1985) *J. Colloid Interface Science*, **105**, 216-234.
23. Verwey, E.J.W. and Overbeek, J.Th.G. (1948) *Theory of Stability of Lyophobic Colloids*, Elsevier, Amsterdam.
24. Hogg, R., Healy, T.W. and Fuerstenau, D.W. (1966) *Trans. Faraday Soc.*, **62**, 1638-1651.
25. Devereux, O.F. and de Bruyn, P.L. (1963) *Interaction of Plane-Parallel Double Layers*, M.I.T. Press, Cambridge, Mass. 1-361.
26. Hough, D.B. and White, L.R. (1980) *Advances in Coll. Int. Sci.*, **14**, 3-41.
27. Ingram, B.T. and Ottewill, R.H. (1991) in Rubingh, D.N. and Holland, P.M. (eds.), *Cationic Surfactants*, Marcel Dekker Inc., New York, 87-140.
28. Churaev, A.P., Ershov, A.P., Esipova, N.E., Iskadarjan, G. A., Madjarova, E.A., Sergeeva, I.P., Sobolev, V.D., Svitova, T.F., Zakarova, M.A., Zorin, Z. M., Poirier, J.E., (1994) *Colloids and Surfaces A*: **91**, 97-112.
29. Napper, D. H. (1983) *Polymeric Stabilisation of Colloidal Dispersions*, Academic Press, London.

INVESTIGATIONS OF THE INTERACTION BETWEEN SUSPENSIONS AND EMULSIONS (SUSPOEMULSIONS)

THARWAT TADROS, JULIA CUTLER, RAMON PONS and PASCALE ROSSI
Zeneca Agrochemicals, Jealott's Hill Research Station, Bracknell, Berkshire RG42 6ET, U.K.

The interaction between suspensions and emulsions (suspoemulsions) was investigated using rheological and sediment volume measurements. Practical suspensions and emulsions of agrochemicals were initially used to study the nature of the interaction. In this case, heteroflocculation between the suspension particles and emulsion was observed and this could be significantly reduced using a strongly "anchored" dispersant for the solid particles. Model suspoemulsions of polystyrene latex (which either contained adsorbed block copolymers or contained grafted poly(ethylene oxide) and hydrocarbon oil emulsions were then investigated using viscoelastic measurement. With suspoemulsions containing adsorbed block copolymers, interaction between the particles and droplets occurred and this resulted in an increase in the relaxation time of the system. When using latex dispersions with grafted PEO chains, no specific interaction could be detected, i.e. the latex-latex, emulsion-emulsion and latex-emulsion interactions are of the same type. However, the viscoelastic properties of the suspensions, emulsions and their mixtures revealed interesting behaviour that could be related to the deformability of the emulsion droplets.

1. Introduction

Suspoemulsions are mixtures of suspensions and emulsions. They occur in many industrial applications, of which the following are worth mentioning: cosmetics, paints, agrochemicals and pharamaceuticals. These systems may produce complex interactions such as particle-particle, droplet-droplet, particle-droplet, as well as some phase transfer of the particles into the droplets followed by recrystallisation. These interactions are governed by adsorption (and "anchoring") of the dispersant/emulsifier system used for their preparation. Wetting of the particles by the oil phase and their solubility in the oil can also affect the physical stability of the system. The particle and

R. H. Ottewill and A.R. Rennie (eds.), Modern Aspects of Colloidal Dispersions, 247–256.

droplet size distributions also play a major role in these interactions.

Due to the above complex interactions, most suspoemulsions are formulated using a simple trial and error approach. In an attempt to understand the interactions in suspoemulsions and control of their stability, we have carried out a systematic study using both practical and model systems. For the practical examples agrochemical suspensions and emulsions were investigated using sediment volume measurements and optical microscopy. The model suspoemulsions consisted of polystyrene latex dispersions (which were stabilised either by polymer adsorption or by grafting polyethylene oxide chains) and hydrocarbon oil-in-water emulsions (which were prepared using block copolymers). The interactions in these model suspoemulsions were investigated using viscoelastic measurements (1,2).

2. Experimental

Chlorothalonil (density ρ = 1.85 g cm^{-3}) and Dichlobutrazol (ρ = 1.25 g cm^{-3}) suspensions were prepared by a bead milling process using Synperonic NPE 1800 (nonyl phenol with 13 moles propylene oxide, PPO, and 27 moles ethylene oxide PEO; this was supplied by ICI surfactants, Wilton, U.K.). Tridemorph (ρ = 0.87 g cm^{-3}) emulsion was prepared using the same surfactant.

Polystyrene latex dispersions (volume mean diameter, VMD = 1.84 μm) were prepared using the surfactant free method (3) and it was stabilised using Synperonic PE P94 (an ABA block copolymer supplied by ICI and containing 47 PPO units and 42 PEO units). Isoparaffinic oil-in-water emulsions (VMD = 0.98 μm) were prepared using Synperonic PE L92 (47.3 PPO units and 15.6 PEO units). The same polymer was used to prepare hexadecane-in-water emulsions (Z-average droplet radius of 280 + 10 nm as measured by PCS).

Polystyrene latex, containing grafted PEO chains, was prepared by dispersion polymerisation using the "Aquersemer" method (4). The Z-average particle radius was 310 + 10 nm (as measured by PCS).

Sediment volume measurements were carried out using measuring cylinders, whereby the suspoemulsion was left standing at room temperature till equilibrium was reached.

Rheology measurements were carried out using a Bohlin VOR (Bohlin Rheologie, Cirencester, U.K.). Three geometries were used depending on the consistency of the sample: concentric cylinder C14, concentric cylinder C25 and concentric cylinder double gap DG24/27.

3. Results and Discussion

The creaming/sedimentation of suspoemulsions prepared using various ratios of suspension : emulsion showed an interesting behaviour, depending on the density difference between the particles and emulsion droplets as well as the total volume fraction of the suspoemulsion. When the particle density was

1.25 (Dichlobutrazol) and the total volume fraction ϕ of the suspoemulsion was < 0.2, the suspoemulsion showed both a creamed and a sedimented layer. When the suspension : emulsion ratio was 2 : 8, the creamed layer was much larger than the sedimented layer, as would be expected. The reverse was true when the suspension : emulsion ratio was 8 : 2. However, at $\phi > 0.2$ for the suspoemulsion, there was no separation into a creamed and sedimented layer. When the ratio of the suspension : emulsion was 2 : 8, all particles and droplets creamed to the top, whereas when the suspension : emulsion ratio was 8 : 2, all particles and droplets sedimented to the bottom. This would be expected from the average density of particles and droplets, which is < 1 in the first case and > 1 in the second case. When the higher density suspension particles (Chlorothalonil) were used only sedimentation could be observed, as would be expected from the average density.

A particularly useful method to illustrate the interactions in suspoemulsions is to compare the total observed sediment plus cream heights with those that would be obtained based on simple additivity. This is illustrated in Figure 1 for Chlorothalonil and Tridemorph suspoemulsions at $\phi = 0.2$.

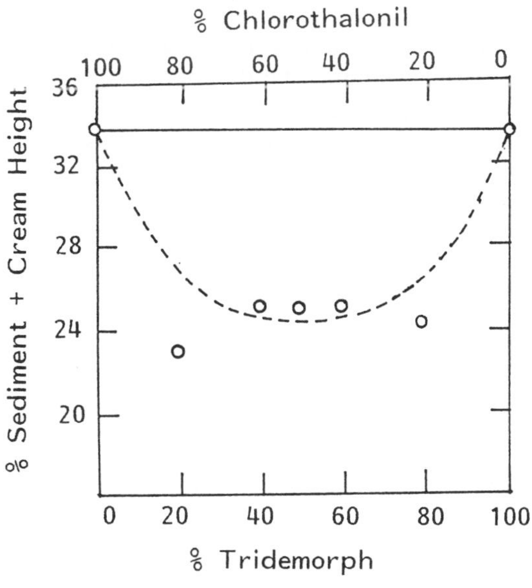

Figure 1: Comparison of predicted, ---- and experimental, - - - finnal sediment + cream heights for Chlorothalonil-Tridemorph suspoemulsions.

It can be seen from the results in Figure 1 that the observed sediment + cream heights are much smaller than would be expected from simple additivity. It is possible that the small suspension particles become trapped in between the larger oil droplets in the cream layer and the small oil droplets become trapped between the larger suspension particles in the sedimented layer. Alternatively, some deformation of the oil droplets may have occurred in the sedimented layer. This phenomenon may have some application for densely packed systems.

Optical microscopic investigations of some suspoemulsions showed some heteroflocculation between the suspension particles and emulsion droplets. In some cases, the emulsion droplets were seen to contain several suspension particles at the droplet periphery. This heteroflocculation was significantly reduced or eliminated by addition of a "comb" type dispersant, namely Atlox 4913 (supplied by ICI), an acrylic graft copolymer consisting of polymethylmethacrylate backbone and PEO chains ("teeth"). Previous studies (5) showed that this polymeric surfactant adsorbs very strongly and irreversibly on suspension particles, thus preventing any attraction with the emulsion droplets, which is believed be the cause of heteroflocculation.

Figure 2 shows typical viscoelastic results for a mixture of 90% isoparaffinic oil-in-water emulsion and 10% polystyrene latex (both stabilised with Synperonic PE block copolymer as described in the experimental section), at a total volume fraction of 0.57 for the suspoemulsion. The results shown in Figure 2 were obtained at a strain amplitude in the linear viscoelastic region. Both G^* (the complex modulus) and G' (the elastic modulus) show a rapid increase with increase in frequency above 0.1 Hz. However, G'' starts to decrease above 1 Hz and a clear cross over point (G' = G'') is observed for the system. The dynamic viscosity show a continuous reduction with increase in frequency as expected. Similar results were obtained for other suspoemulsions containing various emulsion : latex ratios, as well as for the emulsion and latex dispersion alone.

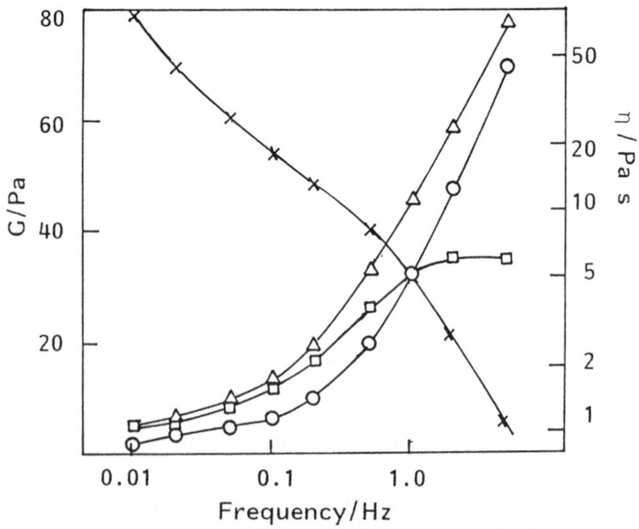

Figure 2: Viscoelastic results for a suspoemulsion of isoparaffinic oil-in-water emulsion and latex; G^*, △; G', ○; G'', □; η, x. $\phi = 0.57$; 90% emulsion + 10% latex.

The characteristic frequency ω^* ($2\pi\nu$, where ν is the frequency in Hz) can be used to calculate the relaxation time of the system ($t^* = 1/\omega^*$). Addition of the latex to the emulsion causes a shift in t^* to higher values, indicating stronger interaction between the latex particles and emulsion droplets.

Figure 3 shows the variation of G*, G' and G" at $\phi = 0.57$ and $\nu = 1$ Hz with % emulsion and latex in the suspoemulsion. The emulsion has much higher moduli than the latex dispersion at the same volume fraction. Although the emulsion has a VMD (0.98 μm) that is close to that of the latex (1.18 μm), the former is much more polydisperse than the latter. The smaller emulsion droplets present may account for

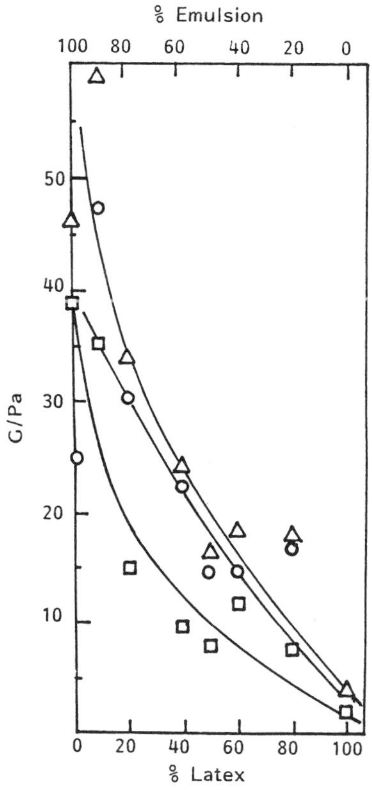

Figure 3 Variation of G*, △; G', ○; and G", □ with % emulsion and latex. $\phi = 0.57$.

the higher modulus of the emulsion when compared with the latex. As the proportion of the latex in the suspoemulsion is increased the moduli decrease. Replacement of emulsion with latex would mean replacing a proportion of small droplets with larger latex particles and the moduli values are reduced. It should be mentioned, however, the mixture of emulsion and latex becomes relatively more elastic than viscous, indicating stronger interaction between emulsion droplets and latex particles.

It is clear from the above discussion that the interaction between suspension and emulsion droplets depends on the nature of the stabiliser used for the particles and droplets. For that reason, we have investigated model systems (6) whereby the latex particles contain grafted PEO chains (with no possible desorption) and the emulsion was based on hexadecane stabilized by Synperonic PE L92. The particle and droplet radii were very similar (315 and 280 nm respectively) in order to avoid any complications arising from the change in particle size distribution on mixing the suspension and emulsion.

Steady state (shear stress-shear rate) measurements were used to obtain the relative viscosity-volume fraction relationship for the latex and the emulsion and the results are shown in Figure 4. The same figure contains the theoretically predicted curve based on the Dougherty-Krieger equation (7),

$$\eta_r = [\,1 - (\phi/\phi_M)\,]^{-[\eta]\phi m} \qquad (1)$$

where ϕ_M is the maximum packing fraction and $[\eta]$ is the intrinsic viscosity (2.5 for hard spheres). Two values for ϕ_M were used, namely 0.60 and 0.61. Reasonable agreement with the latex data is obtained for ϕ_M = 0.61 and $[\eta]$ = 2.5 and this suggests that the latex dispersions behave closely as hard spheres.

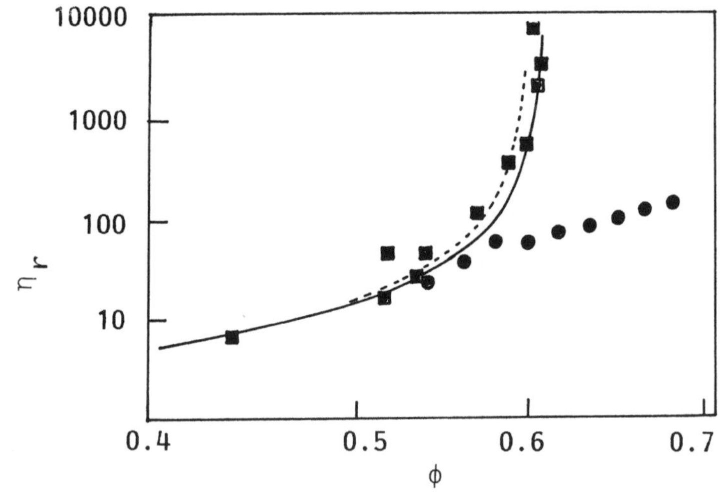

Figure 4: Reduced viscosity, η_r as a function of volume fraction for latices, ■ and emulsions, ●. Lines from equation (1) with $[\eta]$ = 2.5 and - - -, ϕ_{max} = 0.60;----.ϕ_{max} = 0.61.

From the experimental values of the maximum packing fraction, one can calculate the grafted layer thickness, Δ, assuming a maximum packing fraction (a value of 0.68 was used which corresponds to a body-centred cubic lattice),

$$\phi_{max,exp} = \phi_{max,theor}\,[\,1 - (\Delta/R)\,]^3 \qquad (2)$$

The calculated value of Δ is 9 nm which is reasonable since the grafted PEO chain gave a molecular weight of 2000.

The above procedure cannot be applied for the emulsions, since these are deformable. This is indeed the case as illustrated in Figure 4.

Figure 5 shows G' and G" (measured at ν = 1 Hz and in the linear viscoelastic region) as a function of the volume fraction of latex or emulsion (a log-log scale is used). At low volume fractions, G" is higher than G' and both quantities increase with increasing volume fraction. However, G' increases faster with volume fraction than G" and a cross-over point (G' = G") is obtained at ϕ = 0.52 for the latex and at ϕ = 0.58 for the emulsion. Before the cross-over point, the slopes of both curves are little affected by the nature of the particles. However, above the cross-over point, the behaviour is entirely different. The slope of the G' vs ϕ curve (ϕ > 0.6) is much higher for the latex compared with the emulsion. Indeed, with the emulsion G" seems to reach a plateau value at ϕ > 0.63, and the slope of the G' vs ϕ line is greatly reduced. This is consistent with the deformability of the emulsion droplets at high volume fractions. From the volume fraction at the cross-over point and assuming ϕ_M to be 0.61 (the value obtained from high shear measurements) a rough estimation of the adsorbed layer thickness can be obtained. The values obtained were 7 - 15 nm for the latex and 5 - 10 nm for the emulsion. These values are reasonable (at least on a relative basis) considering the molecular weights of the grafted PEO for the latex (2000) and the PEO of the adsorbed Synperonic PE L92 (~ 700).

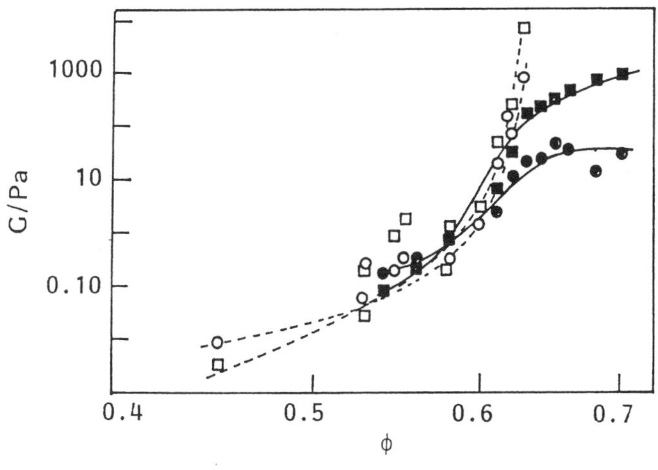

Figure 5: Log-log plots of G', (□,■) and G", (○,●) as a function of volume fraction of latex, ϕ (open symbols) and emulsion (closed symbols).

Figure 6 shows log-log plots of G' vs ϕ for the latices, emulsions and their mixtures. All the curves look very similar for volume fractions below ~ 0.62. This seems to indicate that the interactions inferred from the rheology data are of the same kind. This is what one would expect for particles stabilised in the same way. Thus, below ϕ = 0.62 both latex particles and emulsion droplets behave as near hard spheres. However, at ϕ > 0.62, the behaviour of the latices and the emulsions differ significantly, as indicated by the much reduced slope of the log G' vs log ϕ curve for the emulsion when compared with the latices. Above this volume fraction, the interaction between two emulsion droplets is quite high and the system can reduce this interaction by deformation of the emulsion droplets. This situation is not possible with the latices, where the particles are rigid. Similar behaviour is observed for the suspoemulsion when the percentage of the emulsion in the mixture exceeds 60%. This implies that the behaviour of emulsions and suspoemulsions with more than 60% emulsion are close in their rheological behaviour to concentrated emulsions. In these cases, the elastic modulus is proportional to the volume fraction, following the equation proposed by Princen (8), namely,

$$G_o = a[\,\Gamma/R_{32}\,]\,\phi^{1/3}\,(\,\phi - b\,) \tag{3}$$

where Γ is the interfacial tension, R_{32} is the surface weighted mean droplet radius, and the constants a and b have values of 1.76 and 0.71 according to Princen. G_o is the shear modulus which can be replaced by the elastic modulus of the emulsion (9). The Princen equation is based upon a semiempirical extension of a model system of cylinders arranged in a hexagonal array. When such an arrangement is strained, the total interface is increased; this creates a restoring force that is proportional to the interfacial tension. The origin of the constant b is the value of the maximum packing of undistorted cylinders in the hexagonal array.

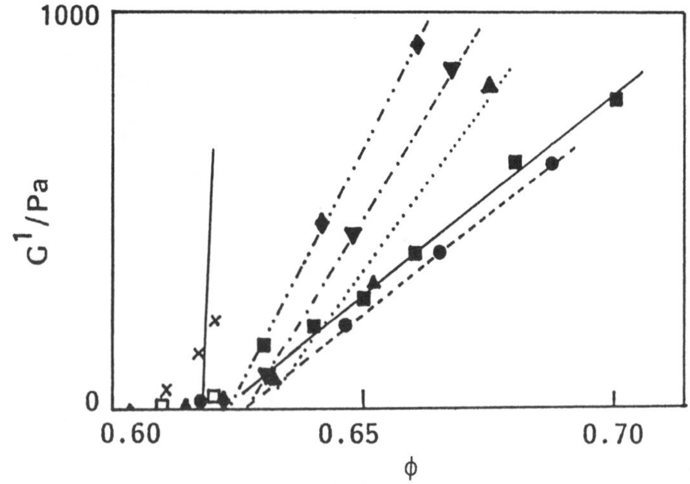

Figure 6: Log-log plots of G' vs ϕ for latices, emulsions and their mixtures:- ■, 100% emulsion; ●, 80% emulsion-20% latex; ,60% emulsion-40% latex; ,50% emulsion-50% latex; ,40% emulsion-60% latex; x, 100% latex. The lines correspond to the linear fits whose parameters are given in Table 1.

For the samples studied, the radius of the emulsion droplets is constant, and due to the excess of the surfactant in the emulsion, the interfacial tension is expected to be little dependent on the composition of the system. The values of the parameters used to fit the data of Figure 6 are given in Table 1.

Table 1
Parameters of the fit of Equation 3 for the Emulsion, Latex and their Mixtures

%Emulsion	$a\Gamma/R_{32}$	b
100	9896 + 690	0.63 + 0.01
80	9900 + 1700	0.63 + 0.01
60	17500 + 20000	0.63 + 0.03
50	20700 + 2200	0.63 + 0.01
40	23700 + 2200	0.62 + 0.02
20	~100000	0.61 + 0.01
0	~1000000	0.62 + 0.02

The first constant of the fit corresponds to the factor $a\Gamma/R_{32}$ and the second corresponds to b in equation (3). The parameter b corresponds to the volume fraction at the onset of elasticity. The values of b given in Table 1 are lower than the values reported by Princen (8) and Pons et al (9). Part of the difference can be accounted for by the role of the adsorbed layer thickness.

The mixture of latex and emulsion may be regarded as two elastic elements in series with the appropriate weight fractions, i.e.

$$1/G_m = f_e/G_e + (1 - f_e)/G_l \qquad (4)$$

where G_m, G_e and G_l are the elastic moduli of the mixture, emulsion and latex respectively. f_e is the weight fraction of the emulsion in the mixture. A plot of the slope of the linear lines of Figure 6 ($a\Gamma/R_{32}$) vs f_e is shown in Figure 7 together with the predicted line based on equation (4), The agreement between experimental results and those based on a simple model of two elastic bodies in series is reasonably good. However, some deviation occurs (in particular for the point with $f_e = 0.2$ and this could be due to the system behaving in an intermediate way between series and parallel elements.

5. Conclusions

Mixtures of suspensions and emulsions (suspoemulsions) may undergo interaction that depends on the nature of the stabiliser, the particle size distribution and the density difference between the particles and droplets. In cases where heteroflocculation (between particles and droplets) occurs, strongly anchored dispersing agents for the suspension are required to prevent this instability. The viscoelastic properties of model systems of polystyrene latex, paraffinic oil-in-water emulsions and their mixtures showed that the deformation of the emulsion droplets at high volume fractions (ϕ) determines the rate of increase of the elastic modulus with ϕ. The elastic modulus of the mixtures of latex and emulsions can be approximated by a simple model in which the elasticity of the mixture arises from the combination in series of two elastic elements that correspond to the latex and emulsion elasticities with the appropriate weight fractions.

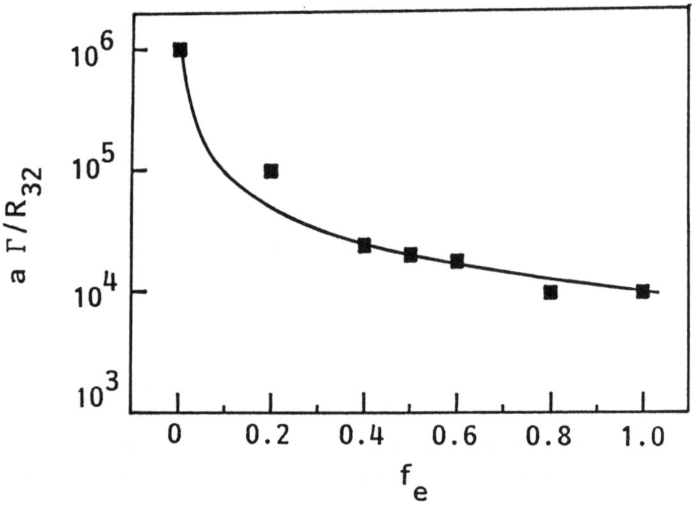

Figure 7: ($a\Gamma/R_{32}$) against weight fraction of the emulsion, f_e; ■, experimental; ---, equation (4).

6. References

1. Tadros, Th. F. (1990) *Langmuir*, **6**, 28.
2. Tadros, Th. F., Liang, W., Costello, B. and Luckham, P. F. (1993), *Colloids and Surfaces*, **79**, 105.
3. Goodwin, J. W., Hearn, J., Ho, C. C. and Ottewill, R. H. (1974), *Colloid & Polymer Sci.*, **252**, 464.
4. Bromley, C. (1986), *Colloids and Surfaces*, **17**, 1.
5. Heath, D., Knott, R. D., Knowles, D. A. and Tadros, Th. F. (1984), *ACS Symposium Series*, **254**, 2.
6. Pons, R., Rossi, P. and Tadros, Th. F. (1995), *J. Phys. Chem.* **99**, 12624.
7. Krieger, I. M. (1972) *Advances Colloid Interface Sci.*, **3**, 111.
8. Princen, H. M. and Kiss, A. D. (1986), *J. Colloid Interface Sci.*, **112**, 427.
9. Pons, R., Erra, P., Solans, C. and Stebe, M. J. (1993), *J. Phys. Chem.*, **97**, 12320.

TOPOLOGICAL ASPECTS OF LIQUID CRYSTALLINE COLLOIDS — EQUILIBRIA AND DYNAMICS

E.M. TERENTJEV

Cavendish Laboratory, University of Cambridge
Madingley Road, Cambridge CB3 0HE, U.K.

This is a brief review of effects arising due to topological defects in liquid crystalline colloids. The distortions around colloid particles in a l.c. matrix, their long-range pair interaction and anisotropic friction drag are analysed theoretically. I then study (experimentally and theoretically) the case of an "inverted" l.c. colloid - the emulsion of l.c. droplets in an isotropic fluid matrix and describe the new effect of topological stability.

1. Introduction

Colloidal suspensions and emulsions are characteristically mesoscopic systems with structure and time scales such that typical shear rates can bring them out of equilibrium and into some exotic states. Even without flow, the structure and properties of colloids pose a number of theoretical and experimental challenges [1,2]. Various novel interactions, for instance, hydrodynamic and polymeric solvent-mediated forces are of much interest. Many important practical applications of colloid systems, ranging from paint to food and detergents industry, add to the fundamental interest in this class of objects. Because of the large amount of inner surfaces in colloids, the use of various surfactants is common and, in case of emulsions - even necessary to stabilise the system. Surfactant molecules, on phase separation in solutions, often form lyotropic liquid crystalline phases, nematic, hexagonal and lamellar [3], and so the concept of a liquid crystal is very common in "colloid science".

In this work, however, we shall be dealing with a different kind of liquid crystalline (l.c.) colloids. Here we study a colloid suspension of solid particles in a thermotropic liquid crystal solvent and an emulsion of l.c. droplets in an isotropic suspending fluid. These will act as a model of systems where the liquid crystal order exists independently, in its own right, and brings new physical properties to heterogeneous materials formed on its base. Surfactants will still remain necessary to control the properties of interfaces and, in particular, the anchoring of the l.c. (nematic) director on these interfaces.

Colloid suspensions in a liquid crystal matrix are qualitatively different from their isotropic analogues because, due to specific anchoring on "inner surfaces", a long-range deformation field $\mathbf{n}(\mathbf{r})$ is created by particles suspended in the liquid crystal. In particular, the principal subject of this article is the effect of topological mismatch between boundary conditions on inner (particles) and outer (container, external fields) surfaces: this leads to

R. H. Ottewill and A.R. Rennie (eds.), Modern Aspects of Colloidal Dispersions, 257–267.

the so-called topological defects in the liquid crystal: point singularities and line disclinations. We shall see that topological defects have a profound effect on macroscopic, equilibrium and kinetic properties of the material. Perhaps two most important applications of l.c. colloids is the moulding of filled l.c. polymers into circuit boards and other tough and light-weight products, and the suspension of abrasive particles in lyotropic mesophases in the production of detergents. In these, and most other practical situations, one is faced with the questions of structure and dynamics of a highly frustrated material, filled with topological defects in equilibrium; here I address some of these questions. There are two principal types of l.c. order relevant for their colloid-related applications: nematics and smectics. Although smectic (lamellar) phases are, perhaps, more common in lyotropic systems, their description is more difficult due to the combination of orientational and translational order of layers, less clear models of surface anchoring and mixture of topological defects and dislocations. In this article I concentrate on the most basic l.c. system - the nematic fluid with no translational order and the unit vector $\mathbf{n}(\mathbf{r})$ field, describing the local uniaxial anisotropy and the axis of birefringence. Several new characteristic features of l.c. colloid system based on a nematic are obtained. These results help to pursue the analogous study of the smectic and lamellar l.c. colloid, which is currently underway.

This paper is organised as follows: in the next section we shall study the director field $\mathbf{n}(\mathbf{r})$ around an isolated colloid particle in the nematic l.c. and the long-range pair interaction between such particles. Then we examine the hydrodynamics of a particle flowing through the nematic matrix and obtain the non-central drag force. Finally, in section 4, the case of an inverted l.c. colloid - the emulsion of a nematic l.c. in an isotropic fluid is discussed, and I describe the effect of topological dereliction - stabilisation of emulsion due to high energy barriers imposed by topological defects.

2. Colloid particles in nematic l.c. - Equilibrium

The reason for frustrations of the director field $\mathbf{n}(\mathbf{r})$ in l.c. colloids is the competition of two conflicting energy factors, specific to the liquid crystal phase. Bulk distortions of the director are penalised by the Frank elasticity, expressed very crudely in the one-constant approximation in Equation (1) as F_e. This elastic energy requires (at least theoretically) that the global equilibrium texture of nematic l.c. is uniform, $\mathbf{n}$=const, although for kinetic reasons or due to thermal fluctuations this may not always be easy to achieve in practice. The second energy in Equation (1), F_s, is surface anchoring - the penalty for the director deviation from the predetermined alignment on the surface (one may consider a particular case of homeotropic anchoring, when $\boldsymbol{k}$ is the surface normal and the constant W>0).

$$F_e = \int \tfrac{1}{2} K \left(\nabla \boldsymbol{n}\right)^2 dV \qquad \text{and} \qquad F_s = -\int \tfrac{1}{2} W \left(\boldsymbol{k}\cdot\boldsymbol{n}\right)^2 dA \tag{1}$$

In some cases these two elastic effects work together, for instance in a cell with uniformly aligning surfaces, and then the resulting texture is indeed uniform. However, in many situations, and in l.c. colloids in particular, they work against each other (see Figure 1). The crude outcome is then determined by the energy balance of these competing terms. For deformations, induced by a colloid particle of the size R, the characteristic magnitude of the Frank elastic energy is $K\cdot(1/R)^2\cdot\text{Volume} \approx K\cdot R$. If, on the other hand, the bulk director is uniform, the surface deformations will contribute the anchoring energy of the magnitude W·Area ≈ W·R². Hence, when (K·R)»(W·R²), the tendency to have the uniform director in the bulk prevails and the director simply breaks away from the predetermined alignment on the particle surface (Figure 1a), remaining largely undistorted in the bulk. In the opposite limiting case, (K·R)«(W·R²), the demand to have the homeotropic surface alignment must be satisfied and the bulk director field suffers from the topological incompatibility between the closed (spherical) inner surface of a particle and the uniform conditions on **n** far away (given, for instance, by the container boundaries). The result is a topological defect, confined to the particle and making the whole construction topologically neutral (Figure 1b,c).

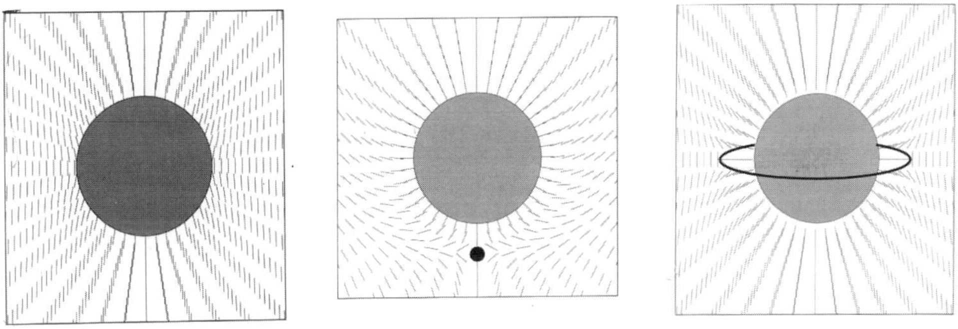

Fig.1 Spherical particle with homeotropic (radial) anchoring in a nematic l.c. (the director is shown by thin lines). Weak anchoring (a) allows director deformations on the surface and causes only small distortions **n**(**r**) in the bulk. Rigid anchoring, WR/K»1, forces topological defects: a disclination ring (b) or a satellite point defect (c).

Therefore, we identify the characteristic parameter, WR/K, which selects between the two regimes: large particles with strong surface anchoring of the director have topological defects around them, while small particles suspended in a fluid with high Frank elastic constants distort the surrounding very little. It is interesting that the crossover between these two regimes for a typical thermotropic nematic l.c. ($K \approx 10^{-11}$ J/m; $W \approx 10^{-5}$ J/m² [4]) takes place for particles with $R^* \sim 10^{-6}$ m, a micron-size. Note also, that both energy constants, K and W, depend on the nematic order parameter Q and vanish in the isotropic phase. Usually one can take $K \sim Q^2$, $W \sim Q$ and, perhaps surprisingly, the topological effects become stronger as the nematic order weakens.

The theoretical analysis of spherical particle in the "weak anchoring" case[†], WR/K«1, is rather simple because the problem of free energy minimisation can be linearised with respect to the small angle of director deviation from its uniform alignment $\mathbf{n}_0$. The solution for **n**(**r**) is exact and takes the form [5]:

$$\beta(\boldsymbol{r}) = \frac{1}{4}\frac{WR}{K}\left(\frac{R}{r}\right)^3 \sin 2\theta\ , \qquad (2)$$

where β(**r**) is the angle between **n** and $\mathbf{n}_0$, r is the distance from the particle centre and θ is the polar angle in the spherical co-ordinate system based on the direction $\mathbf{n}_0$. This weak, but long-range distortion field has quadrupolar symmetry, same as the nematic order itself.

[†] Quotes are put here to remind the reader that one may manipulate the control parameter WR/K by other means than just varying the anchoring energy W.

The analysis of the "strong anchoring" case with topological defects is more difficult. One reason is that the exact solution of non-linear Euler-Lagrange equations is impossible and one has to resort to numerical solutions and approximate ansatz functions. There is another subtle point in choosing the solution: all topological defects have some sort of a core, which rectifies the non-physical singularity in the director field **n**(**r**). To keep the argument simple, let us assume that the core is occupied by the "melted" mesophase with Q=0 (hence K=0), with no elastic penalty, but with the additional thermodynamic energy density of melting E_c. For a linear singularity (in a disclination ring, Figure 1b) the energy balance per unit length is $E_c \approx K/\rho_c^2$, giving the characteristic size of the cylindrical core $\rho_c \approx (K/E_c)^{1/2} \sim 100Å$, the nematic coherence length. For a point defect (like the hyperbolic monopole in Figure 1c) the energy of the melted core is $E_c\rho_c^3 \approx K^{3/2}E_c^{-1/2}$, which does not allow to eliminate the poorly controlled parameter E_c, making the energy balance ambiguous. Therefore, the full contribution of the core energy is not as simple as in the case of linear singularities and the direct comparison of the two possible scenarios, Figure 1b and Figure 1c, is not possible.

The case of the disclination "Saturn ring" around the particle with strong anchoring (Figure 1b) has been considered in detail [5] and the best approximation for the director field takes the form

$$\beta(\boldsymbol{r}) \approx \theta - \tfrac{1}{2}\arctan\frac{\sin 2\theta}{\cos 2\theta + (a/r)^3}, \tag{3}$$

where a is the ring radius. The limit $r >> a$ gives the far-field approximation $\beta \propto r^{-3}\sin 2\theta$, which is identical to the simple quadrupolar field in the weak-anchoring case, Equation (2). The ring radius strongly depends on our key parameter WR/K: in the limit $W \to \infty$ the equilibrium radius is estimated $a^* \approx 1.2R$, only slightly greater than the particle size. Numerical solution of the problem and Monte-Carlo simulations [6] indicate that $a(W)$ rapidly decreases and at WR/K~10 there is a discontinuous transition (without the overall change of quadrupolar symmetry) into the topologically trivial case, Figure 1a.

If the dipolar texture with a satellite point defect (hyperbolic monopole, Figure 1c) is considered [7], the director deformations in the far-field asymptotic limit $r >> R$ are described by

$$\beta(r) \propto 0.3\,(a/r)^2 \sin\theta. \tag{4}$$

Here the equilibrium position of the monopole is estimated as $a^* \approx 1.17R$ [7] and, according to the Monte-Carlo simulations [6], has a similar sharp dependence $a(W)$ with a transition to the quadrupolar weak-anchoring limit at a very close value WR/K~10.

When there are several colloid particles with the director deformation field around each, a long-range pair interaction potential is expected to act between them. The form of this pair potential strongly depends on the type of director deformations - quadrupolar "Saturn ring" as in Equation (3), or dipolar "satellite" as in Equation (4). (Note that at weak anchoring, the quadrupolar solution (2) is exact). The resulting interaction produces a complex non-central force acting on colloid particles, a result of quadrupole-quadrupole (see [8,9]), or dipole-dipole [7] coupling via the director **n**(**r**):

$$U_{qq} \approx \frac{W^2R^8}{K\,\boldsymbol{d}^5}\left(1 - 2.2\cos 2\theta + 3.9\cos 4\theta\right) \qquad \text{at weak anchoring, } WR/K \ll 1 \tag{5}$$

$$U_{qq} \propto Const \cdot K \frac{a^6}{d^5}(1-2.2\cos 2\theta + 3.9\cos 4\theta) \quad \text{disclination ring at WR/K»1}$$

$$U_{dd} \propto Const \cdot K \frac{a^4}{d^3}(2-3\cos 2\theta) \quad \text{satellite defect at WR/K»1 (from [7])}$$

where d is the distance between particles and θ - the angle between the particle separation line and the undistorted director $\mathbf{n}_0$. As it has been argued above, there is no clear theoretical possibility to decide between the two possible scenarios at strong anchoring (the Equation (5) is always valid in the weak anchoring regime). Therefore, an experimental input should be sought. There are several possible effects to look for, and the most dramatic is the dependence of the regime on the nematic order parameter Q. If the particles are small enough or the anchoring sufficiently weak for a given l.c. colloid, by raising the temperature in thermotropic or by diluting the lyotropic l.c. matrix one can often bring the system into the strong anchoring regime - and perhaps see a change in the colloid aggregation due to the above long-range pair interaction. Another option is to look at the phase diagram of such a colloid, taking into account the additional nematic energy of deformations, which couples the colloid concentration and the order parameter Q (the total elastic energy associated with each particle is $F_n \approx 13\,KR \propto Q^2$ in the strong anchoring regime, and $F_n \approx 0.2\,W^2R^3/K \propto const$ at weak anchoring). Such an attempt has been recently made by a French group [10]. Finally, the characteristic time τ of l.c. colloid aggregation due to the long-range forces given above (dynamically controlled by the friction drag force $\sim 6\pi\eta R v$) can be estimated, very qualitatively, as

$$\tau \propto \frac{K\eta}{W^2}\left(\frac{1}{c^{7/3}} - \frac{1}{c_m^{7/3}}\right) \quad \text{at weak anchoring, } \frac{WR}{K} \ll 1 \tag{6}$$

$$\tau_{qq} \propto \frac{\eta R^2}{K}\left(\frac{1}{c^{7/3}} - \frac{1}{c_m^{7/3}}\right) \quad \text{and} \quad \tau_{dd} \propto \frac{\eta R^2}{K}\left(\frac{1}{c^{5/3}} - \frac{1}{c_m^{5/3}}\right), \quad \text{at } \frac{WR}{K} \gg 1$$

where c is the colloid concentration (should be assumed sufficiently small to allow the pair-potential approximation to control the aggregation time, at least approximately) and c_m is the maximum close-packing volume fraction.

3. Colloid particles in nematic l.c. - Friction drag

In the last section we have concluded with a very crude estimate of a characteristic time scale for the motion of colloid particles in a l.c. matrix. However, the question of flow and dissipation around such particles, especially in the topologically non-trivial case WR/K»1, is by no means simple. The nematic flow depends strongly on the director orientation and gradients (see the Leslie-Ericksen theory of nemato-hydrodynamics [4]). In an anisotropic fluid the drag force is not parallel to the line of motion - causing the so-called "lift force" perpendicular to the particle velocity. In general, the linear relation between the velocity vector $\boldsymbol{v}$ and the drag force $\boldsymbol{f}$ is expressed by the mobility tensor $M_{ik}(\mathbf{n})$, which, in turn, is related to the matrix of diffusion coefficients:

$$f_i = M_{ik} v_k, \quad \text{with } M_{ik} = M_\perp \delta_{ik} + (M_\parallel - M_\perp)\, n_i\, n_k \quad \text{and } D_{ik} = k_B T\, M_{ik}^{-1}$$

In the case of isotropic fluid (when the order parameter Q→0) these expressions reduce to the classical Stokes' result: $M_{||} = M_{\perp} = -6\pi\eta R$, whilst Leslie coefficients $\alpha_1 - \alpha_6(Q)$ determine the mobility in the nematic phase. The theoretical analysis of this problem [11] is based on the numerical method of artificial compressibility, when the hydrodynamic variables - the velocity field $\mathbf{v}(\mathbf{r})$ and the pressure $p(\mathbf{r})$ - are relaxing to the correct stationary solution The problem is very difficult due to the presence of a high-energy singularity in the director field. However, in the limit of small Ericksen number ($Er = \eta v R/K \ll 1$, a condition that is often met in l.c. colloids with $K \approx 10^{-11}$ J/m; $\eta \approx 10^{-1}$ Pa.s and $R \approx 1\mu$), the flow may be considered for an approximately fixed static director field, described in the previous section.

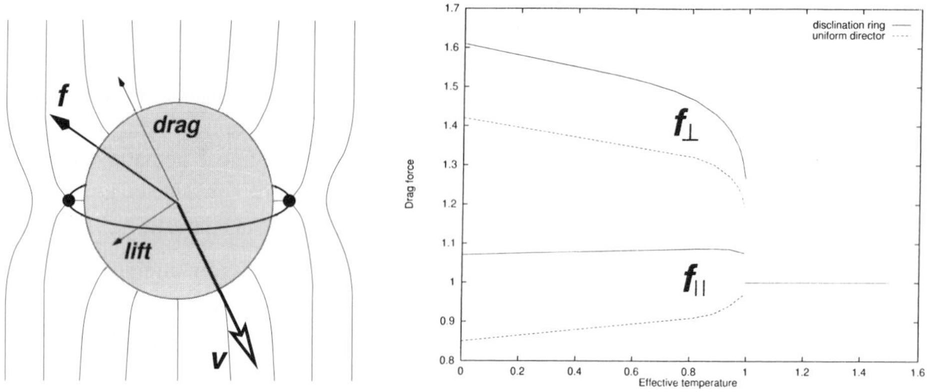

Fig.2 (a) The drag force $\boldsymbol{f}$ on a colloid particle moving with the velocity $\boldsymbol{v}$ at an angle to the distant uniform director $\mathbf{n}_0$ (the quadrupolar case of rigid anchoring with the disclination ring). (b) Temperature dependence of the two principal values of force (in units of f_{iso}), for $\mathbf{v} \perp \mathbf{n}$ and $\mathbf{v} || \mathbf{n}$, obtained from the order-parameter dependence of Leslie coefficients $\alpha_j(Q)$ for thermotropic nematic l.c. (Effective temperature = T/T_{ni})

The results are summarised in the Figure 2. When the particle moves in the direction of far-distant director (assumed uniform), $\mathbf{v} || \mathbf{n}_0$, the drag force is parallel to $\mathbf{v}$ for symmetry reasons. The same applies for the other principal geometry, $\mathbf{v} \perp \mathbf{n}_0$, but the values of these forces can be very different, determined by the principal values of the mobility tensor $M_{||}$ and $M_{\perp}$. The difference depends on the nematic order parameter (for instance, Q=Q(T) in thermotropic l.c. - the case plotted in Figure 2b) and on the topological properties of the particle. In Figure 2b solid lines show the plots $f_{||}(T)$ and $f_{\perp}(T)$ calculated for the MBBA l.c. matrix and the rigid-anchoring (disclination ring) case. Dashed lines represent the opposite limiting case, when $WR/K \to 0$ and the director field is approximately uniform. Above the nematic transition the force has a single value, given by Stokes as $6\pi\eta R v$. Clearly, the topological defect causes a considerable increase in dissipation even at the most basic isolated-particle level.

Another important quantity stemming from this calculation is the anisotropy of mobility and the corresponding drag force. The characteristic ratio $f_{\perp}/f_{||}$ also depends on the nature of the l.c. matrix and on the colloid particle anchoring regime. For example, a typical thermotropic nematic MBBA has this ratio of the order ≈ 1.5. This would also be the value of another observable quantity, the anisotropy of diffusion coefficients $D_{||}/D_{\perp}$, and related kinetic constants. In conclusion of this section, let us emphasise the result that the drag force is not, in general, parallel to the line of particle motion. The perpendicular lift force is non-dissipative (because $\boldsymbol{f}_{\perp}\cdot\boldsymbol{v}=0$) and is similar to the Hall effect: particles travelling in an oblique director field are deflected sideways by this force. One can imagine several applications of this phenomenon, for instance, charged particles sinking in a nematic fluid would accumulate on one edge of the cell, creating an electric potential.

4. Topological stability of l.c. emulsions

When two immiscible fluids are thoroughly blended together, an emulsion is usually formed. The best known examples are, of course, oil-in-water or water-in-oil emulsions. There is an understanding, quite general for colloids and emulsions, that their structural stability is a kinetic concept and not a thermodynamic one [12]. Some emulsions have only a short lifetime before complete phase separation, whereas others remain kinetically stable for years. In order to prepare a stable emulsion there must be a surfactant material present to protect the newly formed droplets from immediate re-coalescence. By aggregating on interfaces, the surfactant molecules reduce the surface tension and, when such a reduction is complete, a microemulsion is formed. This represents a truly thermodynamically stable phase, as opposed to the kinetic stability of macroscopic droplets. The argument is simple: if the local energy of the interface covered with surfactant is very small, the entropy preference of having many small particles in favour of the few big ones makes the molecular micelles a ground state of the total free energy. If, on the other hand, the local interface energy is considerable, the configurational entropy is irrelevant and the ground state is the completely separated phases with the minimal surface area - which the system may take some time to reach, however. It is not easy to give a reference to such a broad and well-studied area, but an interested reader, new to the field, could find useful information in [13].

In this article we study the l.c. order when one of the two emulsified fluids is a liquid crystal, which brings its own energy and entropy contributions into the problem. The effect of l.c. as a majority phase, in which the particles (or isotropic fluid droplets) are suspended, has been examined. Here I shall address the opposite end of the problem, when the isotropic majority phase (water, for instance) has small droplets of a liquid crystal dispersed in it. We shall look at co-operative effects for such droplets in a dense emulsion, focusing on the mechanism of their coalescence, which one would expect to happen by analogy with isotropic macroemulsions[†]. This section presents the experimental result, which is followed by a theoretical argument rationalising the observations [16]. The thermotropic nematic MBBA is blended with water (with a varying proportion of glycerol added to adjust the droplets buoyancy). Surfactant is needed, as usual, to reduce the interface energy and also to impose strong anchoring boundary condition for the nematic director - in this case I used the ionic surfactant C_{16}TAB. The elastic energy of topological

[†] Droplets of a l.c. in isotropic fluid and the related topological defect structures have been extensively studied before [14]. Interesting physical effects here stem from the topological constraint, imposed on the director field in a closed volume by the director anchoring on its surface, leading to topological defects of total point charge (+1) for a nematic and more complex structures in other phases. Since the concept of polymer-dispersed liquid crystals (PDLC) has been formulated [15], much more work have been done in this area. However, in all cases the properties and effects of individual droplets have been addressed.

defects in the confined nematic creates a significant barrier for the droplet coalescence and leads to a new effect of kinetic stabilisation, or "topological dereliction" of the macroemulsion.

As one expects from an effect based on the basic symmetry and topological properties of the system, the particular experimental details are not very significant. One dissolves a fair amount of CTAB in water, 3-5 weight %, taking care not to reach the surfactant aggregation below the Krafft boundary. (One should add at this point that the preparation details appear not to be very demanding: the topologically stable l.c. emulsion is formed in any case, due to the free surfactant remaining in the solution, as described below; however, the microscopic observation of droplets could be difficult if multiple surfactant aggregates are present in the surrounding water matrix). Then a small amount (also ≈5 weight %) of MBBA is added and the mixture is thoroughly blended in an ultrasonic bath. As will be clear below, it is important to keep the "oil" all the time in the nematic phase, below the clearing temperature T_{ni}≈45°C. As a result, a very small sub-micron size emulsion of MBBA in water, with CTAB aggregated on the interfaces, is formed. Due to the initial small size of the droplets, the material scatters light and appears turbid. After a short time these droplets recombine into much bigger ones (≈30-100μ), after which the dynamics stops: most emulsion samples then stayed unchanged for more than two years.

The time of initial recombination can vary between minutes and hours depending on conditions and concentrations, but the physics and the resulting structure of emulsion appear to be fairly robust. For instance, since the density of MBBA is slightly higher than of water (d=1.027), the droplets eventually sink to the bottom of the container. Squeezed in a pile by gravity, the droplets still do not coalesce. In order to improve their buoyancy one may mix water with glycerol (d=1.26) to equilibrate the density with MBBA. One also may significantly alter the concentration of CTAB in the mixture, or blend again at some moderate ultrasonic energy input. None of these factors appear to influence the resulting stable macroemulsion in any qualitative way. If, on the other hand, we raise the temperature of the material above the nematic-isotropic transition point T_{ni}, all nearby droplets coalesce within seconds and form completely phase-separated regions.

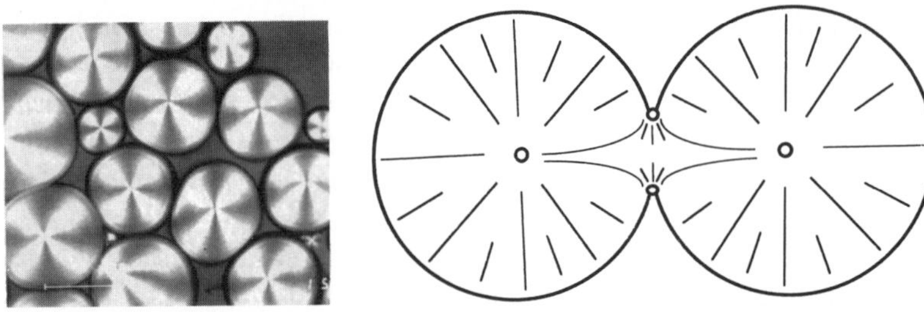

Fig.3 (a) The picture of a group of tightly squeezed MBBA droplets, viewed between crossed polars of a microscope to reveal the radial director distribution; (b) a scheme of the neck formation with the topological ring defect emerging to bring the total point charge of the connected volume back to (+1), from [16].

How can we rationalise these observations? The following argument is based on the assumption that hydrophobic tails of CTAB molecules, densely packed on the surface, impose a sufficiently strong homeotropic (radial) anchoring on the director (this assumption is certainly supported by Figure 3a, where the radial director texture with a monopole topological defect in the middle is apparent). We know from the previous consideration that when the droplet radius increases, the bulk elastic energy of the distorted director is increasing as K·R, while the penalty for violating the surface anchoring is growing faster, as $W \cdot R^2$. Hence, as before, the control parameter WR/K determines the outcome: large droplets with R»K/W are in the rigid anchoring regime and possess the topological charge (this is what we see in Figure 3a). Small droplets rather prefer to have a uniform director in the bulk and concentrate all deformations on the surface, meaning that the system is topologically trivial.

When two topologically charged droplets come into contact, their coalescence must comply with the law of this charge conservation. After the bridge between the droplets is formed, their volume becomes simply connected and, by the Poincare theorem, must have the total topological charge (+1). But each droplet possessed its own point charge of (+1) before their volume became connected! Therefore, in the moment of formation of any neck or bridge between the two volumes, a new topological defect of the charge (-1) must be created. Later this defect may travel towards one of the radial monopoles and annihilate it, to allow the final spherical volume to minimise its elastic energy. However, this cannot happen instantly and the formation of a new topological defect in a closed volume costs an elastic energy ≈K·R (by now the familiar estimate). This energy barrier can be very high and the system may take a long time to overcome it; a Kramers-type estimate would give the characteristic lifetime $\tau \propto \exp\left[KR / k_B T\right]$.

Now one can trace the evolution of the nematic emulsion. Initial droplets, broken to very small sizes by the ultrasonic energy of blending, prefer to violate the surface anchoring energy of the director but have a uniform director field inside. No topological arguments apply in this situation and small droplets recombine without an energy barrier. At a critical size R*≈K/W the surface energy cost becomes too high and droplets form a radial director distribution with a monopole in the middle. Thereafter, for each pair of droplets to coalesce, they have to overcome the topological energy barrier ΔE ≈K·R $\approx K^2/W$. (We can now understand why the conditions *outside* the droplet do not significantly affect the emulsion stability, as long as the strong anchoring constraint imposed by the ordered surfactant is preserved). Near the nematic-isotropic transition the Frank elastic constant and the anchoring energy vanish (see section 2), making the energy barrier to disappear, $\Delta E \sim Q^3$. This is why the complete phase separation takes place so rapidly above T_{ni}.

The same estimate, $\Delta E \approx K^2/W$, controls the parameter of another test on the emulsion stability. We saw already that the weak force of gravity, compressing the droplets together, is not sufficient to overcome this barrier. However, if we place the established emulsion in a high-energy ultrasonic field, the droplets may again break into very small sizes. Note, however, that in order to break the droplet in two parts the same topological analysis demands the formation of a disclination ring on the surface, preferably in the breaking-neck region, in order to provide *two* point defects after the separation. The energy estimate is the same, ΔE≈K·R, and dictates the threshold value for the ultrasonic energy input of a blender (shaking by hand is certainly not enough to provide this energy per droplet).

For a typical thermotropic nematic liquid crystal far from its clearing point (as before, we take $K \sim 10^{-11}$ J/m and $W \sim 10^{-5}$-10^{-6} J/m^2) the characteristic size of stable droplets is $R^* \approx 10$-100μ, in good qualitative agreement with observations, Figure 3a. The corresponding energy barrier is very high, $\Delta E \approx 10^{-16}$ J (compare with room-temperature thermal energy $k_B T \approx 5 \cdot 10^{-21}$ J). In my view this effect of kinetic topological dereliction of nematic emulsions in an isotropic fluid matrix may serve as an explanation of why the nematic droplets continue to stay apart on preparation of PDLC materials [15], and even provide a theoretical estimate of characteristic droplet size, $R^* \approx K/W$, determined by parameters of a given material. Note that at weaker anchoring the characteristic size R^* would increase and the spherical droplets may transform into large flexible vesicles (in practice this happens at $R \sim 1$-10 mm), after which the energy barrier ΔE will decrease again. Thus, when the director anchoring on surfactant-treated interfaces is weak, no stable emulsion should be expected.

5. Conclusions

The main conclusion of this article is that the l.c. order brings an additional and relevant degree of freedom to a colloid material. The most important effect is due to the elastic energy of deformations of the l.c. director field - Frank elasticity in the bulk and surface anchoring on interfaces. Topological defects, inevitable in l.c. colloids in certain circumstances, have very large elastic energy and may dominate the macroscopic properties of the system.

Several effects have been discussed and here, in the end of this article, it may be useful to outline several possibilities to make the effects stand out even more:

• The topological defect around a colloidal particle in the strong anchoring regime (a disclination ring, or a satellite point defect) can be made much more pronounced if a particle is electrically charged (and not immediately screened by a nematic around it). In this case, the electric field emanating, at least initially, in the radial direction will enforce the radial director field near the particle surface: this would make the topological defect separation much greater than $a^* \approx 1.2$ R obtained in section 2.

• Next, analysing the long-range pair interaction potential (5), one may consider a situation when an external electric or magnetic field is applied to the system. The balance of elastic and magnetic energies introduces a new coherence length, $\xi_H \approx (K/\chi_a)^{1/2}/H$ in the case of magnetic field (where χ_a is the diamagnetic anisotropy), and the exponential decay of director distortions far away from a particle. Accordingly, the force of particle interaction also becomes short-range, proportional to $\propto \exp\left[-r/\xi_H\right]$, i.e. particles effectively decouple in a low-concentration colloid (but can be brought into the interaction range again by simply switching off the external H-field).

• The application of an external field should have a similarly distinct effect on the topologically stable l.c. emulsion described in the last section. In a sufficiently strong field the director distribution inside each droplet will become uniform (for the director anchoring, forcing the topological defect, have to be overcome: $\chi_a H^2 \cdot R^3 \gg W \cdot R^2$). Then the topological barrier for the coalescence will no longer be in effect and the droplets should coalesce rapidly, as they do on raising the temperature above T_{ni}. The estimate of the characteristic field required to force the "topologically derelict" emulsion to coalesce should be an alternative way of measuring the surface anchoring energy of the given nematic-isotropic interface: $H_c \approx (W/\chi_a R^*)^{1/2} \approx W \cdot (K\chi_a)^{-1/2}$.

The topological effects in colloid materials involving only the thermotropic nematic liquid crystal have been discussed here. One should emphasise that the conclusions should not be restricted to this system only: the basic topological and elastic effects will manifest themselves in the same way in other liquids with spontaneously broken symmetry. Two such systems should be mentioned in particular, as being more intimately related to colloid science: the lyotropic nematic phase made of cylindrical micelles and the lamellar (smectic) phase. Theoretically, no difference should be expected in lyotropic nematics - apart from different values of Frank constants K and the surface anchoring W, and their primary dependence not on temperature but on the concentration. The quantitative difference in K and W may, however, affect the control parameter of l.c. colloids (WR/K) and have an effect on practical consequences.

The colloid based on the lamellar phase, instead of a nematic, should have some qualitatively different properties. Although the basic philosophy of this article remains in force and various defects will be necessary to match between the topologically incompatible inner and outer surfaces, the additional translational symmetry breaking of smectics allows dislocations, as the other class of defects to be included into the picture. In addition, the elasticity of lamellar phases does not allow easy twisting or splaying of layers and one should expect to find many new qualitative effects in lamellar l.c. colloids. In summary, the topological concepts in liquid crystalline colloidal systems continue to promise a large number of exciting theoretical, experimental and practical problems.

Portions of this work have been supported by the DTI Colloid Technology programme and EPSRC. Special thanks to R. Ruhwandl, who took part in many discussions and practical calculations. Many useful and encouraging discussions with A.R. Rennie, R.P. Townsend, O.D. Lavrentovich, F.M. Leslie, J. Prost, S. Ramaswamy are gratefully appreciated, as well as the invaluable experience of the Polymers & Colloids group of the Cavendish Laboratory.

6. References

1. Russel, W. B., Saville, D. A. and Schowalter, W. R. (1989) *Colloidal Dispersions*, CUP, Cambridge.
2. Lekkerkerker, H. N. W. in: (1992) *Structure and Dynamics of Strongly Interacting Colloids*, S.-H. Chen *et al.* Eds., NATO ASI series C:369, Kluwer, Dordrecht.
3. Funari, S. S. , Holmes, M. C. and Tiddy, G. J. T. (1994) *J. Mater. Chem.* **98**, 3015.
4. de Gennes, P. G. and Prost, J. (1993) *Physics of Liquid Crystals*, Clarendon Press, Oxford.
5. Kuksenok, O. V., Ruhwandl, R. W., Shiyanovskii, S. V. and Terentjev, E. M. (1996) *Phys. Rev.* E**54**, 5198.
6. Ruhwandl, R. W. and Terentjev, E. M. *Phys. Rev.* E - *to be published.*
7. Poulin, P., Stark, H., Lubensky, T. C. and Weitz, D. A. (1997) *Science,* **275**, 1770.
8. Ruhwandl, R. W. and Terentjev, E. M. (1997) *Phys. Rev.* E**55**, 2958.
9. Ramaswamy, S. , Nityananda, R., Raghunathan, V. A. and Prost, J. (1996) *Mol.Cryst.Liq.Cryst.* **288**, 175.
10. Poulin, P. , Raghunathan, V. A., Richetti, P. and Roux, D. (1994) *J Physique II* **4**, 1557.
11. Ruhwandl, R. W. and Terentjev, E. M. (1996) *Phys. Rev.* E**54**, 5204.
12. Bibette, J., Morse, D. C., Witten, T. A. and Weitz, D.A. (1992) *Phys. Rev. Lett.* **69**, 2439.
13. Clint, J. H. (1992) *Surfactant Aggregation*; Blackie, Glasgow-London.
14. Volovik, G. E. and Lavrentovich, O. D. (1983) *Sov. Phys. JETP*, **58**, 1159.
15. Zumer, S., Golemme, A. and Doanne, J.W. (1989) *J. Opt. Soc.* A**6**, 403;
Drzaic, P. S. and Muller, A. (1989) *Liq. Cryst.* **5**, 1467.
16. Terentjev, E. M. (1995) *Europhys. Lett.* **32**, 607.

FILTRATION OF DEFORMABLE PARTICLES IN SUSPENSION

G. H. MEETEN
Schlumberger Cambridge Research
High Cross
Madingley Road
Cambridge CB3 0EL
U.K.

Filtration is described of neutrally–buoyant suspensions of deformable (swollen cross–linked dextran) and rigid (polyethylene) particles. The filtration pressure P and suspension concentration were varied. The rigid particles gave an incompressible filtercake. The deformable particle filtercake was incompressible for $P < 1.5$ kPa, but compressible for $P > 3$ kPa. This behaviour is compared with that of emulsions.

1. Introduction

A filtercake is compressible if its porosity depends on the stress applied to its solids matrix. Colloidal filtercake compressibility originates mainly from interparticle forces; it has been widely studied. Compressibility of deformable–particle suspensions can arise from the pressure–dependence of the packing efficiency; this mechanism has received little attention. Particles in a compressible filtercake will be most concentrated close to the septum (filter). There may be no sharp distinction between cake and suspension, and the pressure–dependence of filtration is weaker than for an incompressible filtercake. Here we use rigid polyethylene particles for the incompressible filtercake, similar to those used by Mackley and Sherman [1] for cross–flow filtration. Cross–linked dextran particles (Sephadex) gave the compressible filtercake. Evans and Lips [2] have studied the effect of particle deformability on the rheology of Sephadex gels.

2. Theory

We briefly review unidirectional filtration. The suspension has void ratio (liquid volume ÷ solids volume) e_0 and forms a filtercake of void ratio e_1. Void ratio e, solids volume fraction (solidosity) v, and liquid volume fraction (porosity) ε are related by $e = (1-v)/v = \varepsilon/(1-\varepsilon) = \varepsilon/v$.

R. H. Ottewill and A.R. Rennie (eds.), Modern Aspects of Colloidal Dispersions, 269–278.

2.1 INCOMPRESSIBLE FILTERCAKE

The quantity G = (cake volume)/(filtrate volume) is constant throughout the process. If v and v_{cake} refer to suspension and cake, respectively, then a volume balance gives

$$G = \frac{1+e_1}{e_0 - e_1} = \frac{v}{v_{\text{cake}} - v}. \tag{1}$$

If a filtercake of unit area acquires thickness dL in a time dt, the filtrate volume is $\mathrm{d}Q = \mathrm{d}L/G$, i.e. $L = GQ$. For cake of permeability k Darcy's law is $\mathrm{d}Q/\mathrm{d}t = kP/\eta L$ for a filtrate of viscosity η. Eliminating dQ gives $\eta L\,\mathrm{d}L = GkP\,\mathrm{d}t$ which integrates from $L = t = 0$ to give $\eta L = 2GkPt^{1/2}$. Using $L = GQ$ the desorptivity S is then given by

$$Q = \left(\frac{2kP}{\eta G}\right)^{1/2} t^{1/2} \equiv S t^{1/2}. \tag{2}$$

2.2. COMPRESSIBLE FILTERCAKE

One–dimensional compressible–cake filtration has been widely–described ([3], [4], [5], [6]). Standard analysis, e.g. Philip & Smiles [4], assumes the cake void ratio e to depend only on the uniaxial stress σ within the solids matrix, and the permeability k to depend only on e. It is then shown that cake growth is governed by a diffusion equation for e, with a diffusivity $D(e) = k(e)/[\eta(1+e)\kappa(e)]$ where η is the viscosity of the filtrate and $\kappa(e) = -\mathrm{d}e/\mathrm{d}\sigma$ is a matrix compressibility. The void ratio e_1 is smallest at the septum where the matrix stress $\sigma = P$ is a maximum, and rises through the cake to e_0 in the suspension. During constant pressure filtration e_1 and e_0 are constant. For a given suspension void ratio e_0, and filtration pressure P (or e_1 at the filter), the liquid flux q_{rel} relative to the solids matrix will be a function of position (i.e. of e) and t, but the flux ratio $F = q_{\text{rel}}(e,t;e_1,e_0)/q_{\text{rel}}(e_1,t;e_1,e_0)$ is t–independent . Philip & Knight [7] show that

$$S^2(e_0,e_1) = 2\int_{e_1}^{e_0} \frac{(e_0 - e)D(e)}{F}\,\mathrm{d}e. \tag{3}$$

By definition $F = 1$ at the septum ($e = e_1$) and $F = 0$ in the suspension ($e = e_0$). Smiles and Harvey [3] assumed $F = [(e_0 - e)/(e_0 - e_1)]^r$ where $0 < r < 1$, and $r = 2 - \pi/2$ makes Equation (3) exact when $D(e)$ is constant. The lower and upper bounds for S are given by $r = 0$ and $r = 1$ respectively (van Duyn and Peletier [8], Sherwood [9]).

2.3. SEPTUM RESISTANCE

Resistance to filtrate flow through the septum will reduce the pressure which forms the filtercake. The filtration pressure can be written $P = P_A - R\,\mathrm{d}Q/\mathrm{d}t$, where P_A is the pressure applied to the filter cell, and R is the hydraulic resistance coefficient of the septum. For the incompressible cake this assumption gives

$$t = Q^2/S^2 + RQ/P_A . \qquad (4)$$

For the incompressible polyethylene particles the filtrate volume $V = QA$ was found to fit $t = a + bV + cV^2$, where A is the filter cell area, as predicted by Equation (4). Term a allows for timing error. For compressible cake filtration, Equation (3) does not give $S(P)$ analytically and a relation such as Equation (4) cannot be derived. However, the $V(t)$ data for compressible filtercakes still fitted $t = a + bV + cV^2$, and S thus obtained.

3. Experimental

3.1. METHOD

The filtration cell was a 40 mm ID acrylic pipe with acetal end–caps. The suspension filtrand and compressed air entered by the upper cap. The lower cap held a 10 micron aperture nylon filter cloth on a rigid support grid. The time–dependence of the filtrate volume was measured with a stop–watch and a graduated cylinder. In filtration experiments, filtercake grew as the filtrate volume was measured. The suspension concentration v and the filtration pressure P were varied. Both suspensions gave filtercake thickness measurable to ±0.3 mm using a rule outside the pipe. In the permeation experiments, filtrate above a pre–formed cake was driven through using a range of pressure.

The rigid particles were polyethylene grains (Hoechst; Hostalen GUR 412). They were polydisperse, having < 8%wt greater than 250 micron, < 16%wt less than 63 micron, and of density 940 kg m^{-3}. Microscopy showed an irregular shape, but not strongly anisometric. Deionised water (64.3%vol) and methanol (35.7%vol) were mixed to make a neutrally–buoyant suspending liquid. Particles were dispersed by adding 0.75%wt Tergitol 7 ($C_{17}H_{35}NaO_4S$) to the water–methanol mixture. The time for settling was long compared with that of filtration.

The deformable particles were water–swollen, cross–linked dextran (Pharmacia; Sephadex G100S), as studied rheometrically by Evans and Lips [2], who describe their swelling and give the particle size moments to be $d_{1,0}$ = 67 micron, and $d_{3,0}$ = 93 micron. The partial specific volume is given (Pharmacia) as 0.65 mL g^{-1}, but owing to the high swelling ratio (11.1 mL g^{-1} [2]) the density of the swollen particle was close to water, and the settling time was long compared with that of filtration.

The 10 micron filter cloth septum retained the polyethylene and Sephadex particles, but its resistance was not negligible and data were analysed accordingly; *vide* §2.3.

4. Results

4.1. RIGID POLYETHYLENE PARTICLES

Desorptivity S was measured as a function of pressure P and solidosity v.

4.1.1. *Desorptivity at fixed v and variable P*

Figure 1 shows S to be linear in $P^{1/2}$ as expected from Equation (2). The filtrate viscosity was η = 1.80 mPa s and the (P–independent) cake solidosity was v_{cake} = 0.495. The gradient of Figure 1, using Equation (2), gave a filtercake permeability k = 8.63 μm^2.

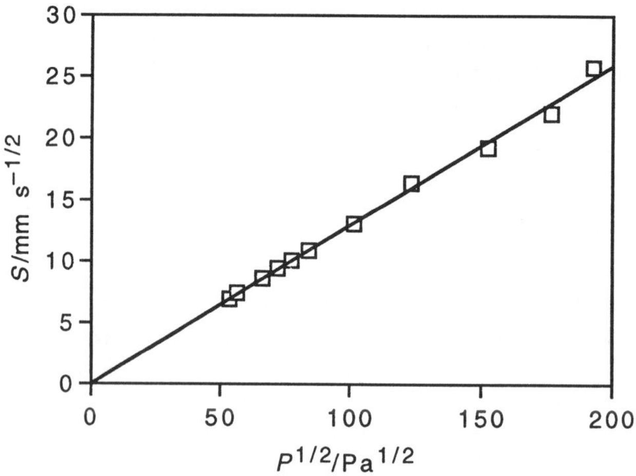

Figure 1. Dependence of S on filtration pressure P for a polyethylene particle suspension of v = 0.18.

4.1.2. *Desorptivity at fixed P and variable v*

Figure 2 shows a plot of S versus $G^{-1/2}$, showing linearity as expected from Equation (2), except for the most concentrated suspensions (small G), where v is close to v_{cake}, and for which S appears to be too large. This is attributed to some structure in the suspension which is transferred to the cake and increases its permeability. The existence of a shear yield stress τ_0 is evidence of suspension structure, and τ_0 of the suspension was measured using a vane attached to the spindle of a rheometer. Figure 2 shows τ_0 becomes non–zero at the suspension concentration where S deviates from the $G^{-1/2}$ dependence. The line through the yield stress data fits τ_0 to $v^{16.5}$; a power law being expected for fractal flocs. The line gradient of Figure 2 gives, using Equation 2, a cake permeability k = 8.05 μm^2, in fair agreement with k = 8.63 μm^2 from the variable–P data (§4.1.1).

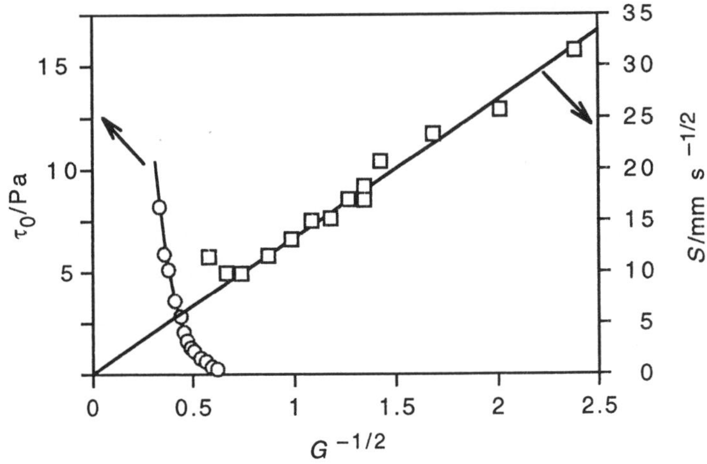

Figure 2. G–dependence of τ_0 (circles) and S (squares) for polyethylene particle suspensions.

4.1.3. *Measurement of filtercake permeability*

The filtrate gravity–permeated a pre–formed cake. If $V(t)$ is the collected permeate volume and V_0 the suspension height above the septum at t = 0, analysis gives $\ln(1-V(t)/V_0) = -k\rho gt/\eta h$. Figure 3 shows $\ln(1-V(t)/V_0)$ versus t. Its gradient gives k = 8.55 μm^2; in good agreement with 8.63 μm^2 in §4.1.1 and 8.05 μm^2 in §4.1.2.

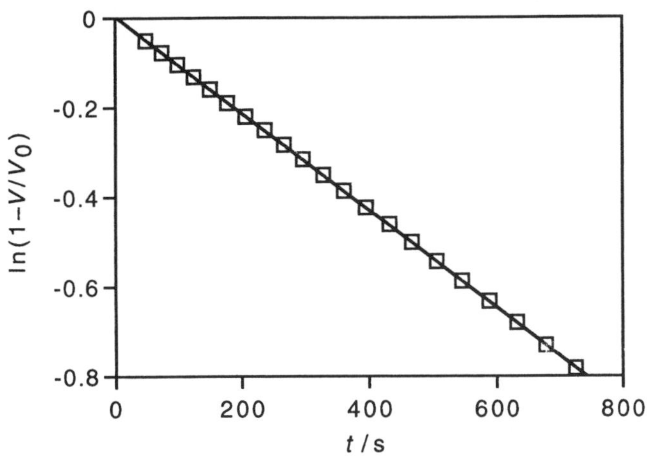

Figure 3. Variable–head permeation of polyethylene filter cake by filtrate. See §4.1.3.

4.2. DEFORMABLE SEPHADEX G100S PARTICLES

We measured the effect of pressure and suspension concentration on desorptivity, and the effect of pressure on the permeance and cake thickness of a pre–formed cake.

4.2.1. *Desorptivity at fixed v and variable P*

Plots (Figure 4) of S versus $P^{1/2}$ for suspension concentrations c of 5 and 20 gL^{-1} show a linear $P^{1/2}$ dependence up to about 1.5 kPa, and S nearly P–independent above about 3 kPa. This behaviour contrasts strongly with that of rigid particles (Figure 1).

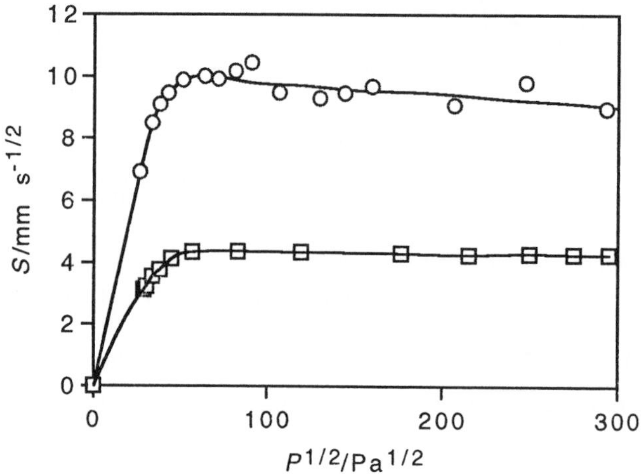

Figure 4. Desorptivity pressure–dependence of Sephadex at c = 5 g/L (circles) and c = 20 g/L (squares).

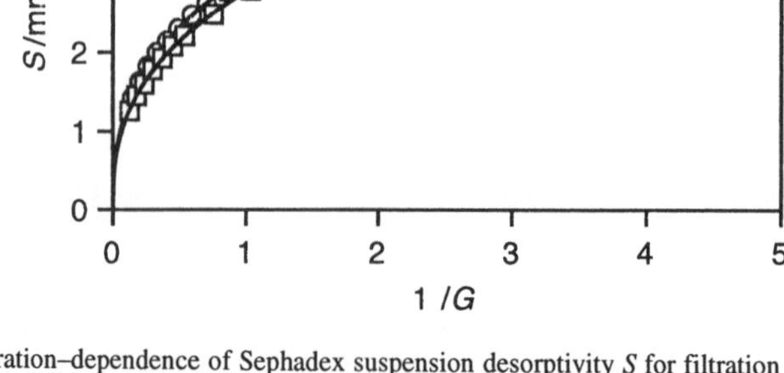

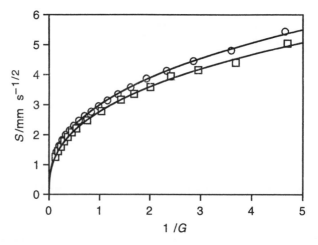

Figure 5. Concentration–dependence of Sephadex suspension desorptivity S for filtration pressure P = 31.37 kPa (circles) and P = 1.37 kPa (squares). Lines show the fit to equation (6).

4.2.2. *Desorptivity at fixed P and variable v*

The desorptivity S was measured over a range of suspension concentration c at pressures each side of the transition region. Linearity between S and $G^{-1/2}$, as for the polyethylene particles, was absent. A good empirical fit was found between S and G_c^{-n}, where

$$G_c = c/(c_{\text{cake}} - c) \tag{5}$$

is the same as G previously defined only if c and c_{cake} are similarly proportional to v and v_{cake}. This may not be so if deformable particles change volume. Figure 5 shows S fitted to power–law plots in G_c for $P = 1370$ Pa and 31370 Pa; c_{cake} being chosen to minimise the fit errors. Table 1 shows the P–dependence of f, n and c_{cake} in the empirical relation

$$S = fG_c^{-n}. \tag{6}$$

Within experimental uncertainty the value of n is shown to be independent of the filtration pressure P, either side of the transition pressure region.

Table 1. Parameters of Equations (5) and (6)

P/Pa	f/mm s$^{-1/2}$	c_{cake}/g L^{-1}	n
1370	2.771	57	0.375
31370	2.997	56.5	0.377

4.2.3. *Sephadex particle bed permeance*

A sedimentary cake was grown from a 20 g L^{-1} suspension, and the clear supernatant driven through it by a pressure of $P = 1200$ Pa, the cake height L and the filtrate volume V subsequently being measured with time. Figure 6 shows that from 0 to about 100 s the cake compacts, and the outflow rate $\mathrm{d}V/\mathrm{d}t$ decreases with time.

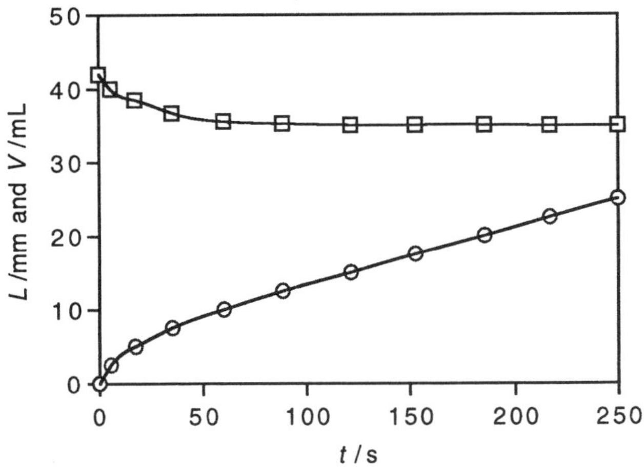

Figure 6. Time–dependence of filtercake height L (squares) and permeate volume V (circles) during permeation of pre-formed filtercake by filtrate at pressure $P = 1200$ Pa. See §4.2.3.

Initially most filtrate is from the compacting cake. After about 50 – 100 s the cake attains equilibrium compaction, dV/dt is time–independent, and the filtrate outflow is mostly from the filtrate supply above a time–stable cake. The cake is not necessarily of uniform concentration as a deformable cake's concentration is greatest at the septum; at the greatest matrix stress. An average cake permeability k_{av} may be obtained from the application of Darcy's law to the cake when L and dV/dt are time–independent, using $dV/dt = k_{av}PA/(\eta L)$. This average permeability was measured as a function of P by repeating the above experiment for various pressures. Figure 7 shows the pressure–dependence of the equilibrium bed height L and equilibrium outflow rate dV/dt.

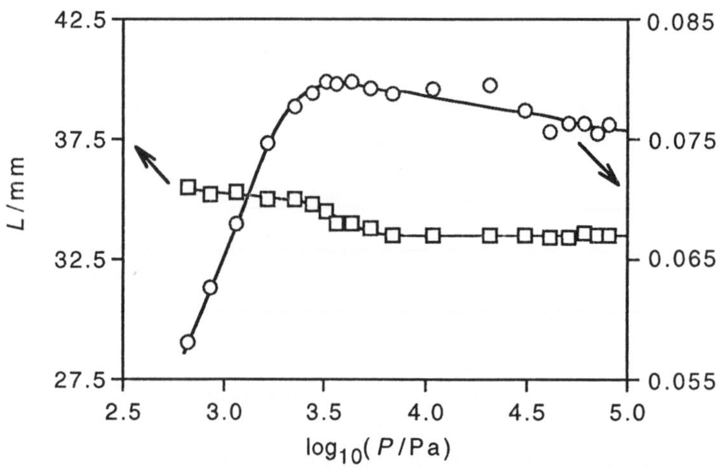

Figure 7. P–dependence of Sephadex filtercake equilibrium height L (squares) and flux dV/dt (circles).

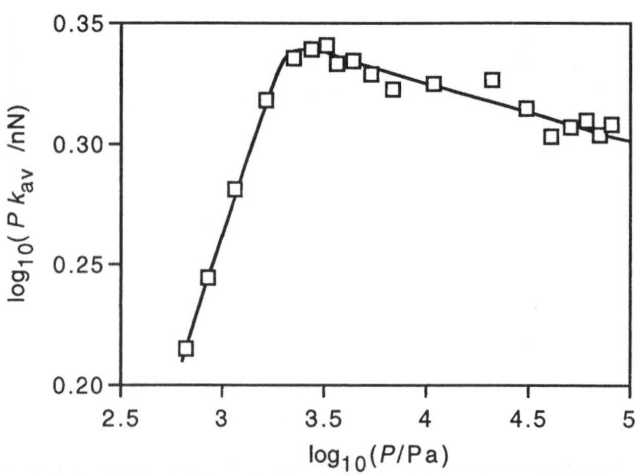

Figure 8. Dependence of Pk_{av} on pressure for Sephadex filtercake, calculated from data shown in Figure 7.

The bed height is shown to decrease mostly between 2 – 6 kPa, by about 4%, and it is also in this region that dV/dt begins to decrease as P increases. The data show that k_{av} decreases with increasing P over the whole pressure range, but with different dependencies below and above the transition region. Figure 8 plots $\log_{10}(k_{av}P)$ versus $\log_{10}P$, showing that $k_{av} \propto P^{-0.74}$ and $k_{av} \propto P^{-1.02}$ on the low and high pressure sides of the region, respectively. Both differ strongly from the pressure–independent k_{av} of the incompressible filtercake.

5. Discussion

The rigid particle results agree largely with textbook theory for incompressible filtercakes but previous experimental confirmation has been sparse.

The results for the crosslinked dextran show a behaviour which is strongly dissimilar to that of the rigid particles. The demonstration of a transition pressure region in filtration and permeation of the Sephadex particles appears to be novel, although Galvin [10] has stated that Sephadex G200 particles are rigid at pressures less than 1600 Pa and deformable above. Andrei *et al* [11] have studied the deformation of single swollen Sephadex particles (G25 and G50) without observing any unexpected variation in the force–deformation behaviour. Their minimum applied force was however 10^{-5} N, so that all measurements were at pressures above the transition pressure region. These workers also found that under a constant applied strain the stress of a Sephadex particle relaxed in a time similar for a filtercake to equilibriate, shown in Figure 6 to be about 50 to 100 s. This relaxation they suggest originates from the migration of liquid inside the particle to the ambient liquid outside, and it is likely that this mechanism of particle deformation relaxation also controls the filtercake equilibriation.

Despite the differences in behaviour of k_{av} below and above the transition pressure, the filtercake concentrations c_{cake} below and above the critical pressure (Table 1) are almost identical; 57 and 56.5 gL^{-1}, respectively. These compare closely with the Sephadex concentration at which shear elasticity first appears; 5.95 wt% found by Evans and Lips [2] (their figure 3), which is 57.4 gL^{-1} in our units. The similarity of c_{cake} below and above the critical pressure, and the strong dependence of k_{av} on P, suggest that the permeability is strongly changed for little change in filtercake concentration. This result is predicted in the emulsion model of Sherwood [12]. He considers the filtration of a suspension of deformable liquid droplets immiscible with a continuous liquid phase, and shows that the permeability is indeed a strong function of filtercake concentration when the droplets close–pack. This model also leads (Sherwood's figure 6) to a pressure–independent desorptivity for sufficiently large pressure, as we find experimentally. In Sherwood's model however the individual fluid droplets are impermeable, with a Laplacian pressure arising from the curvature of their boundary and the difference in surface tension across it. The filtrate flows around and not through the droplets. Thus although some experimental results herein are qualitatively in accord with Sherwood's model, it is not certain whether our water–swollen crosslinked dextran particles are sufficiently impermeable to resemble emulsion droplets, or whether the liquid migration from the particle in stress relaxation (Andrei et al [11]) indicates that most of the liquid transport through the filtercake is in fact mostly through the particles.

The transition pressure behaviour we find in filtration may be compared with the behaviour of the elastic shear modulus G' measured by Evans and Lips [2]; their figures 3 and 4. The connection between rheology and filtration can be made through the observation of Mason *et al.* [13] that the osmotic pressure required to compress an emulsion is nearly the same as its shear modulus G'. The G' data [2] show two distinct domains, separated by a region spanning approximately 2 – 3 kPa, very similar to the filtration transition pressure range 1.5 – 3 kPa. The connection of Mason *et al* [13] shows the matrix compressibility $\kappa(e)$ (*vide* §2.2) to be small for $P < 2$ kPa and large for $P > 3$ kPa; a ratio of about 5. This explains the approximate $P^{1/2}$ dependence found for S at low pressure (Figure 4); the particle matrix behaved as incompressible. Above the transition pressure, both the near–independence of S on P, and the strong inverse dependence of the permeability on pressure ($P^{-1.02}$; Figure 8) are thought to result from the high compressibility of the particle matrix.

This work was carried out within the DTI Colloid Technology Project.

1. Mackley, M. R. and Sherman N. E. (1992) *Chem. Engng Sci.* **47**, 3067.
2. Evans, I. D. and Lips A. (1990) *J. Chem. Soc. Faraday Trans.* **86**, 3413.
3. Smiles, D. E. and Harvey E. G. (1973) *Soil Sci.* **116**, 391.
4. Smiles, D. E. and Kirby J. M. (1987) *Sep. Sci. Technol.* **22**, 1405.
5. Philip, J. R. and Smiles D. E. (1982) *Adv. Colloid Interface Sci.* **17**, 83.
6. Kirby, J. M. and Smiles D. E. (1988) *Aust. J. Soil Sci.* **26**, 561.
7. Philip, J. R. and Knight J. H. (1974) *Soil Sci.* **117**, 1.
8. van Duyn, C. J. and Peletier L. A. (1977) *Nonlinear Anal. Theory Methods Appl.* **3**, 223.
9. Sherwood, J. D. (1993) *Chem. Engng Sci.* **48**, 2767.
10. Galvin, K. P. (1996) *Chem. Engng Sci.* **51**, 3241.
11. Andrei, D. C., Briscoe B. J., Luckham P. F. and Williams D. R. (1996) *J. Chim. Phys.* **93**, 960.
12. Sherwood, J. D. (1993) *Chem. Engng Sci.* **48**, 3355.
13. Mason, T. G., Bibette J. and Weitz D. A. (1995) *Phys. Rev. Lett.* **75**, 2051.

THE RHEOLOGY, PROCESSING AND MICROSTRUCTURE OF COMPLEX FLUIDS

C. BOWER, M. R. MACKLEY, J. B. A. F. SMEULDERS,
D. BARKER AND J. HAYES.
Department of Chemical Engineering
Pembroke Street
Cambridge
CB2 3RA
UK

This paper describes the development of two pieces of equipment that enable precision rheological, processing and microstructure studies to be carried out on rheologically and structurally complex fluids. Using the experimental equipment detailed, we also report microstructure observations on a variety of structured materials which show shear rate dependent aggregation or alignment effects. A droplet chaining effect, in which alignment of the disperse phase of certain droplet dispersions in viscoelastic fluids was also seen.

1. Background

There are many commercial situations where the manner of processing of certain complex fluids can have a significant effect on the quality of the final product. For example the rheology and texture of surfactant-water solutions encountered in detergents and other cleaning products, [1] is intimately connected with the orientation of anisotropic micelles within the water matrix, [2] and this in turn depends on the processing shear history of the solution. The final properties of paints, [3-5] and printing inks, [6-9] are other examples of where pre-processing of the fluid can have a significant influence on resulting properties. It is therefore important that the link between processing and properties can be made and that experimental equipment is available to enable this link to be made. At present high quality rheological apparatus such as the Rheometrics, Bohlin and Carri Med are available but these apparatus do not cover all of the complex processing conditions that are met in many industrial processes, and these instruments do not necessarily provide direct microstructural information. This paper describes two pieces of apparatus where we have attempted to address both these issues, we report both technical specifications for the instruments and results obtained from them.

R. H. Ottewill and A.R. Rennie (eds.), Modern Aspects of Colloidal Dispersions, 279–288.

2. The Multipass Rheometer (MPR).

The multipass rheometer shown schematically in Figure 1, is a double piston apparatus for obtaining controlled shear conditions in a fully constrained environment [10,11]. The apparatus consists of a central test section which is normally a capillary, but can also be a flow visualisation section or a baffled tube. The test material is constrained within the test section by two servo hydraulically driven pistons which can be operated independently or together. By moving one piston alone, the test material can be pressurised to between 1-250 bar. The system is temperature jacketed which also means experiments can be carried out between 5-250° C. After the mean test pressure is set, the two pistons can then be moved synchronously in order to achieve a constant velocity and constant volume capillary measurement. Alternatively the pistons can be oscillated to obtain linear viscoelastic data. A strategic advantage of the apparatus is that the test material is fully enclosed and that the same material can be subjected to many measurements by virtue of it being forced repeatedly within the test section from one cavity to the other.

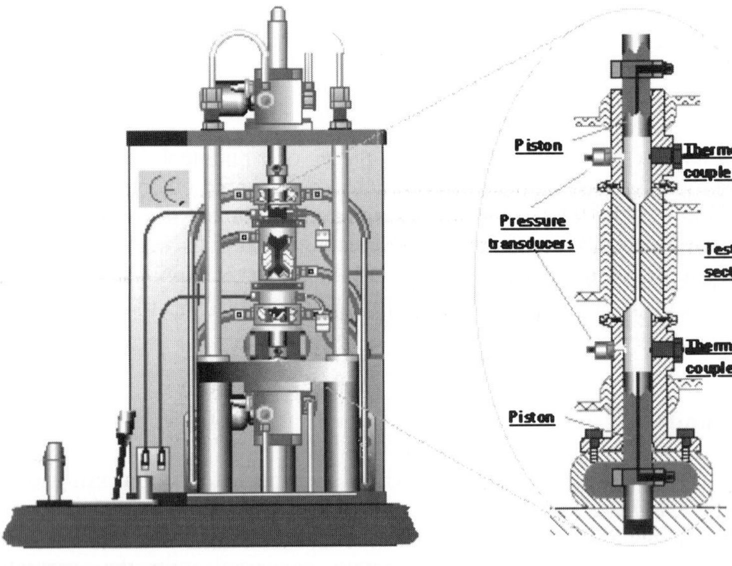

Figure 1 The multipass rheometer with close up of the insert, in this case a capillary, (reproduced with permission from [10]).

An example of the MPR capability is shown in Figure 2, where both the differential pressure across the test section and piston displacement are recorded for a polyisobutylene solution. The data shows a repeatable pressure trace for a multipass steady mode of operation. The two pistons initially move together at a constant upwards velocity. There is a pre-set delay time before they move down again at the same velocity. In this way many successive steady flow measurements can be achieved on the test fluid. The range of shear rates that can be achieved is dependant on the geometry of the test section used, for example, selection of a 40mm capillary with diameter of 8mm allows shear rates between 1-500s^{-1}, whilst using a 1mm diameter capillary of the same length allows shear rates of between 500-2.5x$10^5 s^{-1}$.

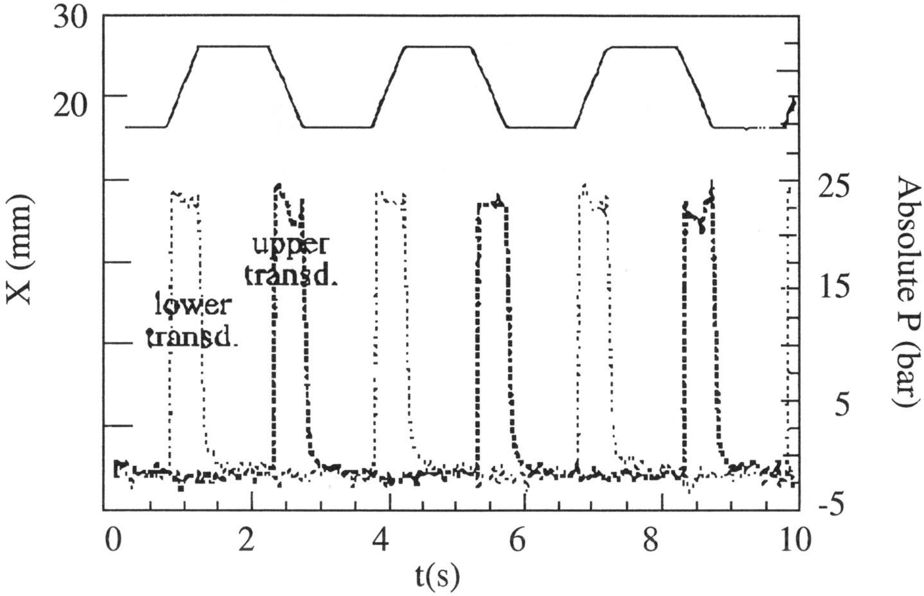

Figure 2 Sample results from MPR showing piston displacement and pressure variation at either end of the capillary, (reproduced with permission from [10])

In Figure 3 data for the storage and loss moduli G' and G" is shown for the polyisobutylene solution. This was obtained by applying a coupled sinusoidal oscillation to the two pistons. The data is compared with that obtained using an RDSII (Rheometrics) dynamic spectrometer. The agreement is reasonable and demonstrates the ability of the machine to make both oscillatory and steady flow capillary measurements. We have found that the fully constrained geometry of the device enables processing and rheological experiments to be carried out on materials that contain a significant volatile component. For example experiments using 20% cellulose acetate in acetone have successfully been carried out for a wide range of process conditions and the use of different diameter capillaries enables shear rates in the range of $0.1\text{-}10^5 s^{-1}$ to be explored. The use of servo hydraulic pistons enables substantial pressures to be achieved and we have found the machine is well suited for the study of soft solids that are intermediate between liquids and solids. Interpretation of the differential pressure data is however not necessarily straight forward as in some cases the wall slip boundary condition is not known.

Recently we have developed optical imaging techniques to study the flow behaviour of certain fluids using the MPR and this, together with the versatile software for driving the pistons means that the apparatus is well suited to a broad range of studies including, rheology, processing, microstructure and reactions.

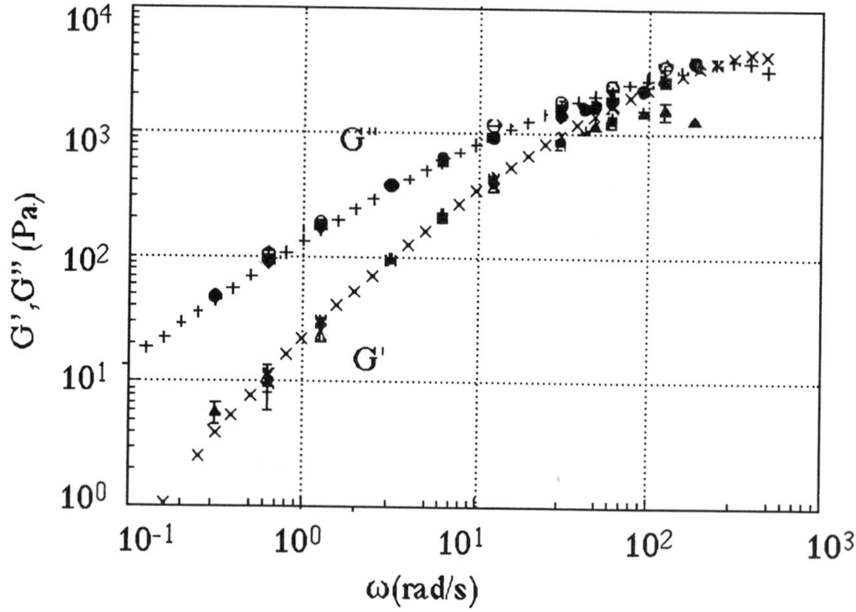

Figure 3 Comparison of G'(×), G'' (+) data from RDSII and G' (▲), G'' (●) data from MPR for 20% PIB in Decalin using an 8mm diameter, 40mm long capillary. (Reproduced with permission from [10])

3. The Cambridge Shear System - CSS450

In addition to the MPR, fluid microstructure observations were performed using a specialised optical shear cell having parallel plate geometry (initially developed in Cambridge and built by Linkam Scientific Instruments, Surrey). Approximately 2cm^3 of sample is placed between two circular quartz discs, with diameters of 32mm and 55mm. The larger bottom disc is rotated by a stepper motor to allow application of a controllable shear rate. A second stepper motor allows the separation of the discs to be varied so that the sample gap could be adjusted from 2500μm down to10μm. Changing the wave-form on the input of the stepper motor allowed the shear applied to the sample to be either continuous, step or oscillatory. With this equipment it was therefore possible to generate shear conditions identical to those on a conventional parallel plate rheometer such as the RDSII (Rheometrics). The range of shear rates attainable were dependant on the gap setting, for example with a 10μm gap it was possible to achieve shear rates of between 0.1-3000s^{-1}. Temperature control was achieved using two silver-block heating elements above and below the quartz discs, to allow precision temperature adjustment between 20 and 450° C. Full control of the shear cell was facilitated by a software interface running on a personal computer. Figure 4 shows elevated and side view schematics of the shear cell. The optical stage was mounted on a standard Olympus optical microscope with a JVC CCD camera fitted so that experiments could be recorded onto S-VHS video tape. Feeding the signal from the CCD into a frame grabber board also allowed capture of digital images facilitating measurements of droplet size, elongation and orientation with standard image analysis software.

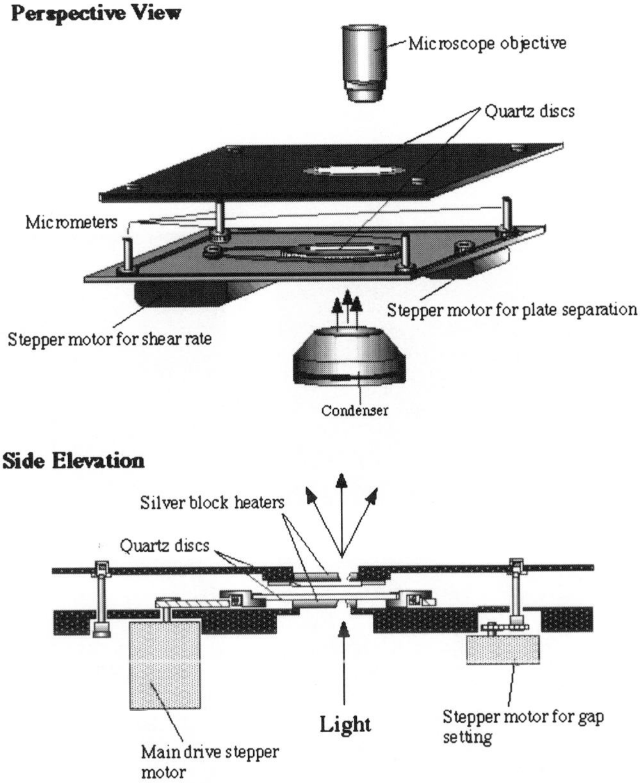

Figure 4 Schematic diagram of the CSS450 optical shearing stage used for detailed microstructure observations.

The optical shear cell facilitated microstructure observations for a wide range of liquid samples and complex fluids such as emulsions, solid suspensions, polymer melts and liquid crystals, all under controlled temperature and shear conditions Some examples of shear rate dependant microstructure changes in various complex fluids are described below.

3.1. Equine blood

The study of blood rheology is a subject in its own right, blood is an extreme example of a complex fluid composed of a large number of biological proteins and liquids, and numerous deformable solid components [12-14]. The most frequently encountered entity in a typical sample of blood is the red blood cell (RBC), a corpuscular solid with a diameter of about 5μm that is responsible for the familiar colouration of the fluid. In a static sample of blood, red blood cells spontaneously aggregate to form stacks of cells or 'rouleaux' [15], as the process continues the rouleaux become longer and join with other rouleaux so that the entire sample becomes one convoluted aggregate. The aggregation behaviour of blood is a highly shear rate dependant process, exhibiting various states of aggregation and disruption. Figure 5 shows images of equine blood which has been sheared at $1000s^{-1}$ to create a uniform dispersion, and then subjected to a steady shear rate of either $5s^{-1}$ or $10s^{-1}$ the RBC's are seen to aggregate into large clusters over a period of 120s. Temperature control allows the sample to be kept at the biological temperature of

37.5°C throughout the experiments.

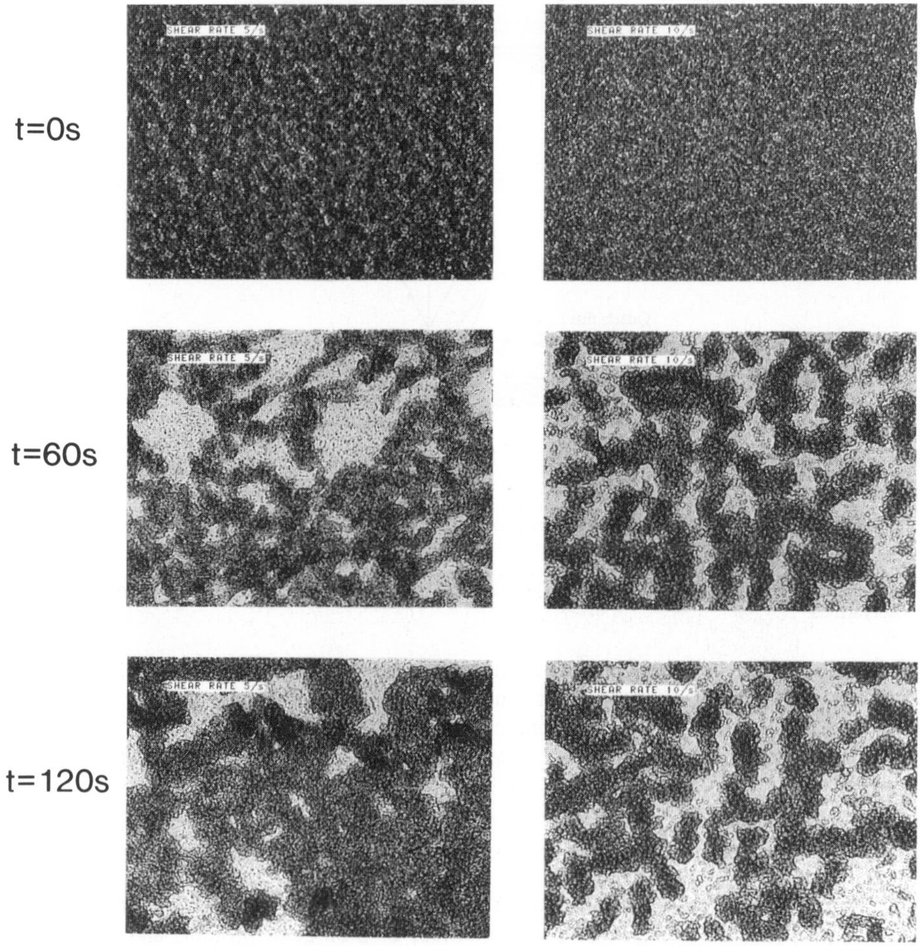

Figure 5. Microstructure changes in equine blood subjected to a steady shear rate. The sample was first sheared at $1000s^{-1}$ before reducing the shear rate to either 5 or $10s^{-1}$ at t=0, aggregation behaviour was seen to be a strong function of the applied shear rate. The shear direction is vertical in all images.

3.2. Shampoo

The formulations of commercial detergents, soaps and other cleaning products are becoming increasingly more sophisticated to meet the demands of the consumer, so that a product may frequently consist of an emulsion or dispersion with many different components, [16]. The shampoo under test is a commercial example of a shear-thinning viscoelastic emulsion, consisting of an aqueous surfactant solvent with a disperse oil phase stabilised by polymer molecules [17. Under conditions of zero shear, or high shear rates exceeding $10^5 s^{-1}$, the oil phase droplets were uniformly dispersed throughout the aqueous surfactant solution. However when the sample was subjected to intermediate shear rates of $40\text{-}80 s^{-1}$ droplet chaining and subsequent coalescence of the oil droplets was observed. Figure 6 shows the effect of increasing steady shear rates on shampoo microstructure. At $40 s^{-1}$ the droplets begin to align into chains oriented along the shear direction (vertically down the page) this behaviour was even more pronounced at a shear rate of $80 s^{-1}$ when droplet coalescence also becomes apparent. Droplet coalescence continued under shearing at $80 s^{-1}$ and after a period of approximately ten minutes the droplet chains started to form elongated filaments that were stable on cessation of the shear. At higher shear rates (> $1000 s^{-1}$) the filaments and droplet chains were disrupted to form discrete droplets.

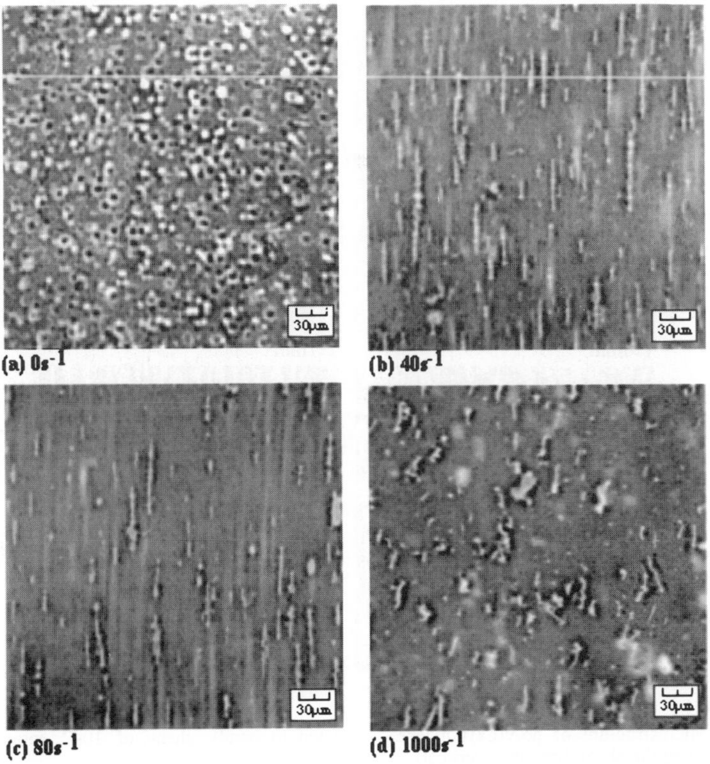

Figure 6. Effect of steady shear rate on shampoo microstructure. Under zero shear the shampoo is a well dispersed emulsion with droplet diameters of approximately 10μm (a). At intermediate shear rates a droplet chaining phenomena was observed along the direction of shear (vertical) (b), with droplets coalescing to form filaments under continued shearing, (c). At higher shear rates droplet chains and filaments were disrupted (d).

3.3. Poly(oxyethylene) solutions

Surfactants are of great importance in both industrial chemical and consumer sectors. They are used in a large variety of applications as stabilising agents and emulsifiers, or as cleaners and detergents. One of the most frequently encountered uses is in aqueous solution, where at low concentrations the surfactant molecules aggregate to form micelles, whilst at high concentrations they adopt a liquid crystalline phase, [18, 19]. Poly(oxyethylene) / water systems form lamella liquid crystals at concentrations of 65-80% w/w, and consequently the rheological response is strongly dependant on the rearrangement of structure during flow. When a liquid crystal surfactant system is subjected to a steady shear rate, it is commonly found that the sample viscosity decreases over a period of several minutes [2], and further repeats of the experiment yield decreasing viscosities. The material properties are therefore strongly dependant on the shear history of the sample, and are intimately connected with the degree of structural alignment of the microstructure. Figure 7 shows the microstructural changes in a sample of lamella liquid crystal surfactant (80% w/w Triton N-101 in aqueous solution) when subjected to a steady shear rate of $10s^{-1}$. Over a period of 10 minutes the material becomes aligned along the shear direction (vertically down the page). Observation of the sample using crossed polarizers further confirms the observed alignment, which persists after cessation of the applied shear.

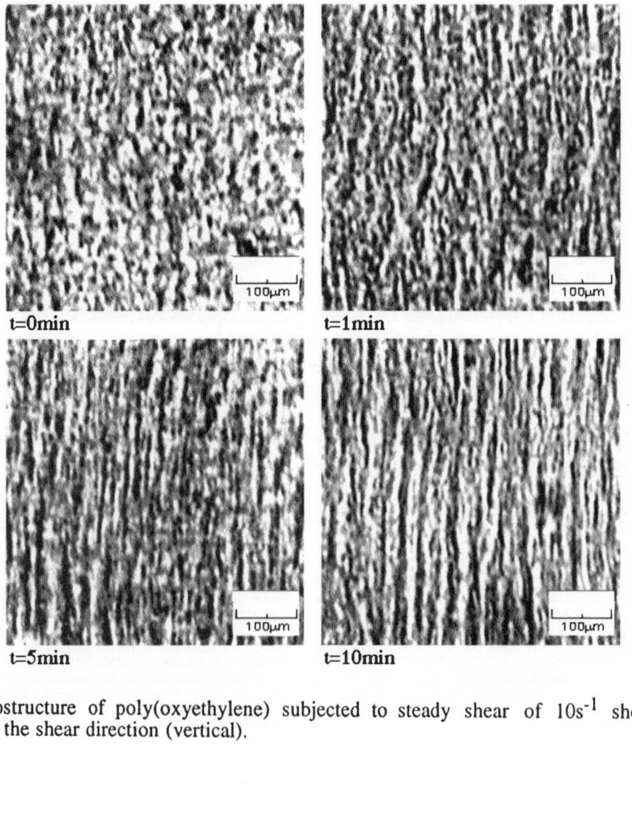

Figure 7 Microstructure of poly(oxyethylene) subjected to steady shear of $10s^{-1}$ showing structural alignment along the shear direction (vertical).

4. Summary and conclusions

The rheology and processing of complex fluids is a difficult issue, and behaviour under controlled shear conditions is often difficult to predict when fluid microstructure changes are taking place. Examples of this behaviour occurring in a variety of materials have been shown in the Figures of this paper. Such microstructure changes must be accounted for if a successful model of the material behaviour is to be developed. Most constitutive equations currently in use assume a continuum model, where the material rheology in arbitrary finite elements is assumed to be a continuum. The microstructural observations presented here suggest that complex fluid rheology may in fact be significantly changed in localised regions of the material, when the sample is subjected to shear stress.

The photomicrographs show the striking shear induced blood cell aggregation effect for equine blood, and the observations demonstrate clearly that rouleaux formation in equine blood is as strong if not stronger than in human blood. At present the mechanism for rouleaux formation is not fully understood in either human or equine blood, however observations of the type made here may help in the future to resolve the issue. The reported alignment of shampoo droplets in a viscoelastic fluid to form chains of aligned droplets is another, as yet unexplained effect which in our opinion deserves closer attention.

The apparatus and techniques described here have allowed exploration of a range of experimental conditions previously unattainable with existing commercial apparatus. The ability to explore both temperature and pressure effects on sample rheology in a constrained environment makes the MPR well suited to the quantitative study of a wide range of processing conditions. Additional optical microstructure facilities have now been installed on the MPR which means both processing rheology and microstructure studies can now be achieved with precision using just one machine.

Acknowledgments

We wish to acknowledge the financial support of the DTI/ICI/Unilever/ Schlumberger/ Zeneca Colloid Technology grant which enabled the prototype MPR to be constructed. We also thank both Eland Test Plant and Linkam Scientific for their skills in constructing the MPR and CSS 450 respectively.

References

1. Paasch, S., Schambil, F. and Schwuger, M. J.(1989) *Langmuir*, **5**, 1344.
2. Franco, J. M., Munoz, J. and Gallegos, C.(1995) *Langmuir*, **11**, 669.
3. Liddell, P. V. and Boger, D. V.(1994) *Industrial & engineering chemistry research*, **33**, 2437-2442.
4. Christ, U. and Bittner, A. (1994) *Progress in organic coatings*, **24**, 29-41.
5. McKay, R. B. (1993) *Progress in organic coatings*, **22**, 211-229.
6. Vanrooden, B. J. M. S. and Annen, T. M. S. (1996) *Wochenblatt für papierfabrikation*, **124**, 854-861.
7. Testa, C. (1993) *Jocca-surface coatings international*, **76**, 40.

8. Sirost, J.C. and Cole, P. (1993) *Jocca-surface coatings international*, **76**, 142.
9. Blayo, A., Noel, N., Lenest, J. F. and Gandini, A. (1993) *Jocca-surface coatings international*, **76**, 164.
10.Mackley, M.R., Marshall, R.T.J. and Smeulders, J.B.A.F. (1995) *J.Rheol.*, **39-6**, 1293-1309.
11.Mackley, M. R. and Spitteler, P. H. J.(1996), *Rheol. Acta*, **35**, 202-209.
12. Kaibara, M. (1996) *Biorheology* , **33**,101-117.
13. Weiss, D.J., Evanson, O. and Geor, R.J. (1994) *Comparative haematology international*, **4**, 11-16.
14. Weng, X., Cloutier, G., Pibarot, P. and Durand, L.G.(1996) *Biorheology*, **33**, 365-377.
15. Copley, A. L., King, R. G., Chien, S., Usami, S., Skaluk, R.and Hannay (1975) *Biorheology*, **12**, 257-263.
16. Rushton, H., Gummer, C. L. and Flasch, H. (1994) *Skin pharmacology*, **7**,78-83.
17. Mackley, M. R., Marshall, R. T. J., Smeulders, J. B. A. F. and Zhao, F. D. (1994) *Chem. Eng. Sci.*, **49**, 2551.
18. Luzatti, V., Mustacchi, H., Skoulios, H. and Husson, (1960) F. *Acta Crystallogr.* **13**, 660.
19. Ringsdorf, H., Schlarb, B., Venzmer, J. (1988) *J. Angew. Chem.* **100**, 117.

DEPLETION EFFECTS IN AQUEOUS DISPERSIONS OF HYDROPHILIC PARTICLES IN THE PRESENCE OF SODIUM POLYACRYLATE

G.C.ALLAN[1], M.J. GARVEY[3], J.W.GOODWIN[1], R.W.HUGHES[2], R.MacMILLAN[1] and B.VINCENT[1,*]

1) School of Chemistry, University of Bristol, BS8 1TS, UK
2) Bristol Colloid Centre, University of Bristol, BS8 1TS, UK
3) Unilever Research Port Sunlight Laboratory, Quarry Road East, Bebington, Merseyside, L63 3JW, UK
* to whom correspondence should be addressed

We have extended some previous studies, carried out in this laboratory, of the colloidal stability of polystyrene latex particles (carrying terminally-anchored polyethylene oxide chains) in the presence of sodium polyacrylate, to similar studies with *Ludox* silica particles plus sodium polyacrylate. In particular, we show that the stability / instability depletion flocculation boundaries depend on the way in which the system is assembled, as does the nature of the flocs formed. The observed behaviour also depends strongly on the presence of background electrolyte. We have also extend the earlier latex work, on depletion effects in the presence of sodium polyacrylate, to investigate two rheological parameters as a function of particle volume fraction: the dynamic shear stress and the high-frequency plateau modulus. Reasonable fits to the experimental data, for both these parameters, could only be obtained if the background visco-elastic nature of the polyelectrolyte solution, continuous phase is taken into account.

1. Introduction

Interactions between colloidal particles are mediated by the addition of polymers [1-6], including polyelectrolytes. There are many applications, e.g. in personal care products, waste water treatment and food emulsions [2-7]. The added polymer may adsorb to the particle surface. In turn, this may result in an enhancement of the particle stability, known as steric stabilisation [2-4,7], or it may result in flocculation of the dispersion via the bridging mechanism [2-4,6,8]. Alternatively, if the polymer does not adsorb onto the particles, this may lead to depletion flocculation [1-5,9]. This situation was first discussed by Asakura and Oosawa [10,11]. These authors concluded that a polyelectrolyte would have a stronger depletion attraction than a neutral polymer [11]. Since that time, there has been a large amount of both experimental work [3,12-18] and theoretical work [1-3,9-12,16,17,19-27] directed towards understanding depletion interactions.

In this paper we discuss the effects of adding a polyelectrolyte, sodium polyacrylate, to dispersions of two types of hydrophilic particles: (i) *Ludox* silica particles; (ii)

R. H. Ottewill and A.R. Rennie (eds.), Modern Aspects of Colloidal Dispersions, 289–303.

polystyrene latex particles, carrying short, terminally-anchored poly(ethylene oxide) chains [PS-g-PEO particles].

In the case of the silica particles, stability "maps" (plots of polymer concentration versus particle concentration) have been constructed (at pH 9.8), which delineate regions of dispersion stability and instability. Differences, not only in the positions of the stability/instability boundaries, but also in the nature of the aggregates (as observed microscopically) occur, depending on the method used to "transverse" the stability map: (i) simple mixing of the polyelectrolyte solution and the silica dispersions; or (ii) slow evaporation of water from a silica dispersion containing polyelectrolyte, premixed in the stable region. The effect of adding background electrolyte (NaCl) has also been investigated.

Adsorption isotherms for poly(acrylic acid) onto silica particles have been determined [28,29], as a function of pH. At pH 9.8 there is no adsorption (within experimental error) of the polymer onto the silica particles, so the aggregation we are reporting in this work is due to depletion attraction between the silica particles.

Corresponding adsorption isotherms and stability maps have previously been established in this laboratory [13,30] for PS-g-PEO particles in the presence of poly(acrylic acid), as a function of pH. At low pH (< 4) adsorption of (essentially undissociated) poly(acrylic acid) chains onto the PS-g-PEO particles occurs, corresponding to a zone in the low pH region of the stability map where bridging flocculation is observed. At pH 9, the polymer exists in the neutralised sodium polyacrylate form, and no adsorption occurs onto the PS-g-PEO particles; in this region of the stability map depletion flocculation is found to occur over a certain range of polyelectrolyte concentration s [13,30].

Rheology has been extensively used to investigate systems exhibiting depletion attraction [7,14,15,18,31-35]. In most of these studies an uncharged polymer has been used as the free polymer. Comparable studies using a polyelectrolyte as the free polymer are less common [18]. In this study, we probe the viscoelastic nature of dispersions that have undergone weak flocculation, originating from depletion attraction. Shear wave propagation measurements have been used to probe the elastic nature of flocculated, high volume fraction dispersions of PS-g-PEO particles, in the presence of non-adsorbing poly(acrylic acid). In addition, some controlled-stress viscometry experiments are also reported.

2. Experimental

2.1 MATERIALS

Styrene (B.D.H Chemicals) and methanol were distilled under vacuum prior to use. 2,2'-azobisisobutyronitrile [ABIN] (Waco Chemical Industries) was used as received. All water used was filtered through a *Milli-Q* purification system. Methoxy-PEO-methacrylate (M_N = 2000 g/mol) was kindly supplied by I.C.I Paints, Slough, U.K. Sodium hydroxide, hydrochloric acid, sodium chloride and methanol (all B.D.H Chemicals) were used without further purification.

Two samples of sodium polyacrylate were used: (i) for use with the PS-g-PEO latex systems, a sample (Polymer Laboratories) having an M_n of 2.2 x 10^5 g mol^{-1} and a polydispersity (M_w/M_n) of 1.5, was used; (ii) for use with the *Ludox* silica particles, a somewhat polydisperse sample (B.D.H.) was used, having a nominal M_v of 2.3 x 10^5 g mol^{-1}. Some of this polymer was fractionated using a non-solvent, sequential precipitation method [29]. One of the fractions, with a nominal M_v of 38000, was used in some of the stability studies.

2.2 *LUDOX* SILICA DISPERSIONS

These were kindly supplied by DuPont Ltd. Some details are given in Table 1 below.

TABLE 1: Details of *Ludox* silicas used.

Type :	*Ludox SM*	*Ludox HS-40*	*Ludox TM*
diameter /nm [quoted by supplier]	7	12	22
diameter /nm [determined by PCS]	12.9 ± 1.4	20.8 ± 2.1	30.1 ± 1.1
specific surface area / m^2g^{-1}	360	230	140
pH	10.1	9.7	9.0

2.3 PS-PEO LATEXES

The polystyrene latex [PS-g-PEO] particles, with terminally-attached poly(ethylene oxide) chains, were prepared by dispersion polymerisation, following the method described by Bromley [36], Cowell and Vincent [37] and Cawdrey and Vincent [30]. They were found to be essentially monodisperse, singlet particles and to have a mean diameter of 259 nm, as determined by transmission electron microscopy.

2.4 STABILITY MAPS

The method for establishing stability (polymer concentration versus particle concentration) maps in the case of PS-g-PEO latex plus poly(acrylic acid) [PAA] systems, has been described previously[13, 36]. Two methods were used to establish similar stability maps for the silica plus sodium polyacrylate systems:

(a) *simple mixing*: in this case samples were prepared by mixing, in the following order, water (or sodium chloride solution), a solution of sodium polyacrylate and finally the silica dispersion, the pH of each of these solution having been previously adjusted to pH 9.8 by the addition of sodium hydroxide solution. It was found in practice that the exact order of mixing did not matter, unless the (final) concentration of sodium polyacrylate was significantly greater than that required to produce an unstable dispersion.

(b) *evaporation method*: in this method the silica particles and sodium polyacrylate solutions were mixed, at pH 9.8, in a beaker, so as to give initial concentrations in the *stable* region of the stability map (as determined by method (a) above. The beakers used were selected so as to give a relatively high solution surface area to volume ratio. Each was placed on a magnetic stirrer. Evaporation of water was allowed to proceed slowly, whilst maintaining slow magnetic stirring for periods up to 30 hours. The

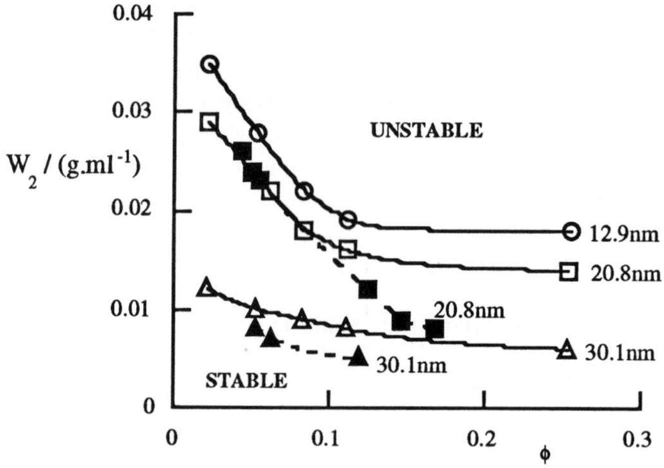

FIGURE 1. Stability maps: polymer (w/v) concentration (W_2) versus particle volume fraction (ϕ), for the three *Ludox* silica dispersions [see TABLE 1], in the presence of the unfractionated sample of poly(acrylic) acid (M_v=2.3 x 10^5 s mol^{-1}) at pH 9.8. No added NaCl . — "simple mixing" method; ---- "evaporation" method.

described in section 2.4, are shown. It may be seen that the "evaporation" method and the "simple mixing" method lead to different stability / instability boundaries, for a given silica particle size. In general, the minimum (w/v) concentration of polymer (W_2) required for the onset of flocculation, is lower, at a given particle volume fraction (ϕ), using the "evaporation" method, particularly at higher ϕ values. This is not surprising, since we have recently shown [44] that non-equilibrium effects are readily encountered, unless the method of mixing is carefully controlled. In the present work, we believe that the "slow evaporation" method corresponds more closely to equilibrium conditions. The fact that higher polymer concentrations are required to induce flocculation using the "single mixing" method, would suggest that some "supersaturation" occurs under these conditions : the system is *metastable* until some polymer concentration is reached (greater than that corresponding to the binodal value), beyond which nucleation and growth of flocs occurs rapidly.

It may also be seen from figure 1 that the minimum polymer concentration ($W_2^\dagger$) required to induce flocculation decreases with increasing particle radius. This trend has been observed previously [44-46]. It has also been discussed theoretically by Napper [47].

3.2 OPTICAL MICROGRAPHS : SILICA DISPERSIONS IN THE ABSENCE OF BACKGROUND ELECTROLYTE:

Figures 2 and 3 show optical micrographs of aggregates of *Ludox HS* particles, in the presence of sodium polyacrylate, at pH 9.8, resulting from the "simple mixing" and percentage light transmission (T) of the mixture was recorded as a function of time (t) using a Pye-Unicam *S.P.1800* Spectrometer, as was that for a control experiment *without* polymer present. The point at which these two T(t) plots deviated was taken to indicate the onset of flocculation in the mixture. By also monitoring the weight of the evaporating mixture, as a function of time, the exact particle and polymer concentrations could be readily calculated.

2.5 OPTICAL MICROSCOPY

Various of the silica plus sodium polyacrylate samples, in which flocculation had been observed, were examined using a Nikon *Optiphot* optical microscope, by placing a small sample on a dimple microslide and covering with a glass slip.

2.6 RHEOLOGY

The PS-g-PEO dispersions, in the presence of added sodium polyacrylate at pH 9.0, were examined rheologically, over the particle volume fraction (ϕ) range: $0.19 < \phi < 0.53$. The following methods were employed:
(a) *Controlled stress viscometry* measurements were carried out in order to determine the dynamic yield stress [39]. This was determined by the standard procedure [40] of extrapolating the linear shear stress-shear rate plot to zero shear rate. A Bohlin *CS* Rheometer, employing a cone and plate geometry, was used for these purposes. The applied stress was varied between 0.06 and 20 Pa, depending on the viscosity of the material under study. All experiments were carried out at 25°C.
(b) *A pulse shearometer* (Rank Bros. Cambridge, U.K.) was used to determine the dynamic rigidity (also referred to as wave rigidity) modulus of the dispersions. This instrument was originally developed by van Olphen [41]. The apparatus and analysis used in this study has been described in detail before [42,43]. The dispersion was placed between two parallel plates, each connected to a piezoelectric crystal. A square shear wave of frequency 200 Hz was generated by the rotation of one of the plates ($< 10^{-4}$ rad) and the time taken to propagate through the medium to the other plate was measured. By carrying out the experiment at a variety of plate separations, the (u) could be determined. The dynamic rigidity modulus (G) could be calculated using equation (1), where ρ is the density of the material and u is the velocity of the shear wave. If the attenuation of the shear wave is sufficiently small, G may be equated with $G(\infty)$, the high frequency limit of the storage modulus.

3. Results and Discussion

3.1 STABILITY MAPS : SILICA DISPERSIONS IN THE ABSENCE OF BACKGROUND ELECTROLYTE.

Figure 1 shows stability / instability boundaries for the three *Ludox* silica dispersions, in the presence of sodium polyacrylate ($M_V = 2.3 \times 10^5$ g mol^{-1}), at pH 9.8, in the *absence* of any additional background electrolyte. Results for both of the "mixing" methods,

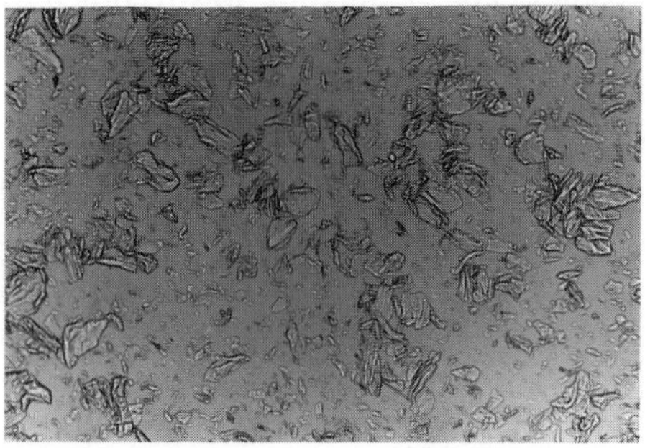

FIGURE 2. Optical micrograph (x 1800) for *Ludox-HS40* silica (ϕ = 0.138) particles, plus poly(acrylic acid) (M_V=38,000, 5.5%), at pH 9.8, 10 minutes after mixing by the "simple mixing" method (see text).

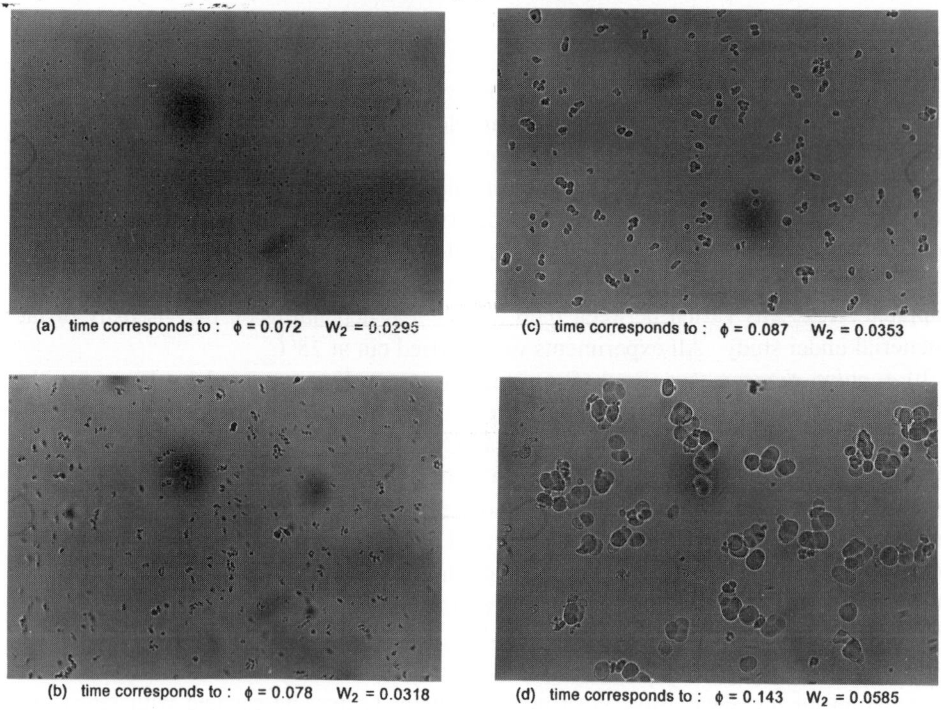

(a) time corresponds to : ϕ = 0.072 W2 = 0.0295
(c) time corresponds to : ϕ = 0.087 W2 = 0.0353
(b) time corresponds to : ϕ = 0.078 W2 = 0.0318
(d) time corresponds to : ϕ = 0.143 W2 = 0.0585

FIGURE 3. Optical micrographs (x 7200) for *Ludox-HS40* silica particles, plus poly(acrylic acid) (PAA; M_V=38,000) at various times after the onset of flocculation, using the slow evaporation method. The corresponding concentrations in the system at these times are: (a) 7.2% silica, 2.95% PAA; (b) 7.8% silica, 3.18% PAA; (c) 8.7% silica, 3.53% PAA; (d) 14.3% silica, 5.85% PAA.

"slow evaporation" methods, respectively. Clearly, the mechanism of flocculation is very different in the two cases. It would seem that "simple mixing" leads to relatively large, polydisperse aggregates (see figure 2; the largest aggregates shown are ~100 μm). As discussed in the previous section, these may well be the result of rapid nucleation and growth (and subsequent aggregation) of flocs from an initially supersaturated, metastable system. Evidence for supersaturation in weakly flocculating systems, in which colloidal phase separation occurs, has been seen before [48,49].

Slow evaporation of the mixture , on the other hand, would seem to lead to much tighter, close-packed aggregates, as may be seen in figure 3. Moreover, these aggregates, after a given time and, hence, at certain particle and free polymer concentrations, seem to be reasonably uniform in size, the mean size increasing with time. This suggests that, unlike the "simple mixing" method, in this case, after the initial nucleation of flocs (i.e. when W_2 just exceeds $W_2{}^{\dagger)}$, further growth is slow and floc aggregation is absent. In this case flocs presumably grow by addition of singlet particles onto existing flocs, as the free polymer concentration slowly increases. The individual silica particles are too small to observe whether any crystalline packing occurs within the flocs in this case. However, in some recent studies carried out in this laboratory [50], in which we studied depletion flocculation of much larger (~ 200 nm) polystyrene latex particles with added (non-adsorbing) sodium carboxymethylcellulose, it was found that, if a slow evaporation method, similar to that described here, was used to induce depletion flocculation, then the systems became iridescent on standing. Scanning electron microscopy revealed that the aggregates did indeed have crystalline order. In that work, we were attempting to model the iridescence observed in the exines of living and fossil megaspores, produced by *Selaginella* (Lycopsida) [51].

3.3 SILICA DISPERSIONS IN THE PRESENCE OF ADDED ELECTROLYTE.

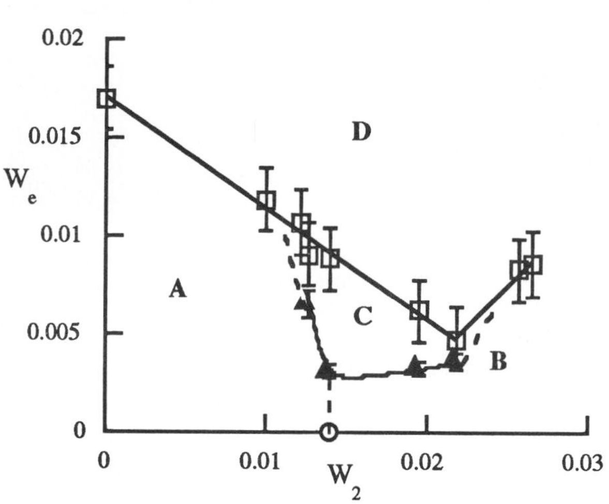

FIGURE 4. Stability map : NaCl (w/v) concentration (W_e) versus poly(acrylic acid) (M_v=2.3 x 10^5) (w/v) concentration (W_2). *Ludox HS-40*, at pH 9.8 at a fixed particle volume fraction (ϕ = 0.25). For description of the regions A, B, C and D see the text.

Figure 4 shows the "stability map", plotted in this case as electrolyte (NaCl) concentration versus sodium polyacrylate concentration (fixed particle concentration), for the system *Ludox HS-40* plus sodium polyacrylate and sodium chloride, at pH 9.8. This was produced using the "simple-mixing" method described earlier. Four regions may be identified. Region A corresponds to the stable dispersion. Region B corresponds to an unstable region which is similar to that observed in the *absence* of added electrolyte. Here fairly large, polydisperse flocs (c.f. figure 3) form. However, an interesting feature, observed in this particular study, is that these flocs form *reversibly*, at least initially. This was monitored as follows. A small aliquot of the dispersion was placed in the optical microscope, some 5-10 minutes after mixing the particle dispersion and the polymer solution. Dilution with water, such that $W_2 < W_2^{\dagger}$, led to break-up of the flocs. Moreover, if these directly-mixed samples were allowed to stand, then a lower white sedimented colloidal phase, and an upper clear liquid phase, separated within a few hours. On "tipping" the sample vial, the lower white sediment flowed freely, and readily "broke-up" on addition of water (or salt solution). However, on prolonged standing, over several days, the sediment no longer flowed-freely, nor was it broken-up on the addition of water, suggesting some floc consolidation had occurred on prolonged standing.

Region C is similar to region B, i.e. reversible flocculation occurred, but in this case the flocs that formed remained rather small (< 1 μm) and were, therefore, difficult to observe in the optical microscope (see figure 5a).

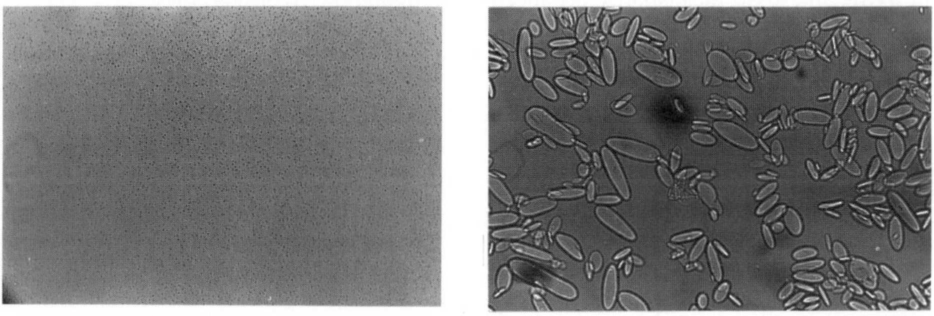

FIGURE 5. Optical micrographs (x 7200) for *Ludox HS-40* silica (ϕ = 0.25) plus poly(acrylic acid) (M_v 38,000) in: (a) region C of figure 5 (W_2 = 0.0256, W_e = 0.0036); (b) region D of figure 5 (W_2 = 0.0191, W_e = 0.0072). Note: the flocs seen in region B of figure 5 are similar to those shown in figure 3.

In region D (relatively high electrolyte concentration) flocs formed *irreversibly*, i.e. they could not be broken-up on diluting the system below $W_2^{\dagger}$. Interestingly, the flocs that formed with *Ludox HS-40* in this region, although polydisperse, are *ellipsoidal* in shape (figure 5b). Those formed with *Ludox SM* were also slightly ellipsoidal, whereas those formed with *Ludox TM* were spherical, but polydisperse.

Clearly, the flocculation behaviour observed with *Ludox* silica particles in the presence of varying concentrations of sodium polyacrylate and sodium chloride is complex. However, it can probably be interpreted in terms of the subtle interplay of the two, relatively long-range, interparticle interactions operating in these particular systems: depletion and electrical double layer overlap. Unfortunately, this is not a situation which is easy to model since there are too many unknown parameters.

3.4 STABILITY MAP: PS-g-PEO DISPERSIONS PLUS POLYELECTROLYTE AND SALT.

Stability maps have been reported previously by us [13,30] for aqueous dispersions of PS-g-PEO particles, in the presence of poly(acrylic acid) and NaCl, as a function of pH. The polymer used in the rheology results, presented below, had a similar molecular weight ($M_v = 2.2 \times 10^5 g\ mol^{-1}$) to that used in this earlier work, and the PS-g-PEO particles were also similar. In contract to the silica particles used in this work, the (negative) surface charge density of the PS-g-PEO particles is rather small [30], and so the dominant long-range interaction between the particles is the depletion interaction. The nature of the depletion flocculation observed with the PS-g-PEO particles is again weak and reversible. Although no optical micrographs were taken for these systems, the floc phase which separated with time was amorphous and reversible to dilution below $W_2^{\dagger}$, that is, similar in behaviour to the silica dispersions in region B in figure 4.

3.5 SHEAR-WAVE PROPAGATION EXPERIMENTS ON PS-g-PEO DISPERSIONS.

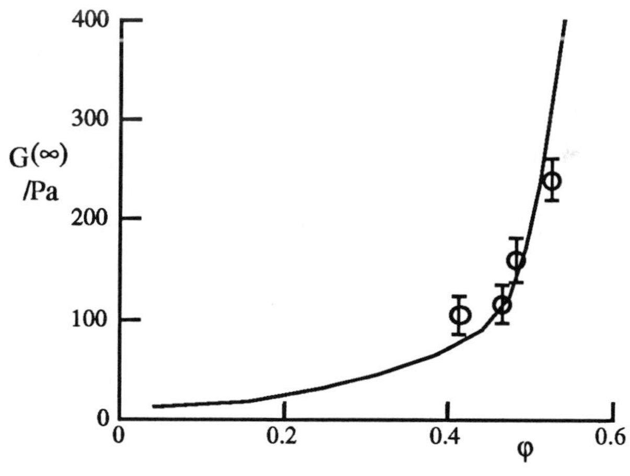

FIGURE 6. $G(\infty)$ versus PS-g-PEO particle volume fraction (ϕ) for $W_2 = 0.0007$ at $C_c = 2 \times 10^{-4}$ mol dm^{-3}. The points are experimental and the line is a theoretical fit (see text).

Figure 6 shows the $G(\infty)$ results which were obtained for the PS-g-PEO dispersions, in the presence of sodium polyacrylate and NaCl, at pH 9, as a function of the particle volume fraction, ϕ. Although reliable values for $G(\infty)$ could only be obtained for $\phi > 0.4$, the elasticity of the flocculated network would appear to be significant at these higher particle volume fractions, with $G(\infty)$ approaching 0.25 kPa for $\phi \sim 0.53$.

In an earlier paper, Goodwin *et al.* [43] described a model for calculating values of $G(\infty)$. In this present work equation (14) of reference 43 was used to calculate the contribution to $G(\infty)$ due to depletion attraction. That equation is reproduced here as equation (2),

$$G(\infty) = \frac{3\phi}{4\pi a^3} + \frac{3}{40\pi a^6}\phi^2 \int_0^\infty g(r)\frac{d}{dr}\left(r^4 \frac{dV_T}{dr}\right)dr \qquad (2)$$

where V_T is the total interaction potential between two particles of radius a, with their centres r apart, and g(r) is the pair distribution function.

In this work it was found that the value for $G(\infty)$, calculated using equation (2), making reasonable assumptions in calculating V_T, was very much less than the value of $G(\infty)$ that had been measured, at all values of ϕ. However, it has to be remembered that the polyelectrolyte chains in bulk solution will also contribute to the overall elasticity of the system, and not just its viscosity [52]. Therefore, we set out below a new approach for calculating $G(\infty)$. We propose to consider the system as a matrix of charged spheres (polyelectrolyte coils), with the particles present as an inert filler. By using a lattice structure for the polyelectrolyte coils, we are able to calculate the plateau modulus (G_0) due to the polyelectrolyte coils alone. The expression has been derived before [53] and is reproduced below, as equation 3

$$G_0 = \left(0.83 / r_p\right)\left(d^2 V_e / dr_p^2\right) \qquad (3)$$

Here r_p is the centre-of-mass to centre-of-mass separation of the polyelectrolyte coils; V_e is their pair electrostatic interaction potential; 0.83 is a geometric factor and applies for a lattice which is face-centred cubic or hexagonally close-packed.

To calculate r_p we must first know ϕ_p, the effective volume fraction of the polyelectrolyte coils, and to calculate V_e we must first know σ_0, the surface charge density of the polyelectrolyte coils. ϕ_p may be estimated using equation (4) below,

$$\phi_p = 4\pi c_p N_A r_g^3 / 3M \qquad (4)$$

where c_p has units of mass/volume, N_A is the Avogadro constant, M is the molecular weight of a polyelectrolyte chain and r_g is the radius of gyration of the coil. r_p is then given by

$$r_p = 2r_g(0.74 / \phi_p)^{1/3} \qquad (5)$$

where 0.74 is the volume fraction for touching, hexagonally close-packed, monodisperse spheres. In order to calculate σ_0 we must know the amount of charge per polyelectrolyte coil. At pH 9, the polymer is ~ 3.5 pH units above its pK_a value of 5.5. Therefore, we may assume that it is fully ionised. The number of charges per molecule, X, is then given by equation (6).

$$X = \frac{M}{M_{mon}} \qquad (6)$$

here M_{mon} is the molecular weight of each acrylate group in the chain. Assuming an even distribution of charge over the surface of each polyelectrolyte coil,

$$\sigma_0=\left(\frac{3X}{4\pi}\right)^{2/3}\frac{e}{r_g^2} \tag{7}$$

where e is the charge on an electron. The corresponding surface potential, ψ_0 is given by Equation (8)

$$\psi_0=\frac{1}{\varepsilon\left(1+\kappa r_g\right)}\left(\frac{3X}{4\pi}\right)^{2/3}\frac{e}{r_g^2} \tag{8}$$

where ε is the dielectric permittivity and κ is the Debye-Hückel parameter. The value of r_g is then iterated in equations (4), (5) and (8) until the value of G_0 in equation (9) below gives a good fit to the experimental data (in this case the data displayed in figure 6),

$$G(\infty)=G_0\left(1-\frac{\phi}{0.62}\right)^{-1.55} \tag{9}$$

Using a value for r_g of 32 nm gives an excellent fit to the data, as shown in Figure 6.

Takahashi and Nagasawa [54] have reported intrinsic viscosity data for sodium polyacrylate in aqueous NaBr solutions. Cawdrey [55] has calculated corresponding r_g values under these conditions. Interpolating Cawdrey's data, for the polymer molecular weight and continuous phase NaCl concentration used in the shear wave propagation experiments reported here, yields a value of 46 nm for r_g.

3.6 CONTROLLED-STRESS VISCOMETRY EXPERIMENTS ON PS-g-PEO DISPERSIONS.

Plots of shear stress against shear rate revealed only slight shear-thinning. However, a Bingham yield stress, σ_B, could be determined for the various PS-g-PEO dispersions investigated here. The values obtained were small (less than 0.5 Pa), but this is reasonable for systems showing only weak flocculation.

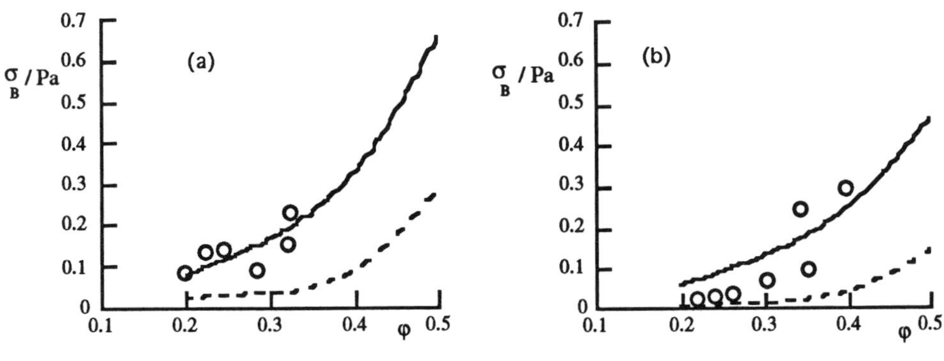

FIGURE 7. Bingham yield stress versus (σ_B) versus PS-g-PEO particle volume fraction (ϕ) for (a) $W_2 = 0.0007$ at $C_e = 2 \times 10^{-4}$ mol dm^{-3}, and (b) $W_2 = 0.0014$ at $c_e = 0.023$ mol dm^{-3}. The points are experimental. Theoretical fits (see text) : ---- electroviscous contribution; — electroviscous *plus* depletion contributions.

Figure 7 shows the yield stress values, as a function of ϕ, for the PS-g-PEO dispersions at two polyelectrolyte (and NaCl) concentrations. It can be seen that σ_B increases with ϕ, which is the expected trend.

We may calculate σ_B from the energy dissipation model of Michaels and Bolger [56],

$$\sigma_B = \frac{24\phi^2}{\pi^2 (2a)^3} \cdot V_{min} \tag{10}$$

where V_{min} is the minimum in the interparticle pair potential, resulting in these systems from depletion attraction. V_{min} may be estimated, to a first approximation, using the original equations of Asakura and Oosawa [10,11] for the depletion interaction. However, there is a problem , as discussed in the previous section (3.5), that the continuous phase, which contains a high concentration of polyelectrolyte, is viscoelastic. Therefore, we must consider the contribution to σ_B from the polyelectrolyte coils in the continuous phase. This may be achieved by considering this contribution to be analogous to the "second electroviscous" effect and then to make use of the analysis given by Russel [57,58] for this effect. In this approach, one calculates an effective particle (in this case, polyelectrolyte coil) "collision" diameter (L), in the low-shear limit, by equating the thermal and electrostatic terms. L is then given by,

$$L = \kappa^{-1} \ell n\left[\alpha / \ell n\left(\alpha / \ell n\left\{\alpha / \ell n\alpha\right\}\right)\right] \tag{11}$$

where

$$\alpha = 4\pi\varepsilon\psi_o r_g^2 \kappa \exp(2\kappa r_g) / kT \tag{12}$$

The high and low shear rate viscosities for effective hard-sphere systems are given by the Dougherty-Krieger equations [59] :

$$\text{low shear:} \quad \eta(o) = \eta_o\left[1 - \frac{\phi}{\phi_{mo}}\right]^{-2.5\phi_{mo}} \tag{13}$$

$$\text{high shear:} \quad \eta(\infty) = \eta_o\left[1 - \frac{\phi}{0.605}\right]^{-1.51} \tag{14}$$

$$\phi_{mo} = 0.54(2r_g / L)^3$$

where the maximum packing volume fraction at zero shear, $\phi_{mo} = 0.54 \left(\frac{2r_g}{L}\right)^3 = 0.407$,

and L is given by equation (11).

The viscosity of the system, as a function of σ, may be estimated using Krieger's equation [60] below, which incorporates the high and low shear limiting viscosity values,

$$\eta(\sigma)=\eta(\infty)+\left(\frac{\eta(o)-\eta(\infty)}{1+\frac{2.55\sigma r_g^3}{kT}}\right) \tag{15}$$

This enables a shear stress / shear rate plot for the continuous phase to be constructed, and, hence, the contribution to the dynamic yield stress from the matrix to be estimated. The results of the theoretical calculation are shown in figure 7. As may be seen that, if both the depletion effect from the particles (equation 10) and the "electroviscous" effect from the matrix are taken into account, then the theoretical calculations fit the (rather scattered) experimental data reasonably well.

4. Conclusions

We have shown that stability / depletion flocculation boundaries in stability phase maps for the *Ludox* silica particles in the presence of sodium polyacrylate, depend strongly on the way in which the systems are assembled, as does the nature of the flocs which form, as observed from optical microscopy. The addition of NaCl to the system leads to more complex behaviour. In that case, both reversible (low NaCl concentration) and irreversible flocculation (high NaCl concentration) may be observed.

Stability maps had been established previously in this laboratory [13,30] for PS-g-PEO latex particles in the presence of sodium polyacrylate. Similar studies have been reported by Liang *et al.* [18] However, the polymer used in that work had a much wider molecular weight distribution than that used in the current work, and the effect of added electrolyte was not investigated. In this paper we have extended our own work with these particular systems, by studying the Bingham yield stress and the high frequency plateau modulus, as a function of particle volume fraction. However, reasonable fits to the experimental data for both these rheological parameters could only be obtained, if the viscoelastic nature of the continuous phase, which contains significant concentration of polyelectrolyte chains, is taken into account, in addition to the (relatively weak) interparticle depletion forces.

5. Acknowledgements

We wish to acknowledge the support of the D.T.I. "Colloid Technology" Programme and Unilever Research, Port Sunlight Laboratory, for financial support for this work.

6. References

1. Fleer, G.J., Scheutjens, J.H.M.H., and Vincent, B. (1984) *ACS Symp.Ser. Polymer Adsorption and Dispersion Stability* **240**, 245.
2 Fleer, G.J., Scheutjens, J.H.M.H., Cohen-Stuart, M.A., Cosgrove, T. and Vincent, B. (1993) *Polymers at Interfaces*, Chapman and Hall, London.

3. Napper, D.H. (1983) *Polymeric Stabilization of Colloidal Dispersions*, Academic Press, London .
4. Hunter R.J. (1989) *Foundations of Colloid Science,* Oxford University Press, Oxford, Vols. I and II.
5. Milling, A.J. (1996) *J.Phys.Chem.,* **100** 8986.
6. de L. Costello, B.A., Luckham, P.F. and Tadros, Th.F. (1992) *Langmuir*, **8** 464.
7. Tadros, Th.F. (1994) *Colloids Surf. A*, **91** 39.
8. Grover, G.S. and Bike S.G. (1995) *Langmuir*, **11** 1807.
9. Vincent, B. (1990) *Colloids Surf.*, **50** 241.
10. Asakura, A. and Oosawa, F. (1954) *J.Chem.Phys.*, **22** 1255.
11. Asakura, A. and Oosawa, F. (1958) *J.Polymer.Sci.* **33** 183.
12. Vincent,B., Luckham, P.F. and Waite, F.A. (1980) *J.Collloid Interface Sci.,* **73** 508.
13. N.F. Cawdrey, A.J. Milling and B. Vincent (1994) *Colloids Surf. A,* **86** 239.
14. Sakellariou, P., Strivens, T.A. and Petit, F. (1995) *Colloid Polym.Sci.* **273** 279.
15. Ogden, A.L. and Lewis, J.A. (1996) *Langmuir*, **12** 3413.
16. Seebergh, J.E. and Berg, J.C. (1994) *Langmuir,* **10** 454.
17. Ilett, S.M., Orrock, A., Poon, W.C.K. and Pusey, P.N. (1995) *Phys.Rev.E*, **51** 1344.
18. Liang,W., Tadros, Th.F. and Luckham, P.F. (1994) *Langmuir,* **10** 441.
19. De Gennes, P.G. *Macromolecules*, (1981) **14** 1637.
20. De Hek, H and. Vrij, A. (1981), *J.Colloid Interface Sci.*, **84** 409.
21. Sperry, P.R. (1981) *J.Colloid Interface Sci.,* **87** 375.
22. Gast, A.P., Hall, C.K. and Russel, W.B. (1983), *Faraday Discuss.Chem.Soc.* **76** 189.
23. Gast, A.P., Hall, C.K. and Russel, W.B. (1983) *J.Colloid Interface Sci.* **96** 251.
24. Shukla, K. and Rajagopalan, R., (1993), *Colloids Surf. A* **79** 249.
25. Walz, J.Y. and Sharma, A. (1994), *J.Colloid Interface Sci.* **168** 485.
26. Dickman, R. and Yethiraj, A. (1994) *J.Chem.Phys.* **100** 4683.
27. Dahlgren, M.A.G. and Leermakers, F.A.M. (1995), *Langmuir* **11** 2996.
28. Milling, A.J. and Vincent, B. (1997) J Chem.Soc. Faraday Trans. in press.
29. MacMillan, R. and Vincent, B. to be published
30. Cawdrey, N. and Vincent, B. in *Colloid Polymer Particles* (1995), ed. Goodwin, J.W. and Buscall, R. Academic Press, London
31. Reynolds, P.A. and Reid, C.A. (1991), *Langmuir*, **7** 89.
32. Prestidge, C. and Tadros, Th.F. (1988), *Colloids Surf.,* **31** 325.
33. Nashima, T. and Furusawa, K. (1991), *Colloids Surf.,* **55** 149.
34. Liang, W., Tadros, Th.F. and Luckham, P.F. (1993), *J.Colloid Interface Sci.* **155** 156.
35. Liang, W., Tadros, Th.F. and Luckham, P.F. (1993), *J.Colloid Interface Sci.,* **158** 152.
36. Bromley, C.W.A., (1986), *Colloids Surf.,* **17** 1.
37. Cowell, C. and Vincent, B. (1982), *J.Colloid Interface Sci.,* **87** 518.
38. Allan, G.C. and Vincent, B. in preparation.
39. Bonnecaze, R.T. and. Brady, J.F (1992), *J.Rheol.* **36** 73.
40. Reference 4, pp. 87-88.
41. Van Olphen, H.(1956), *Clays Clay Miner.*, **4** 204.
42. Buscall, R., Goodwin, J.W. Hawkins, M.W. and Ottewill, R.H., (1982), *J.Chem.Soc. Faraday Trans.1* **78** 2873.
43. Goodwin, J.W., Hughes, R.W., Partridge, S.J. and Zukoski, C.F. (1986), *J.Chem.Phys.*, **85** 559
44. Jenkins,.P.D. and Vincent, B. (1996), *Langmuir,* **12** 3107.
45. Clarke, J. and Vincent, B. (1981), *J.Colloid Interface Sci.*, **82** 208.
46. de Hek, H. and Vrij, A. (1981), *J. Colloid Interface Sci.*, **84** 409.
47. Napper, D.H. (1983), *Polymeric Stabilisation of Colloidal Dispersions*, Academic Press, London.
48. Emmett, S. and Vincent, B. (1990), *Phase Transitions*, **21** 197.
49 Verhaegh, N.A.M., van Duijneveldt, J.S., Dhont, J.K.G. and Lekkerkerker, H.N.W., (1996), *Physica* , **A230** 409.
50. Hemsley A.R., Jenkins, P.D., Collinson, M.E. and Vincent, B. (1996), *Botannical J. Linnean Soc.,* **121** 177.
51. Hemsley, A.R., Collinson, M.E., Kovach, W.I., Vincent, B. and Williams, T. (1994), *Phil. Trans. Royal Soc.* , London, **345** 163
52. Boger, D.V. (1977), *Nature*, **265** 126.
53. Buscall, R., Goodwin, J.W., Hawkins, M.W. and Ottewill, R.H., (1982), *J.Chem.Soc. Faraday Trans.1* **78** 2889.
54. Takahashi, A. and Nagasawa, M., (1964), *J.Amer.Chem.Soc.* **86** 543.
55. Cawdrey, N.F. (1992) Ph.D. thesis, University of Bristol .

56. Michaels, A.S. and Bolger, J.C. (1966), *Ind. Eng. Chem. Fund.*, **1** 153.
57. Russel, W.B. (1978), *J. Fluid Mech.*, **85** 209.
58. Russel, W.B. (1980), *J. Soc. Rheology*, **24** 287.
59. Krieger, I.M. and Dougherty, T.J. (1959), *Trans. Soc. Rheology*, **3** 137.
60. Krieger, I.M. (1977), *Adv. Colloid Interface Sci.*, **3** 111.

BIBLIOGRAPHY

Publications on Concentrated Colloidal Dispersions

The following pages list the publications that have appeared as a result of work under the Colloid Technology Programme. It contains those 300 papers that have been published up to the end of May 1997. Moreover, a considerable amount of work is still in preparation. The papers have been listed alphabetically by author and not separated by topic as many projects have overlapping boundaries and involved collaboration between several partners within the Programme and outside. This bibliography supplements the references provided to the individual Chapters of this book as it was not possible to cover every topic described in this list in the space available in the present volume.

Albiston, L., Franklin, K. R., Lee, E. and Smeulders, J. B. A. F., (1996), 'Rheology and microstructure of aqueous layered double hydroxide dispersions', *J. Mater. Chem.*, **6**, 871-877.

Annable, T., Buscall, R., Ettelaie, R., Shepherd, P. and Whittlestone, D., (1993), 'The rheology of solutions of associating polymers: comparison of experimental behaviour with transient network theory', *J. Rheol.*, **37**, 695-726.

Annable, T., Buscall, R., Ettelaie, R., Shepherd, P. and Whittlestone, D., (1994), 'The influence of surfactants on the rheology of asociating polymers in solution', *Langmuir*, **10**, 1060-1070.

Annable, T. and Ettelaie, R., (1994), 'Thermodynamics of phase separation in mixtures of associating polymers and homopolymers in solution', *Macromolecules*, **27**, 5616-5622.

Annable, T., Buscall, R. and Ettelaie, R., (1996), 'Network formation and its consequences for the physical behaviour of associating polymers in solutions', *Coll. Surf. A*, **111**, 97-116.

Annable, T. and Ettelaie, R., (1996), 'Effects of association on the phase behaviour of mixed solutions of hydrophobically modified and similar unmodified polymers', *J. Chim. Phys.* **93**, 899-919.

Aubony, M., Harden, J. L. and Cates, M. E., (1996), 'Shear-induced deformation and desorption of grafted polymer layers', *J. Phys. II France*, **6**, 969-984.

Bailey, L., Denis, J., Goldsmith, G., Hall, P. L., and Sherwood, J. D., (1994), 'A wellbore simulator for mud-shale interaction studies', *J. Petroleum Sci. & Eng.*, **11**, 195-211.

Bailey, L., Reid, P. and Sherwood, J. D., (1994), 'Interaction between wellbore fluids and shale during drilling hydraulic and hydrochemical characterisation of Argillaceous rocks: Proceedings of an International Workshop, Nottingham, June 7-9, 179-188. Paris: Organisation for Economic Co-operation and Development and the Nuclear Energy Agency, 1995.

Bailey, L. Reid, P. and Sherwood, J., (1994), 'Mechanisms and solutions for chemical inhibition of shale swelling and failure', pp. 13-28 in *Recent Advances in Oilfield Chemistry*, Ogden, P. (Ed.), RSC Spec. Pub. 159 , Cambridge.

Ball, R. C. and Melrose, J. R., (1995), 'Lubrication breakdown in hydrodynamic simulations of concentrated colloids', *Adv. Coll. Interf. Sci.*, **59**, 19-30.

Bartlett, P. and Ottewill, R. H., (1992), 'Geometric interactions in binary colloidal dispersions', *Langmuir*, **8**, 1919-1925.

Bartlett, P., Ottewill, R. H. and Pusey, P. N., (1992), 'Superlattice formation in binary mixtures of hard-spheres', *Phys. Rev. Letts.*, **68**, 3801-3804.

Bartlett, P. and Ottewill, R. H., (1992), 'A neutron scattering study of the structure of a bimodal colloidal crystal', *J. Chem. Phys.*, **96**, 3306-3318.

Barlett, P. and Ottewill, R. H., (1994), 'Gravitational effects on the phase behaviour of dispersions', *Adv. Coll. Interf. Sci.*, **50**, 39-50.

Biggs, S. and Vincent, B., (1992), 'Poly(styrene-b-2-vinylpyridine-1-oxide) and poly(dimethylsiloxane-1-2-vinylpyridine-1-oxide) diblock copolymers', *Coll. Polymer Sci.*, **270**, 505-510.

Biggs, S. and Vincent, B., (1992), 'Poly(styrene-b-2-vinylpyridine-1-oxide) and Poly(dimethylsiloxane-water/propan-2-ol mixtures', *Coll. Polymer Sci.*, **270**, 511-517.

Biggs, S. and Vincent, B., (1992), 'Poly(styrene-b-2-vinylpyridine-1-oxide) diblock copolymers. (3) micro-emulsions in water/propan-1-ol/toluene mixtures', *Coll. Polymer Sci.*, **270**, 563-573.
Billingham, J. and Needham, D., (1993), 'Mathematical modelling of chemical clock reactions II. A class of autocatalytic clock reaction schemes', *J. Eng. Math.*, **27**, 113-145.
Billingham, J. and Ferguson, J., (1993), 'Laminar, unidirectional flow of a thixotropic fluid in a circular pipe', *J. Non-Newtonian Fluid Mech.*, **47**, 21-55.
Billingham, J. and Coveney, P. V., (1993), 'Simple chemical clock reactions: an application cement hydration', *J. Chem. Soc. Farad. Trans.*, **89**, 3021-3028.
Billingham, J. and Coveney, P. V., (1994), 'Kinetics of self-replicating micelles', *J. Chem. Soc. Farad. Trans.* **90**, 1953-1959.
Bladon, P., Liu, H. and Warner, M. (1992) 'Biaxial effects in nematic comb-like polymers', *Macromolecules*, **25**, 4329-4338.
Bladon, P., Warner, M. and Cates, M. E., (1993), 'Transesterficiation in nematic polymers', *Macromolecules*, **26**, 4499-4505 .
Bladon, P., Terentjev, E. M. and Warner, M., (1993), 'Transitions and instabilities in liquid crystalline elastomers', *Phys. Rev. E.*, **47**, R3838-R3840.
Bladon, P., Warner, M. and Terentjev, E. M., (1994), 'Orientational order in strained nematic networks', *Macromolecules*, **27**, 7067-7075.
Boek E. S., Coveney P. V. and Lekkerkerker H. N. W., (1996), 'Computer simulations of rheological phenomena in dense colloidal suspensions with dissipative particle dynamics', *J. Phys. Cond. Mat.*, **8**, 9509-9512.
Boek, E. S., Coveney, P. V., Lekkerkerker, H. N. W. and van der Schoot, P., (1997) 'Simulating the rheology of dense colloidal suspensions using dissipative particle dynamics', *Phys. Rev. E*, **55**, 3124-3133.
Boghosian, B. M., Coveney, P. V., and Emerton, A. N., (1996), 'A lattice-gas model of microemulsions', *Proc. Roy. Soc. Lond. A*, **452**, 1221-1250.
Bongiovanni, R., Ottewill, R. H., Rennie, A. R. and Laughlin, R. G., (1996), 'The interaction of small ions with the micellar surface of an ultra long-chain Zwitterionic surfactant', *Langmuir*, **12**, 4681-4690.
Bouchaud, J-P., Cates, M. E., Prakash, J. R. and Edwards, S. F., (1994), 'A model for the dynamics of sandpile surfaces', *J. Phys. I France*, **4**, 1383-1410.
Bouchaud, J-P., Cates, M. E., Prakash, J. R. and Edwards, S. F., (1995), 'Hysteresis and metastability in a continuum sandpile model', *Phys. Rev. Letts.* **74**, 1982-1985.
Bouchaud, J-P., Cates, M. E. and Claudin, P., (1995), 'Stress distributions in granular media and nonlinear wave equations', *J. Phys., I France*, **5**, 639-656.
Braithwaite, G. J. C., Luckham, P. F. and Howe, A. M., (1996), 'Interactions between poly(ethylene oxide) layers adsorbed to glass surfaces probed by using a modified atomic force microscope', *Langmuir*, **12**, 4224-4237.
Braithwaite, G. J. C. and Luckham, P. F., (1997), 'A study of attractive interactions between poly(ethylenoxide) coated surfaces using AFM', *NATO ASI Series E. Applied Science*, **330**, 121-127 Kluwer, Dordrecht.
Braithwaite, G. J. C. and Luckham, P. F., (1997), 'Effect of molecular weight on the interactions between poly(ethylenoxide) layers adsorbed to glass surfaces', *J. Chem. Soc. Farad. Trans.*, **93**, 1409-1415.
Briscoe, B. J., Andrei, D. C., Luckham, P. F. and Williams, D. R., (1996), 'The deformation of microscopic gels', *J. Chimie Phys.*, **93**, 960-976.
Briscoe, B. J., Liu, K. K. and Williams, D. R., (1996), 'The compressive deformation of a single microcapsule', *Phys. Rev. E.*, **54**, 6673-6680.
Brown, A. B. D., Ball, R. C., Clarke, S. M., Melrose, J. R. and Rennie, A. R., (1995), 'Model hard sphere sytems under flow: a novel experimental approach', *Progr. Coll. Polymer Sci.*, **98**, 99-102.
Bucknall, D. G., Penfold, J., Webster, J. R. P., Zarbakhsh, A., Richardson, R. M., Rennie, A., Higgins, J. S., Jones, R. A. L., Fletcher, P., Thomas, R. K., Roser, S. and Dickinson, E., (1995), 'SURF - A second generation neutron reflectometer', *ICANS* XIII, 13th Meeting of the International Collaboration on Advanced Neutron Sources, Switzerland, **95**-02, 123-129.
Buscall, R., (1993), 'The Hamaker coefficient for titania (rutile) in liquid media', *Coll. Surf. A.*, **75**, 269-272.
Buscall, R., McGowan, J. I. and Morton-Jones, A. J., (1993), 'The rheology of concentrated dispersions of weakly-attracting colloidal particles with and without wall slip', *J. Rheol.*, **37**, 621-641.
Buscall, R., D'Haene, P. and Mewis, J., (1994), 'The maximum density for flow dispersions of near monodisperse, spherical particles', *Langmuir*, **10**, 1439-1441.

Buscall, R., (1994), 'An effective hard-sphere model for the non-Newtonian viscosity of soft-sphere dispersions: Comparison with further data for microgels and sterically stablised latices', *Coll. Surf. A.*, **83**, 33-42.

Buzza, D. M. A. and Cates, M. E., (1994) 'Uniaxial modulus of concentrated emulsions', *Langmuir* **10**, 4503-4508.

Buzza, D. M. A., Lu, C-Y. D. and Cates, M. E., (1995), 'Linear shear rheology of incompressible foams', *J. Phys. II France*, **5** 37-52.

Callaghan, P. T., Terentjev, E. M. and Warner, M, (1995), 'Pulsed gradient spin-echo NMR of confined Brownian particles', *J. Chem. Phys.* **102**, 4619-4624.

Callaghan, P. T., Cates, M. E., Rofe, C. J. and Smeulders, J. B. A. F., (1996), 'A study of the 'spurt effect' in worm-like micelles using nuclear magnetic resonance microscopy', *J. Phys. II France*, **6**, 375-393.

Cameron, R. E. and Donald, A. M., (1994), 'Minimizing sample evaporation in the environmental scanning electron microscope', *J. Microscopy*, **173**, 227-237.

Cameron, R. E., (1994), 'Environmental scanning electron microscopy in polymer science', *Trends in Polymer Science*, **2**, 116-120.

Cameron, R. E., (1994), 'Environmental scanning electron microscopy: principles and application', *Microscopy and Analysis*, **41**, 11-13.

Cates, M. E., van der Schoot, P. and Lu, C-Y. D., (1995), 'Giant dielectric response of the sponge phase', *Europhys. Letts.* **29**, 669-674.

Clapperton, R. M., Ottewill, R. H. and Ingram, B. T., (1994), 'A study of fluorocarbon-hydrocarbon surface active agent mixtures by NMR spectroscopy', *Langmuir*, **10**, 51-56.

Clarke, C. J., Jones, R. A. L., Edwards, J. L., Clough, A. S. and Penfold, J., (1994), 'Kinetics of formation of physically end-adsorbed polystyrene layers from the melt', *Polymer*, **35**, 4065-4071.

Clarke, S. M., Melrose, J. R., Rennie, A. R., Ottewill, R. H., Heyes, D., Mitchell, P. J. Hanley, H. J. M. and Straty, G. C., (1994), 'The structure and rheology of hard-sphere systems', *J. Phys. Cond. Mat.*, **6**, A333-A337.

Clarke, S. M., Ottewill, R. H. and Rennie, A. R., (1995), 'Light scattering studies of dispersions under shear', *Adv. Coll. Interf. Sci.*, **60**, 95-118.

Clarke, S. M., Rennie, A. R. and Convert, P., (1996), 'A diffraction technique to investigate the orientational alignment of anisotropic particles: Studies of clay under flow', *Europhys. Letts.*, **35**, 233-238.

Clarke, S. M., Rennie, A. R. and Ottewill, R. H., (1997), 'The stacking of hexagonal layers of colloidal particles. 'A study by small angle diffraction', *Langmuir*, **13**, 1964-1969.

Clegg, S. M., Williams, P. A., Warren, P. B. and Robb, I. D., (1994), 'Phase behaviour of polymers with concentrated dispersions of surfactants', *Langmuir*, **10**, 3390-3394.

Cooke, W. C., Warr, S., Huntley, J. M. and Ball, R. C., (1996), 'Particle size segregation in a 2-dimensional bed undergoing vertical vibration', *Phys. Rev. E*, **53**, 2812-2822.

Cosgrove, T., Phipps, J. S. and Richardson, R. M., (1992), 'Neutron reflection from a liquid/liquid interface', *Coll. Surf.*, **62**, 199-206.

Cosgrove, T., Eaglesham, A., Horne, D., Phipps J. S. and Richardson, R. M., (1992), 'Neutron reflection from liquid-liquid interfaces', Proc. 2nd Intnl. Conf. on Surface X-ray and Neutron Scattering, Springer Verlag **61**, 159-165.

Cosgrove, T., Phipps J. S. and Richardson, R. M., (1992), 'Neutron reflection from polymers adsorbed at the solid-liquid interface', Proc. 2nd Intnl. Conf. on Surface X-ray and Neutron Scattering, Springer Verlag **61**, 169-172.

Cosgrove, T., (1992), 'Volume fraction profiles for terminally attached polymers', *Macromolecular Reports* **A29**, 125-130.

Cosgrove, T. and Griffiths, P. C., (1992), 'Nuclear magnetic resonance studies of adsorbed polymer layers', *Adv. Coll. Interf. Sci.*, **42**, 175-203.

Cosgrove, T. and Griffiths, P. C., (1994), 'Characterisation of adsorbed polymer layers using NMR solvent diffusion studies', *Coll. Surf. A*, **84**, 249-258.

Cosgrove, T. and Griffiths, P. C., (1995), 'Diffusion in bi-modal and polydisperse polymer solutions I', *Polymer*, **36**, 3335-3342.

Cosgrove, T. and Griffiths, .P. C., (1995), 'Diffusion in polydisperse polymer solutions II', *Polymer*, **36**, 3342-3347.

Cosgrove, T., Griffiths, P. C. and Lloyd, P. M., (1995), 'Polymer adsorption: the effect of the relative size of the polymer and the substrate particle', *Langmuir*, **11**, 1457-1463.

Cosgrove, T., Heath, T. G., and Ryan, K., (1994), 'Terminally attached polystyrene chains on modified silicas', *Langmuir*, **10**, 3500-3506.

Cosgrove, T., Mears, S. J., Thompsen, L. and Howells, I., (1996), 'Adsorption in mixed silica polymer surfactant systems', *ACS Symposium Series*, **615**, 196-204.
Cosgrove, T., Patel, A., Webster, J. R. P. and Semlyen, A. J., Zarbakhsh, A., (1993), 'A study of a chemisorption of poly(hydrogen methylsiloxane) using neutron reflectometry and small-angle neutron scattering', *Langmuir*, **9**, 2326-2329.
Cosgrove, T., Phipps J. S. and Richardson, R. M., (1994), 'The structure of copolymers at the liquid-liquid interface as studied by neutron reflectometry', *Langmuir*, **9**, 3530-3537.
Cosgrove, T., Phipps, J. S., Richardson, R. M., Hair, M. L. and Guzonas, D. A., (1993), 'The adsorption of block copolymers of poly-2-vinylpyridine and polystyrene by neutron and surface forces techniques', *Macromolecules*, **26**, 4363-4367.
Cosgrove, T., Phipps, J. S., Richardson, R. M, Hair, M. L. and Guzonas, D. A., (1994), 'Surface force and neutron scattering studies on adsorbed poly(2-vinylpyridine) b-polystyrene', *Coll. Surf.*, **86**, 91-101.
Cosgrove, T., Richards, R. D. C., Semlyen A. J. and Webster, J. R. P., (1993), 'The adsorption of end-functionalised and cyclic polymers', *ACS Symposium Series*, **532**, 111-120.
Cosgrove, T., Turner, M. J., Griffiths, P. C., Hollingshurst, J., Shenton M. and Semlyen, J. A., (1996) 'Self diffusion and spin-spin relaxation in blends of linear and cyclic polydimethylsiloxane', *Polymer* **37**, 1535-1540.
Cosgrove, T., Vincent, B., Fleer, G. J., Cohen Stuart, M. and Scheutjens, J. M. H. M., (1993), *'Polymers at Interfaces'* Chapman and Hall, London
Cosgrove, T., White, S., Heenan, R. H., Howe A. and Zarbakhsh, A., (1995), 'SANS studies of gelatin-surfactant interactions', *Langmuir*, **11**, 744-749.
Cosgrove, T., White, S., Zarbakhsh, A., Heenan, R. K. and Howe, A., (1996), 'SANS studies of gelatin surfactant interaction II. The effect of pH', *J. Chem. S oc. Farad. Trans.*, **92**, 595-599.
Cosgrove, T., Zarbakhsh, A., Luckham, P. F., Hair, M. L. and Webster, J. R. P., (1995), 'The adsorption of PEO-PS block copolymers on quartz using a parallel plate surface force apparatus and simultaneous neutron reflection', *Faraday Discuss.*, **89**, 189-201.
Cosgrove, T., Zarbakhsh, A., Luckham, P. F., Richardson, R. M., and Webster, J. R. P., (1994), 'The measurement of volume fraction profiles for adsorbed polymers under compression using neutron reflectometry', *Coll. Surf. A*, **86**, 103-110.
Coveney, P., (1994), 'Chemical oscillations and non-linear chemical kinetics in "Self-production of supramolecular structures"', *NATO. ASI. Ser. C: Math. and Phys. Sci.*, **146**, 157-176. Fleischaker, G., Colonna, S. and Luisi, P., (Eds.) Kluwer Academic Publishers, Dodrecht.
Coveney, P. V. and Humphries, W., (1996) 'Molecular modelling of the mechanism of action of phosphonate retarders on hydrating cements', *J. Chem. Soc. Farad. Trans.*, **92**, 831-841.
Coveney, P. V. Emerton, A. N. and Boghosian, B. M., (1996), 'Simulation of self-reproducing micelles using lattice-gas automation', *J. Am. Chem. Soc.*, **118**, 10719-10724.
Coveney, P. V. and Wattis, J. A. D., (1996) 'Analysis of a generalized Becker-Doring model of self-reproducing micelles', *Proc. Roy. Soc. Lond. A*, **452**, 2079-2102.
Cronin, D. W., Terentjev, E. M., Sones, R. A. and Petschek, R. G., (1994), 'Twisting transition in a capillary filled with chiral smectic C liquid crystal', *Mol. Cryst. Liq. Cryst.*, **238**, 167-177.
D'Angelo, R., Plona, T. J., Schwartz, L. M., and Coveney, P.V., (1995), 'Ultrasonic measurements on hydrating cement slurries - onset of shear-wave propagation', *Adv. Cem. Based Mater.*, **2**, 8-14.
Debacher, N. and Ottewill, R. H., (1992), 'An electrokinetic examination of mica surfaces in aqueous media', *Coll. Surf.*, **65**, 51-59.
Donath, E., Krabi, A., Allan, G. and Vincent, B., (1996), 'A study of polymer depletion layers by electrophoresis: The influence of viscosity profiles and the non-linearity of the Poisson-Boltzmann equation', *Langmuir*, **12**, 3425-3430.
Donath, E., Walther, D., Krabi, A., Allan, G. C. and Vincent, B., (1996), 'A new relaxation effect with polymer depletion layers', *Langmuir*, **12**, 623-626.
Downes, N., Ottewill, G. A. and Ottewill, R. H., (1995), 'An investigation of the behaviour of ammonium perfluoro-octanoate at the air/water interface in the absence and presence of salts', *Coll. Surf. A.*, **102**, 203-211.
Edwards, S. F. and Mounfield, C. C., (1996), 'A theoretical model for the stress distribution in granular matter. I. Basic equations', *Physica A*, **226**, 1-11.
Edwards, S. F. and Mounfield, C. C., (1996), 'A theoretical model for the stress distribution in granular matter. II. Forces in pipes', *Physica A*, **226**, 12-24.
Edwards, S. F. and Mounfield, C. C., (1996), 'A theoretical model for the stress distribution in granular matter. III. Forces in sandpiles', *Physica A*, **226**, 25-33.
Español, P. and Warren, P. B., (1995), 'Statistical mechanics of dissipative particle dynamics', *Europhys. Lett.*, **30,** 191-196.

Ettelaie, R., (1993), 'On the Helmholtz free-energy of the highly charged double plates in an electrolytes', *Langmuir*, **9**, 1888-1892.
Ettelaie, R., (1995), 'Solutions of the linearized Poisson-Boltzmann equation through the use of random walk simultion methods', *J. Chem. Phys.*, **103**, 3657-3667.
Ettelaie, R. and Buscall, R., (1995), 'Electrical double-layer interactions for spherical charge regulating colloidal particles', *Adv. Coll. Interf. Sci.*, **61**, 131-160.
Evans, I.D. and Lips, A., 'Influence of soluble polymers on the elasticity of concentrated dispersions of deformable food microgel particles', (1993) , pp. 214-223, in *Food Colloids and Polymers: Stability and Mechanical Properties*, Eds. Dickinson, E. and Walstra, P. , RSC Special Publication 113, Cambridge.
Fragneto, G., Thomas, R. K., Rennie, A. R. and Penfold, J., (1995), 'Neutron reflection study of bovine ß-casein adsorbed on OTS self-assembled monolayers', *Science*, **267**, 657-660.
Fragneto, G., Li, Z. I., Thomas, R. K., Rennie, A. R. and Penfold, J., (1996), 'A neutron reflectivity study of the adsorption of aerosol-OT on self-assembled monolayers on silicon', *J. Coll. Interf. Sci.*, **178**, 531-537.
Fragneto, G., Lu, J. R., McDermott, D. C., Thomas, R. K., Rennie, A. R., Gallagher, P. D. and Satija, S. K., (1996), 'Structure of monolayers of tetraethylene glycol monododecyl ether adsorbed on self-assembled monolayers on silicon: a neutron reflectivity study', *Langmuir*, **12**, 477-486.
Frith, W. J., D'Haene, P., Buscall, R. and Mewis, J., (1996), 'Shear thickening in model suspensions of sterically stabilized particles', *J. Rheol.*, **40**, 531-548.
Frith, W. J. and Lips, A., (1995), 'The rheology of concentrated suspensions of deformable particles', *Adv. Coll. Interf. Sci.*, **61**, 161-189.
Frith, W. J. and Lips, A., (1996), 'Rheology of gel suspensions', pp. 306-318 in *Solid-Solid Interactions*, Adams, M. J., Biswas, S. K., Briscoe, B. J. (Eds.), Imperial College Press, London.
Goodwin, J. W., Hughes, R. W. and Bradbury, A. J., (1992), 'Deformation, melting and relaxation of structured colloidal dispersions', *Langmuir*, **8**, 2863-2872.
Goodwin, J. W. and Hughes, R. W., (1992), 'The dynamics and phase of concentrated dispersions', *Adv. Coll. Interf. Sci.*, **42**, 303-351.
Goodwin, J. W., (1996), 'Rheological characterisation of water soluble polymers', pp. 28-41 in *Industrial Water Soluble Polymers*, Ed. Finch, C. A., Roy. Soc. Chem., Cambridge.
Grassia, P. S., Hinch, E. J. and Nitsche, L. C., (1995), 'Computer simulation of Brownian motion of complex systems', *J. Fluid. Mech.*, **282**, 373-403.
Grassia, P. S. and Hinch, E. J., (1996), 'Computer simulations of polymer chain relaxation via Brownian motion', *J. Fluid Mech.*, **308**, 255-288.
Griffiths, P. C., Stilbs, P., Cosgrove, T. and Howe, A. H., (1996), 'A self-diffusion study of the complex formed by sodium dodecyl sulphate and the biopolymer and polyampholyte gelatin', *Langmuir*, **12**, 2884-2893.
Hall, C., Hoff, W. D., Taylor, S. C., Wilson, M. A., Yoon, B-G., Reinhardt, H-W., Sororo, M., Meredith, P. and Donald, A. M., (1995), 'Water anomaly in capillary liquid absorption by cement-based materials', *J. Mat. Sci. Letts.*, **14**, 1178-1181.
Harden, J. L. and Cates, M. E., (1995), 'Extension and compression of grafted polymer layers in strong normal flows', *J. Phys. II France*, **5**, 1093-1103.
Harden, J. L. and Cates, M. E., (1996), 'Deformation of grafted polymer layers in strong shear flows', *Phys. Rev. E*, **53**, 5063-5074.
Hardwick, A. J. and Walton, A. J., (1994), 'Forced oscillations of a bubble in a liquid', *Eur. J. Phys.* **15**, 325-328.
Haw, M. D., Poon, W. C. K. and Pusey, P. N., (1994), 'Structure factors from cluster-cluster aggregation simulation at high concentration', *Physica A*, **208**, 8-17.
Haw, M. D., Sievwright, M, Poon, W. C. K. and Pusey, P. N., (1995), 'Cluster-cluster gelation with finite bond-energy', *Adv. Coll. Interf. Sci.*, **62**, 1-16.
Haw, M. D., Sievwright, M., Poon, W. C. K. and Pusey, P. N., (1995), 'Structure and characteristic length scales in cluster-cluster aggregation simulation', *Physica A*, **217**, 231-260.
Herzog, B., Huber, K. and Rennie, A. R., (1994), 'Characterization of worm-like micelles containing solublized dye-molecules by light scattering techniques', *J. Coll. Interf. Sci.*, **164**, 3870-381.
Heyes, D. M. and Melrose, J. R., (1993), 'Brownian dynamics simulations of model hard sphere suspensions.', J. Non-Newtonian Fluid Mech., **46**, 1-28.
Heyes, D. M., Mitchell, P. J., Visscher, P. B. and Melrose, J. R., (1994), 'Brownian dynamic simulations of concentrated dispersions: visoelasticity and near-Newtonian behaviour', *J. Chem. Soc. Farad. Trans.*, **90**, 1133-1141.
Ho, C. C., Keller, A., Odell, J. A. and Ottewill, R. H., (1993), 'Preparation of monodisperse ellipsoidal polystyrene particles', *Coll. Polymer Sci.*, **271**, 469-479.

Ho, C. C., Ottewill, R. H. and Yu, L. P., (1997), 'Examination of ellipsoidal polystyrene particles by electrophoresis', *Langmuir*, **13**, 1925-1930.
Hinch, E. J. and Nitche, L. C., (1993), 'Nonlinear drift interactions between fluctuating colloid particles: oscillatory and stochastic motions', *J. Fluid Mech.*, **256**, 343-403.
Hinch, E. J., (1994), 'Brownian motion with stiff bonds and rigid constraints', *J. Fluid. Mech.*, **271**, 219-234 .
Huntley, J. M., (1993), 'Vacancy effects on the force distribution in a two-dimensional granular pile', *Phys. Rev. E*, **48**, 4099-4101.
Ilett, S.M., Orrock, A., Poon, W.C. K. and Pusey, P.N. (1995), 'Phase-behaviour of a model colloid-polymer mixture', *Phys. Rev. E*, **51**, 1344-1352.
Ito, M. and Cosgrove, T. (1994), 'Atomistic simulation of short-chain polymers at the liquid/liquid interface', *Coll. Surf. A*, **86**, 125-131.
Johnson, S. A., Gorman, D. M., Adams, M. J. and Briscoe, B. J., (1993), 'The friction and lubrication of human stratum corneum', pp. 663-672 in *Thin Films in Tribology*, Ed. Dowson, D., Elsevier, Amsterdam.
Jones, R. A. L., (1994), 'The wetting transition for polymer mixtures', *Polymer*, **10**, No. 35, 2160-2166.
Jones, R. A. L., (1995), 'A superficial look at polymers', *Phyics World* , March, 47-51.
Keddie, J. L., Jones, R. A. L. and Cory, R. A., (1994), 'Size-dependent depression of the glass transition temperature in polymer films', *Europhys. Letts.*, **27**, 59-64.
Keddie, J. L., Jones, R. A. L. and Cory, R. A., (1994), 'Interface and surface effects on the glass transition temperature in thin polymer films', *Faraday Discuss.*, **98**, 219-230.
Keddie, J. L., Meredith, P., Jones, R. A. L. and Donald, A. M., (1995), 'Kinetics of film formation in acrylic latices studied with multiple-angle-of-incidence ellipsometry and environmental scanning electron microscopy', *Macromolecules* , **28**, 2673-2682.
Keddie, J. L. and Jones, R. A. L., (1995), 'Glass transition behaviour in ultra-thin polystyrene films', *Israel Journal of Chemistry*, **35**, 21-26.
King, S. M., Cosgrove, T. and Eaglesham, A., (1996), 'The adsorption of polystyrene saturated-polydiene block copolymers on silica substrates', *Coll. Surf. A*, **108**, 159-171.
Kuksenok, O. V., Ruhwandl, R. W., Shiyanovskii, S. V. and Terentjev, E. M., (1996), 'Director structure around a colloid particle suspended in a nematic liquid crystal', *Phys. Rev. E*, 53, 5198-5203.
Lee, E. M., Kanelleas, D., Milnes, J. E., Smith, K., Warren, N. and Rennie, A. R., (1996), 'Adsorption of a polyelectrolyte at a charged liquid interface', *Langmuir*, **12**, 1270-1277.
Leong, Y. K., Scales, P. J., Healy, T. W., Boger, D. V. and Buscall, R., (1993), 'Control of the rheology of concentrated aqueous colloidal systems by steric and hydrophobic forces', *J. Chem. Soc. Chem., Comm.*, **7**, 639-664.
Leong, Y. K., Scales, P. J., Healy, T. W., Boger, D. V. and Buscall, R., (1993), 'Rheological evidence of adsorbate-mediated short-range steric forces in concentrated dispersions', *J. Chem. Soc. Farad. Trans.*, **89**, 2473-2478.
Li, Z. I., Lu, J. R., Thomas, R. K., Rennie, A. R. and Penfold, J., (1996), 'Neutron reflection study of butanol and hexanol adsorbed at the surface of their aqueous solutions', *J. Chem. Soc. Farad. Trans.*, **92**, 565-572.
Lips, A., Westbury, T., Hart, P. M., Evans, I. D. and Campbell, I. J., (1993) 'On the physics of shear-induced aggregation in concentrated food emulsions', pp. 31-44 in *Food Colloids and Polymers: Stability and Mechanical Properties*, Eds. Dickinson, E. and Walstra, P., RSC Special Publication 113, Cambridge.
Lips, A., Evans, T., Evans, I. D. and Underwood, D., (1993), 'Linear viscoelastic study of effects of SDS on the gelation behaviour of gelatins', pp. 331-334 in *Food Proteins: Structure and Funtionality*, Eds. Schwenke, K. D. and Mothes, R., VCH, Weinheim, Germany.
Li, Z. X., Lu, J. R., Styrkas, D. A., Thomas, R. K., Rennie, A. R. and Penfold, J., (1993), 'The structure of the surface of ethanol/water mixtures', *Mol. Phys.*, **80**, 925-939.
Lu, C-Y. D. and Cates, M. E., (1994), 'Viscoelasticity of lyotropic smectics', *J. Chem. Phys.* **101**, 5219-5288.
Lu, C-Y. D. and Cates, M. E., (1995), 'Hydrodynamics of soluble surfactant films', *Langmuir*, **11**, 4225-4233.
Lu, J. R., Purcell, I. P., Lee, E. M., Simister, E. A., Thomas, R. K., Rennie, A. R. and Penfold, J., (1995), 'The composition and structure of sodium dodecyl sulfate-dodecanol mixtures adsorbed at the air-water interface. A neutron reflection study', *J. Coll. Interf. Sci.*, **174**, 441-455.
Lubensky, T. C., Terentjev, E. M. and Warner, M. (1994), 'Layer-network coupling in smectic elastomers', *J. Phys. II France*, **4**, 1457-1459.

McCarney, J., Lu, J. R., Thomas, R. K. and Rennie, A. R., (1994), 'Neutron reflection from polystyrene adsorbed on silica from cyclohexane at temperatures at and below the theta temperature', *Coll. Surf. A*, **86**, 185-192.
McDermott, D. C., Kanelleas, D., Thomas, R. K., Rennie, A. R., Satija, S. K. and Majkrzak, C. F., (1993), 'Study of the adsorption from aqueous solution of mixtures of nonionic and cationic surfactants on crystalline quartz using the technique of neutron reflection', *Langmuir*, **9**, 2404-2407.
McDermott, D. C., McCarney, J., Thomas, R. K. and Rennie, A. R., (1994) 'Study of the adsorption from aqueous solution of hexadecyl trimethyl-ammonium bromide on quartz using the technique of neutron reflection', *J. Coll. Interf. Sci.*, **162**, 304-310.
McDonald, J. A., Butler, S. A. and Rennie, A. R., (1995), 'The synthesis and surface activity of some alkanediyl bismorpholines', *Coll. Surf. A*, **102**, 137-141.
McDonald, J. A. and Rennie, A. R., (1995), 'Scattering studies of mixed micelles formed from C_{16}TAB and $C_{12}E_6$ surfactants', *Prog. Coll. Polymer Sci.*, **98**, 75-78.
McDonald, J. A. and Rennie, A. R., (1995), 'A structural study of mixed micelles containing C_{16}TAB and $C_{12}E_6$', *Langmuir*, **11**, 1493-1499.
Mackley, M. R., Marshall, R. T. J., Smeulders, J. B. A. F. and Zhao, F. D., (1994), 'The rheological characterisation of polymeric and colloidal fluids', *Chem. Eng. Sci.*, **49**, 2551-2565.
Mackley, M. R., Marshall, R. T. J. and Smeulders, J. B. A. F., (1994), 'Multi-pass rheometry. *Prog. in Trends in Rheo. IV.*, Proceedings of the 4th European Rheology Conference, Ed. Gallegos, C. 503-505.
Mackley, M. R., Marshall, R. T. J. and Smeulders, J. B. A. F., (1995), 'The multipass rheometer', *J. Rheol.*, **39**, 1293-1309.
Mao, Y., Cates, M. E. and Lekkerkerker, H. N. W., (1995), 'Depletion force in colloidal systems', *Physica A*, **222**, 10-24
Mao, Y., Cates, M. E. and Lekkerkerker, H. N. W., (1995), 'Depletion stabilization by semidilute rods', *Phys. Rev. Letts.*, **75**, 4548-4551.
Meeten, G. H., (1993), 'A dissection method for analysing filter cakes', *Chem. Eng. Sci.*, **48**, 2391-2398.
Meeten, G. H., (1994), 'Shear and compressive yield in the filtration of a bentonite suspension', *Coll. & Surf. A.*, **82**, 77-83.
Meeten, G. H., (1995), 'Anomalous diffraction model of linear optical dichroism in shear-thickening polymer solutions', *Rheologica. Acta*, **34**, 160-162
Meeten, G. H., (1997), 'Refraction by spherical particles in the intermediate scattering region', *Optics. Comms*, **134**, 233-240.
Meeten, G. H. and Lebreton, C., (1993), 'A filtration model for the capillary suction time method', *J. Petrol Sci. & Eng.*, **9**, 155-162.
Meeten, G. H. and North, A., (1995), 'Refractive index measurement of absorbing and turbid fluids by reflection near the critical angle', *Measurement Sci. Tech.*, **6**, 214-221.
Meeten, G. H. and Pafitis, D., (1993), 'Granular fluids rheometry', Flow and microstructure of dense suspensions', *Mat. Res. Co. Symp. Proc.* **289**, 239-244. Struble, I., Zukoski, C and Maitland G, (eds.), MRS, Pittsburgh.
Meeten, G. H., and Sherman, N. E., (1993), 'Ultrasonic veolcity and attenuation of glass ballotini in viscous and viscoelastic fluids', *Ultrasonics*, **31**, 192-199.
Meeten, G. and Sherwood, J., (1994), 'The hydraulic permeability of bentonite suspensions with granular inclusions', *Chem. Eng. Sci.*, **49**, 3249-3256.
Meeten, G. H. and Smeulders, J. B. A. F., (1995), 'Interpretation of filterability measured by the capillary suction time method', *Chem. Eng. Sci.*, **50**, 1273-1279.
Meeten, G. H. and Smeulders, J. B. A. F., (1996), 'Filterability measured by the capillary suction time method-reply', *Chem. Eng. Sci.*, **51**, 1355-1356.
Meeten, G. H. and Smeulders, J. B. A. F., (1996), 'Comments on specific cake resistance determination with a new capillary suction time apparatus', *Ind. Eng. Chem. Res.*, **35**, 4810-4812.
Meeten, G.H. and Wood, P., (1993), 'Optical fibre methods for measuring the diffuse reflectance of fluids', *Meas. Sci. and Tech.*, **4**, 643-648.
Melrose, J. R., (1993) 'The simulation of aggregated colloids under flow', *MRS Pub.* Vol. **289** *'Flow and Microstructure in dense suspensions*, p. 33.
Melrose, J. R. and Ball, R. C., (1994), 'Moving mesh simulations of complex fluids', *Prog. in Trends in Rheo. IV.*, Proceedings of the 4th European Rheology Conference, Eds. Gallegos, C., Steinkopff, Darmstadt (1994) 675-677.
Melrose, J. R. and Ball, R. C., (1995), 'The pathological behaviour of sheared hard spheres with hydrodynamic interactions', *Europhys. Letts.*, **32**, 535-540.

Melrose, J. R. and Heyes, D. M., (1993) 'Simulations of electrorheological and particle mixture suspensions: agglomerate and layer structures', *J. Chem. Phys.*, **98**, 5873.

Melrose, J. R. and Heyes, D. M., (1993), 'Rheology of weakly flocculated suspensions: simulation of agglomerates under shear', *J. Coll. Interf. Sci.*, **157**, 227-234.

Meredith, P. and Donald, A. M., (1995), 'Imaging of "wet" polymer latices in ESEM, some contrast considerations', *Scanning*, **17**, 54.

Meredith, P., Donald, A. M. and Luke, K. (1995), 'Pre-induction and induction hydration of tricalcium silicate: an environmental scanning electron microscopy study', *J. Mat. Sci.*, **35**, 1921-1930.

Meredith, P., Donald, A. M. and Luke, K., (1995), 'In situ hydration studies of the components of ordinary Portland cement by ESEM', *Scanning*, **17**, 57.

Meredith, P. and Payne, R. S., (1996), 'Freeze drying: In situ observations using cryo-environmental SEM + DSC', *J. Pharm. Sci.*, **85**, 631-637.

Meredith, P. and Donald, A. M., (1996), 'Study of 'wet' polymer latex systems in environmental scanning electron microscopy: Some imaging considerations', *J. Microscopy*, **181**, 23-35.

Mills, M. F., Gilbert, R. G., Napper, D. H., Rennie, A. R. and Ottewill, R. H., (1993), 'Small-angle neutron scattering studies of inhomogeneities in latex particles from emulsion homopolymerisations', *Macromolecules*, **26**, 3553-3562.

Mitchell, P. J., Heyes, D. M. and Melrose, J. R., (1995), 'Brownian-dynamics simulations of model stabilised colloidal dispersions under shear', *J. Chem. Soc. Farad. Trans.*, **91**, 1975-1989.

Nicolai, H., Herzhaft, B., Hinch, E. J., Oger, L. and Guazzelli, E., (1995), 'Particle velocity fluctuations and hydrodynamic self-diffusion of sedimenting non-Brownian spheres', *Phys. Fluids*, **7**, 12-23.

Norton, L. J., Kramer, E. J., Bates, F. S., Gehlsen, M. D., Jones, R. A. L., Karim, A., Flecher, G. P. and Kleg, R., (1995), 'Neutron reflectometry study of surface segregation in an isotopic poly (ethylene propylene) blend: Deviation from mean-field theory', *Macromolecules*, **28**, 8621-8628.

Olmsted, P. D. and Terentjev, E. M., (1996), 'Mean field nematic-smectic A transition in a random polymer network', *Phys. Rev. E*, **53**, 2444-2453.

Olmsted, P. D. and Terentjev, E. M., (1996), 'Some properties of membranes in nematic solvents', *J. Phys. II France*, **6**, 49-56.

Osipov, M. A., Sluckin, T. J. and Terentjev, E. M., (1995), 'Viscosity coefficients of smectics C' *Liq. Cryst.*, **19**, 197-205.

Ottewill, R. H., (1992), 'Experimental methods of particle characterization', *Pure & Applied. Chem.*, **64**, 1697-1702.

Ottewill, R. H., (1992), 'Adsorption of surfactants at interfaces - studies by small angle neutron scattering', *Prog. Coll. Polymer Sci.*, **88**, 49-57.

Ottewill, R. H., (1995), 'Polymer colloids in nonaqueous media', pp. 1-25 in *Colloidal Polymer Particles*, Eds. Goodwin, J. W. and Buscall, R., Academic Press, London.

Ottewill, R. H., (1997), 'Stabilization of polymer colloid dispersions', pp. 59-121 in *Emulsion Polymerisation and Emulsion Polymers*, , Eds. Lovell, P. A. and El-Aasser, M. S., J. Wiley, Chichester.

Ottewill, R. H., (1997), 'Stability of polymer colloids', pp. 31-48 in., *Polymer Disperions: Principles and Applications*, Ed. Asua, J. M., NATO ASI Series E, **335**, Kluwer Academic Publishers, Dordrecht.

Ottewill, R. H., (1997), 'Scattering techniques - fundamentals', pp. 217-228 in *Polymeric Dispersions: Principles and Applications*, Ed. Asua, J. M., NATO ASI Series *E*, **335**, Kluwer Academic Publishers, Dordrecht.

Ottewill, R. H., (1997), 'Application of scattering techniques to polymer-colloid dispersions', pp. 229-242 in *Polymeric Dispersions: Principles and Applications*, Ed. Asua, J. M., *NATO ASI Series E*, **335**, Kluwer Academic Publishers, Dordrecht, The Netherlands.

Ottewill, R. H. and Bartlett, P., (1992), 'The study of binary colloidal mixtures by neutron scattering' pp. 795-800 *in 'Structure and dynamics of strongly interacting colloids and supramolecular aggregates in solution'*, Ed. Chen, S. H., Huang, J. S. and Tartaglia, P., Kluwer Academic Publishers, Dordrecht.

Ottewill, R. H., Bee, R. D. and Hoogland, J., (1993), 'Food surfactants at triglyceride-water and decane-water interfaces' pp. 341-353 in '*Food Colloids and Polymers: Stability and mechanical properties*', Ed. Dickinson, E. and Walstra, P., Roy. Soc. Chem., Cambridge.

Ottewill, R. H., Cole, S. J. and Waters, J. A., (1995), 'Characterisation of particle morphology by scattering techniques', *Macromol. Symp.*, **92**, 97-107.

Ottewill, R. H., Hanley, H. J. M., Rennie, A. R. and Straty, G. C., (1995), 'Small-angle neutron scattering studies on binary mixtures of charged particles', *Langmuir*, **11**, 3757-3765.

Ottewill, R. H., Ho, C. C., Keller, A. and Odell, J. A., (1993), 'Monodisperse ellipsoidal polystyrene latex particles: preparation and characterisation', *Polymer International*, **30**, 207-211.

Ottewill, R. H. and Rennie, A. R., (1994), 'Small angle neutron scattering studies on association polymers - polymeric materials', *Sci. & Eng.*, **71**, 569-570.

Ottewill, R. H. and Rennie, A. R., (1996), 'Interaction behaviour in a binary mixture of polymer particles', *Prog. Coll. Polymer Sci.*, **100**, 60-63.
Ottewill, R. H., Rennie, A. R., Laughlin, R. G. and Bunke, G. M., (1994), 'Micellar structure in solutions of an ultralong chain Zwitterionic surfactant', *Langmuir*, **10**, 3493-3499.
Ottewill, R. H. and Satgurunathan, R, (1995), 'Nonionic latices in aqueous media, Part 4, Preparation and characterisation of electrosterically stabilised particles', *Coll. Polym. Sci.*, **273**, 379-386.
Ottewill, R. H., Schofield, A. B. and Waters, J. A., (1996), 'Preparation of composite latex particles by engulfment', *Coll. Polymer Sci.*, **274**, 763-771.
Ottewill, R. H., Schofield, A. B., Waters, J. A. and Williams, N. St. J., (1997), 'Preparation of core-shell polymer colloid particles by encapsulation', *Coll. Polymer Sci.*, **275**, 274-283.
Ottewill, R. H. and Taylor, P., (1994), 'The formation and ageing rates of oil-in-water miniemulsions', *Coll. Surf. A.*, **88**, 303-316.
Pafitis, D. G., and Meeten, G. H., (1993), 'Low strain shear behaviour of cement slurries at quasi-static rates', *Flow & Microstructure of Dense Suspensions*, Eds. Struble, L. J., Zukoski, C. F., and Maitland, G. C., Ch.38, 179-183, MRS Pittsburgh.
Panizza, P., Roux, D., Vuillaume, V., Lu, C-Y. D. and Cates, M. E., (1996), 'Viscoelasticity of the onion phase', *Langmuir*, **12**, 248-252.
Patel, A., Cosgrove, T., Semlyen, J . A., Webster J. R. P. and Scheutjens, J. M. H. M., (1994), 'Adsorption studies of different end functionalised linear PDMS', *Coll, Surf. A*, **87**, 15-24.
Poon, W. C. K., Ilett, S. M. and Pusey, P. N., (1994), 'Phase-behavior of colloid-polymer mixtures', *Nuovo Cimento*, **16**, 1127-1139.
Poon, W. C. K., Meeker, S. P., Pusey, P. N. and Segre, P. N. (1996), 'Viscosity and structural relaxation in concentrated hard-sphere colloids', *J. Non-Newtonian Fl. Mech.*, **67**, 179-189.
Poon, W. C. K, Pirie,.A. D., Haw, M. D. and Pusey, P. N., (1997), 'Non equilibrium behaviour of colloid-polymer mixture', *Physica A*, **235**, 110-119.
Poon, W. C. K., Pirie, A. D. and Pusey, P. N., (1995), 'Gelation in colloid-polymer mixtures', (1995), *Farad. Discuss.*, **101**, 65-77.
Poon, W. C. K. and Warren, P. B., (1994), 'Phase behaviour of hard sphere mixtures', *Europhys. Lett.*, **28**, 513-518.
Poon, W. C. K., Selfe, J. S., Robertson, M. B., Ilett, S. M., Pirie, A. D. and Pusey, P. N., (1993), 'An experimental-study of a model colloid-polymer mixture', *J. Phys. II France*, **3**, 1075-1086.
Pusey, P. N., Pirie, A. D. and Poon, W. C. K., (1993), 'Dynamics of colloid polymer mixtures', *Physica A*, **201**, 322-331.
Pusey, P. N., Poon, W. C. K., Ilett. S. M and Bartlett, P., (1994), 'Phase behavior and structure of colloidal suspensions', *J. Phys. Cond. Matter*, **6**, A29-A36.
Pusey, P. N., Segre, P. N., Behrend, O. P., Meeker, S. P. and Poon, W. C. K., (1997), 'Dynamics of concentrated colloidal suspensions', *Physica A*, **235**, 1-8.
Rallison, J. M. and Hinch, E. J., (1995), 'Instability of a high speed submerged elastic jet', *J.Fluid. Mech.*, **288**, 311-324.
Rennie, A. R., (1995), 'Reduction of data from SANS instruments', pp. 93-105 in *'Modern Aspects of Small-Angle Scattering'*, Ed. Brumberger, H., NATO ASI Series **451**, Kluwer Academic Publishers, Dordrecht.
Rennie, A. R., (1995), 'Bulk polymers', pp. 433-449 in *'Modern Aspects of Small-Angle Scattering'*, Ed. Brumberger, H., NATO ASI Series **451**, Kluwer Academic Publishers, Dordrecht.
Rennie, A. R., (1996), 'Neutron scattering from polymers', pp. 325-345 in *Polymer Spectroscopy*, Ed. Fawcett, A. H., Academic Press, London.
Rennie, A. R., (1996), 'Neutron scattering techniques', pp. 123-165 in *Characterisation of solid polymers*, Ed. Spells, S. J., Chapman & Hall, London.
Rennie, A. R. and Clarke, S. M., (1996), 'Scattering by complex fluids under shear', *Curr. Op. Coll. Interf. Sci.*, **1**, 34-38.
Ruhwandl, R. W. and Terentjev, E. M. (1995), 'Friction drag on a cylinder moving in a nematic liquid crystal', *Z. Naturforsch A*, **50**, 1023-1030.
Ruhwandl, R. W. and Terentjev, E. M., (1996), 'Friction drag on a particle moving in a nematic liquid crystal', *Phys. Rev. E*, **54**, 5204-5210.
Ruhwandl, R. W. and Terentjev, E. M., (1997), 'Long-range forces and aggregation of colloid particles in a nematic liquid crystal', *Phys. Rev. E.*, **55**, 2958-2961.
Russel, W. B. and Goodwin, J. W., (1996), 'Rheology and rheological techniques', *Curr. Opin. Coll. Interf. Sci.*, **1**, 447-449.
Segre, P. N., Behrend, O. P. and Pusey, P. N. (1995), 'Short-time Brownian-motion in colloidal suspensions - Experiment and Simulation', *Phys. Rev. E*, **52**, 5070-5083.

Segre, P. N., Meeker, S. P., Pusey, P. N. and Poon, W. C. K., (1995), 'Viscosity and structural relaxation in suspensions of hard-sphere colloids', *Phys. Rev. Letts.*, **75**, 958-961.

Segre, P. N. and Pusey, P. N., (1994), 'Multiple-scattering suppression in dilute and concentrated colloidal dispersions', *Nuovo Cimento*, **16**, 1181-1191.

Segre, P. N. and Pusey, P. N., (1996), 'Scaling of the dynamic scattering function of concentrated colloidal suspensions', *Phys. Rev. Letts.*, **77**, 771-774.

Segre, P. N. and Pusey, P. N., (1997), 'Dynamics and scaling in hard-sphere colloidal suspensions', *Physica A*, **235**, 9-18.

Segre, P. N., van Megen, W., Pusey, P. N., Schatzel, K and Peters, W. (1995), '2-color dynamic light-scattering', *J. Modern Optics*, **42**, 1929-1952

Sherman, N. E., and Sherwood, J. D., (1993), 'Crossflow filtration: cakes with variable resistance and capture efficiency', *Chem. Eng. Sci.*, **48**, 2913-2918.

Sherwood, J. D., (1992), 'Ionic motion in a compacting filtercake', *Proc. Roy. Soc. Lond. A*, **437**, 607.

Sherwood, J. D., (1993), 'Biot poroelasticity of a chemically active shale', *Proc. Roy. Soc. Lond. A*, **440**, 365-377.

Sherwood, J. D., (1993), 'Formation of a cement filtercake above a compactible mudcake', *Chem. Eng. Sci.*, **48**, 2767-2775.

Sherwood, J. D., (1993), 'A model for static filtration of emulsions and foams', *Chem. Eng. Sci.*, **48**, 3355-3361.

Sherwood, J. D., (1994), 'A model for the flow of water and ions into swelling shale', *Langmuir*, **10**, 2480-2486.

Sherwood, J. D., (1994), 'A model for hindered transport of solute in poroelastic shale', *Proc. Roy. Soc. Lond.*, **A445**, 679-692.

Sherwood, J. D., (1994), 'Thixotropic mud, compacted cake and swelling shale', *Les. Cah. de Rheol.*, **13**, 111-119.

Sherwood, J. D., (1995), 'Ionic transport in swelling shale', *Adv. Coll. Interf. Sci.*, **61**, 51-64.

Sherwood, J. D. and Bailey, L., (1994), 'Swelling of shale around a cylindrical wellbore', *Proc. Roy. Soc. Lond. A*, **444**, 161-184.

Sherwood, J. D. and Durban D., (1996), 'Squeeze flow of a power-law viscoplastic solid', *J. Non-Newtonian Fl. Mech.*, **62**, 35-54.

Sherwood, J. D. and Meeten, G., (1996), 'Friction in drilling fluid filtercakes', pp. 355-363 in *Solid-Solid Interactions*, Adams, M. J., Biswas, S. K.and Briscoe, B. J., (eds.) Imperial College Press, London.

Sherwood, J. D. and Stone, H., (1995), 'Electrophoresis of a thin charged disk', *Phys. Fluids*, **7**, 697-705.

Sherwood, J. D. and van Damme, H., (1994), 'Non-linear compaction of an assembly of highly deformable plate-like particles', *Phys. Rev. E.*, **50**, 3834-3840.

Sones, R. A., Petschek, R. G., Cronin, D. W. and Terentjev, E. M., (1996), 'Twisting transition in a fiber composed of chiral smectic-C liquid crystal polymers', *Phys. Rev. E.*, **53**, 3611-3617.

Spenley, N. A., Yuan, X-F. and Cates, M., (1996), 'Non monotonic constitutive laws and the formation of shear-banded flows', *J. Phys. II France*, **6**, 375-394.

Stevens, J.D., Jones, I. L., Warner, M., Lavin, M. J. and Leaver, P. K., (1992), 'Mathematical modelling of retinal tear formation: Implications for the use of heavy liquids', *The Eye*, **6**, 69-74.

Taylor, P. and Ottewill, R. H., (1994), 'Ostwald ripening in o/w miniemulsions formed by the dilution of o/w microemulsions', *Prog. Coll. Polymer Sci.*, **97**, 199-203.

Taylor, P., (1996), 'Effect of an Anionic Surfactant on the rheology and stability of concentrated O/W emulsions stabilized by PVA', *Coll. Polymer Sci.*, **274**, 1061-1071.

Terentjev, E. M., (1995), 'Stability of liquid crystalline macroemulsions', *Europhys. Letts.*, **32**, 607-612.

Terentjev, E. M., (1995), 'Disclination loops, standing alone and around solid particles in nematic liquid crystals', *Phys. Rev. E.*, **51**, 1330-1337.

Terentjev, E. M., (1995), 'Density functional model of anchoring energy at a liquid crystalline polymer - solid interface', *J. Phys. II France*, **5**, 159-170.

Terentjev, E. M. Callaghan, P. T. and Warner, M., (1995), 'Pulsed gradient spin-echo NMR of confined Brownian particles', *J. Chem. Phys.*, **102**, 4619-4624.

Terentjev, E. M., Osipov, M. A. and Sluckin, T. J., (1994), 'Ferroelectric instability in semi-flexible liquid crystalline dipolar polymers', *J. Phys. A* **27**, 7047-7059.

Terentjev, E. M. and Warner, M. (1994), 'Continuum theory of elasticity and piezoelectric effects in smectic A elastomers', *J. Phys. II France*, **4**, 111-126.

Terentjev, E. M. and Warner, M., (1994), 'Continuum theory of ferroelectric smectic C* elastomers', *J. Phys. II France*, **4**, 849-864.

Terentjev, E. M., Warner, M. and Bladon P., (1994), 'Orientation of nematic elastomers and gels by an electric field', *J. Phys. II France*, **4**, 667-676.

Terentjev, E. M., Warner, M. and Lubensky, T. C., (1995), 'Fluctuations and long-range order in smectic elastomers', *Europhys. Letts.* **30**, 343-348.

Terentjev, E. M., Warner, M. and Verwey, G. C., (1996), 'Non-uniform deformations in liquid crystalline elastomers', *J. Phys. II France*, **6**, 1049-1060.

van der Schoot, P, McDonald, J. A. and Rennie, A. R., (1996), 'Static scattering by linear micelles', *Langmuir*, **11**, 4614-4616.

Verwey, G. C., Warner, M. and Terentjev, E. M., (1996), 'Elastic instability and strip domains in liquid crystalline elastomers', *J. Phys. II France*, **6**, 1273-1290.

Vincent, B., (1992), 'Dispersion stability in mixed solvent (aqueous/organic) media', *Adv. Coll. Interf. Sci.*, **42**, 279-302.

Vincent, B., (1993), 'The preparation of colloidal particles having (post-grafted) terminally-attached polymer chains', *J. Chem. Eng. Sci.*, **48**, 429-437.

Vincent, B., (1995), 'Electrically-conducting polymer colloids and composites', *Polym Adv. Techn.*, **6**, 356-361.

Vincent, B., Cawdery, N. and Milling, A., (1994), 'Instabilities in dispersions of hairy particles on adding solvent - miscible polymers', *Coll. Surf. A*, **86**, 239-249.

Vincent, B., Leon, A. and Cawdery, N., (1994), 'The synthesis and characterisation of monodisperse poly(acrylic acid) and poly(methacylic acid)', *Coll. Polymer Sci.*, **86**, 239-249.

Vincent, B., Snowden, M. and Marston, N. J., (1994), 'The effect of surface modifications on the stability characteristics of poly(N-isopropyl acrylamide) latices under Brownian and flow conditions', *Coll. Polymer Sci.* **272**, 1273-1280.

Wang, X-J, and Warner, M., (1992), 'Molecularly non-homogenous nematic polymers', Chapter 25, *NATO ASI on Quantum Electronics*, Kluwer Academic Publishers, Dordrecht.

Wang, X-J, and Warner, M., (1992), 'Theory of main chain nematic polymers with spacers of varying degree of flexibility', *Liquid Crystals* **12**, 385-401.

Wang, X-J. and Warner, M., (1996), 'The swelling of nematic elastomers by nematogenic solvents', *Macromol. Theory Simul.*, **6**, 37-52.

Warner, M. and Wang, X-J., (1992), 'The rubber elasticity of nematic solids', Chapter 6 of *'Elastomeric Polymer Networks'*, pp.63-92 ed. J E Mark and B Erman, Prentice Hall, New Jersey.

Warner, M. and Wang, X-J., (1992), 'Phase equilibria of swollen nematic elastomers' *Macromolecules*, **25**, 445-449.

Warner, M. and Williams. D. M. R., (1992), 'NMR spin-lattice relaxation from molecular defects in nematic polymer liquid crystals', *J. Phys.II France*, **2**, 471-486.

Warner, M. and Wang. X-J., (1992), 'Discrete and continuum models of nematic polymers', *J. Phys. A*, **25**, 2831-2841.

Warner, M. and Cates, M. E., (1993), 'Non local dielectric response to dipolar polymers', *J. Phys., II France*, **3**, 829-849.

Warner, M. and Terentjev, E. M., (1996), 'Nematic elastomers - a new state of matter?', *Prog. Polym. Sci.*, **21**, 853-891,

Warr, S., Jacques, G. T. H. and Huntley, J. M., (1994), 'Tracking the translational and rotational motion of granular particles: use of high-speed photography and image processing', *Powder Tech.*, **81**, 41-56.

Warr, S., Huntley, J. M. and Jacques, G. T. H., (1995), 'Fluidisation of a two-dimensional granular systems. Experimental study and scaling behaviour', *Phys. Rev. E*, **52**, 5583-5595.

Warr, S. and Huntley, J. M., (1995), 'Energy input and scaling laws for a single particle vibrating in one dimension', *Phys. Rev. E* **52**, 5596-5601.

Warren, P. B., (1994), 'Hydrodynamics of fractal aggregates', *Il Nuovo Cimento*, **16**, 1231-1236.

Warren, P. B., (1994), 'Depletion effect in a model lyotropic liquid crystal - theory', *J. Phys. I France*, **4**, 237-244.

Warren, P. B., (1997), 'Simplified mean field theory for polyelectrolyte phase behaviour', *J. Phys. II France*, **7**, 343-361.

Warren, P. B., Ball, R. C. and Boelle, A., (1995), 'Convection limited aggregation', *Europhys. Lett.*, **29**, 339-344.

Warren, P. B., Ilett, S. M. and Poon, W. C. K., (1995), 'The effect of polymer non-ideality in a colloid + polymer mixture', *Phys. Rev. E.*, **52**, 5205-5213.

West, A. H. L., Melrose, J. R. and Ball, R. C., (1994), 'Computer simulations of the breakup of colloid aggregates', *Phys. Rev. E.*, **49**, 4237-4249.

Wilson, L. M., Stühn, B. and Rennie, A. R., (1995), 'Temperature dependent smectic mesophase ordering in fluorocarbon side chain polyesters', *Liquid Crystals*, **18**, 923-926.

Wittmer, J. P., Claudin, P., Cates, M. E. and Bouchaud, J-P., (1996), 'An explanation of the central stress minimum in sandpiles', *Nature*, **382**, 336-338.
Wittmer, J. P., Cates, M. E. and Claudin, P., (1997), 'Stress propagation and arching in static sandpiles', *J. Phys. I France*, **7**, 39-80.
Wu, D. T. and Cates, M. E., (1993) 'Hydrodynamics of adsorbed polymer layers: a new approach', *Phys. Rev. Letts.*, **71**, 4142-414.
Wu, D. T. and Cates, M. E., (1994), 'Hydrodynamic response of adsorbed polymer layers', *Coll. Surf.A*, **86**, 274.
Wu, D. T. and Cates, M. E., (1996), 'Nonlocal hydrodynamic theory of flow in polymer layers', *Macromolecules*, **29**, 4417-4431.
Wyatt, D. and Vincent, B., (1993), 'Electrical effects in non-aqueous systems', *J. Biopharm. Sci.*, **3**, 27-31.
Young, A. M., Higgins, J. S., Peiffer, D. G. and Rennie, A. R., (1995), 'Effect of sulphonation level on the single chain dimensions and aggregation of sulphonated polystyrene ionomers in xylene', *Polymer*, **36**, 691-697.
Young, A. M., Higgins, J. S., Peiffer, D. G., Rennie, A. R., (1996), 'Effect of aggregation on the single-chain dimensions of sulfonated polystyrene ionomers in xylene', *Polymer*, **37**, 2125-2130.
Yuan, J., (1993), 'Relationship between electron energy loss and interparticle dispersion force', *Institute of Physics Conf. Series*, **138**, 63-68.

INDEX